The Evolutionary Biology of Species

TIMOTHY G. BARRACLOUGH

Imperial College London, UK

OXFORD

UNIVERSITY PRESS

OXFORD

UNIVERSITY PRESS

Great Clarendon Street, Oxford, OX2 6DP,
United Kingdom

Oxford University Press is a department of the University of Oxford.
It furthers the University's objective of excellence in research, scholarship,
and education by publishing worldwide. Oxford is a registered trade mark of
Oxford University Press in the UK and in certain other countries

Published in the United States of America by Oxford University Press
198 Madison Avenue, New York, NY 10016, United States of America

British Library Cataloguing in Publication Data
Data available

Library of Congress Control Number: 2019933999

ISBN 978–0–19–874974–5 (hbk)
ISBN 978–0–19–874975–2 (pbk)

DOI: 10.1093/oso/9780198749745.001.0001

To my family

Acknowledgements

I would like to thank the many friends, colleagues, and family who made this book possible. First to my supervisors Sean Nee, Paul Harvey, and Alfried Vogler, who taught me how to do my own science and set high standards that I have aspired to ever since. A particular thanks to Paul for nudging me into writing this book. I have been fortunate to have had many excellent colleagues during my long residence at the Silwood Park campus who contributed greatly to ideas presented in this book or to its completion in other ways; in particular Austin Burt, Tom Bell, Martin Bidartondo, Lauren Cator, Magda Charalambous, Steve Cook, Mick Crawley, Julia Day, Pat Evans, Rob Ewers, Diego Fontaneto, Richard Gill, Ivana Gudelj, Charles Godfray, Mike Hassell, Vasso Koufopanou, Russ Lande, John Lawton, Simon Leather, Georgina Mace, Claire de Mazancourt, Shorok Mombrikotb, David Orme, Ian Owens, Samraat Pawar, Ally Phillimore, Andy Purvis, Carsten Rahbek, Dan Reuman, Damian Rivett, James Rosindell, Vincent Savolainen, Julia Schroeder, Ibi Wallbank, Chris Wilson, Guy Woodward, and Denis Wright. Not least, many thanks to members of my lab group over the years: Gail Reeves, Jonathan Davies, Joseph Butlin, Emma Barrett, Elisabeth Herniou, Richard Waterman, Jan Schnitzler, Luis Valente, Yael Kisel, Martine Claremont, Tomochika Fujisawa, Isobel Eyres, Diane Lawrence, Francesca Fiegna, Gabriel Perron, Kevin Balbi, Amr Aswad, Alex Lee, Meirion Hopkins, Aelys Humphreys, Alejandra Moreno-Letelier, Cuong Tang, Chris Culbert, Laura Johnson, Thomas Scheuerl, Sina Omosowon, Rowan Schley, Michael Schmutzer, Thomas Smith, Bruce Murphy, Reuben Nowell, Pedro Almeida, Richard Sheppard, and Lily Peck.

The work evolved through many collaborations and I particularly acknowledge Alfried Vogler, Joan Pons, Vincent Savolainen, Mark Chase, Bill Birky Jr., Alan Tunnacliffe, Claudia Ricci, Chiara Boschetti, Diego Fontaneto, Chris Wilson, Tom Bell, Gary Frost, Glenn Gibson, and Gemma Walton. Many others provided advice, feedback, and encouragement during aspects of work found here, including Graham Bell, Jerry Coyne, Trevor Price, Dolph Schluter, Sally Otto, Loren Rieseberg, Richard Cowling, Michael Turelli, Roger Butlin, Lacey Knowles, Peter Linder, Angus Buckling, Mike Brockhurst, and Ziheng Yang. Bill Birky Jr., Austin Burt, Diego Fontaneto, Tomochika Fujisawa, Aelys Humphreys, Meng Lu, Bruce Murphy, Reuben Nowell, Siobhan O'Brien, Harrison Ostridge, Ayush Pathak, Lily Peck, Loren Rieseberg, James Rosindell, Vincent Savolainen, Thomas Scheuerl, Dolph Schluter, Alfried Vogler, Chris Wilson, and Jiqiu Wu provided useful feedback on earlier drafts of chapters.

The book was mainly written 'in and among' my day job in the Department of Life Sciences, Imperial College London. Chapters 3 and 4 were developed in part for the ForBio graduate course on 'species delimitation' at the Norwegian University of Science and Technology (NTNU), Trondheim in October 2016. Chapter 7 was completed while on a short visit to Wissenschaftskolleg zu Berlin in December 2018. Many thanks to these institutions for support, and to Ian Sherman, Bethany Kershaw, and Lucy Nash at OUP for their sensitive handling of the book's gestation, to Julie Musk for copy-editing and to Keerthana Sundaramoorthy for production. Thanks to Diego Fontaneto and Giulio Melone for the use of their amazing scanning electron micrographs of bdelloid rotifers for the cover. A special thanks goes to David Barraclough for indexing the book, and to Jo, Roan and Callum Barraclough for their help with the figures.

Finally, many thanks and much appreciation to my family, Jo, Callum, Roan, my parents David and Christine who encouraged my interest in science from an early age, Melanie, Paul, Alexander, Toby, and Valerie and Colin who supported me in various ways, and my grandparents Harold, Eileen, Harry, and Ida, who I am fortunate to have known into adulthood.

Contents

Introduction

1.1 Why species?

This book is about species. Indeed, every ecology and evolutionary biology book is about species, because all life is classified into units of diversity that we call species. But this book is about the units themselves—what species are, how they form, the consequences of species boundaries and diversity for evolution, and patterns of species accumulation over time. Finding a title was hard because 'species' is used as a catchall term for organisms and life. This is not a book about the whole of evolutionary biology and I was painfully aware that an earlier author had first dibs on a similar title for a more general account of the evolution of life. Underline the word species on the front cover and I hope the aim of the book is clear.

Species are central for understanding the origin and dynamics of biological diversity. Explaining why lineages split into multiple distinct species is one of the main goals of evolutionary biology. Yet, we often take the existence of species and their properties for granted. Precisely what we mean by species and whether they really exist as a property of nature has been widely discussed, but rarely modelled or tested critically with data. Approaches for understanding the origins of diversity differ markedly within species (the realm of microevolution) versus between species (the realm of macroevolution). Does this reflect a true discontinuity in biological processes or simply an artefact of how different scientific fields developed? In turn, genetic and ecological interactions between species should play a dominant role in structuring evolutionary dynamics. Yet, most studies of contemporary evolution focus on single populations, and do not consider explicitly the effects of multiple coexisting species.

The time is ripe to revisit the concept of species and its consequences for how organisms evolve. Description of the diversity of life has been revolutionized by the use of molecular markers and increasingly by whole-genome sequencing. With the power to reconstruct the tree (or web) of life for all organisms, do we still need species? Maybe it would be better to abandon them altogether and portray diversity as a branching (sometimes fusing) hierarchy?

The central thesis of this book is that species represent more than just a unit of taxonomy; they are a model of how diversity is structured and how groups of related organisms evolve. The 'species hypothesis' is that natural processes act in a way that generates units of diversity, called species, which then determine evolutionary dynamics: organisms interact in a qualitatively different way within species than between

The Evolutionary Biology of Species. Timothy G. Barraclough, Oxford University Press (2019).
© Timothy G. Barraclough 2019. DOI: 10.1093/oso/9780198749745.001.0001

species. In theory, diversity could be structured in other ways; for example, organisms might interact with each other via reproduction, sharing of genes, competition, and so on in ways that just decline gradually with increasing evolutionary divergence. What is exciting is that tools are now available to test alternative models for the structure of diversity and to estimate the role of alternative processes such as selection and gene flow. Species are no longer the focus of philosophical debate, but they represent a theory amenable to empirical tests and estimation. The answer is important both for understanding where diversity comes from and for predicting contemporary evolution.

1.2 The evolutionary dynamics of species

The scope of the book can be summarized as follows. Try to visualize all life on earth tracing back over time to its origins and forward into a hazy future (Fig. 1.1). Myriad strands are visible that constitute lineages of genes, which come together in individual organisms visible as dots on the plane of the present. In some parts of life, these strands shuffle each generation through sexual reproduction; in others they associate over longer periods through clonal inheritance and only rarely transfer genes. Zooming out a little, dots of contemporary individuals do not form a starry sky on the plane of the present, but group in clusters that share similar genetic composition and biological characteristics (Fig. 1.1). Tracking backwards and forwards in time, it can be seen that the strands of gene lineages group within these clusters, whereas there are few exchanges between them. These clusters are our hypothetical species units. They are evolutionarily independent, because interactions such as gene exchange and competition are stronger within species, and weaker, rare, or absent between them. We draw ellipses round them to highlight our hypothesis that this is the structure of diversity.

What then describes the evolutionary dynamics of these species entities? And what processes control those dynamics and cause them to vary among different types of organisms? Species originate by speciation. Tubes formed by the time-integration of species circles occasionally split and diverge into two separate species (Fig. 1.1). Biological attributes of individuals and environmental conditions around these divergence events reveal the causes of speciation, such as geographical isolation and the availability of new ecological niches. In turn, species are lost through extinction, when the final individual from a species dies with no descendants.

Speciation operates over long timescales and most of our understanding comes from retrospective studies. The snapshot of time since pioneering naturalists of the 1700s and 1800s until the present has seen few species origins across the tree of multi-cellular life (although plenty of extinctions). But over these contemporary timescales (Fig. 1.1), species continue to evolve as they encounter new conditions and shift their geographical locations. A great deal is known about the genetic and environmental determinants of evolution within species and populations. Often these accounts focus on a single species at a time. Yet, species can still interact with each other either ecologically or by occasional exchange of genes through hybridization or other

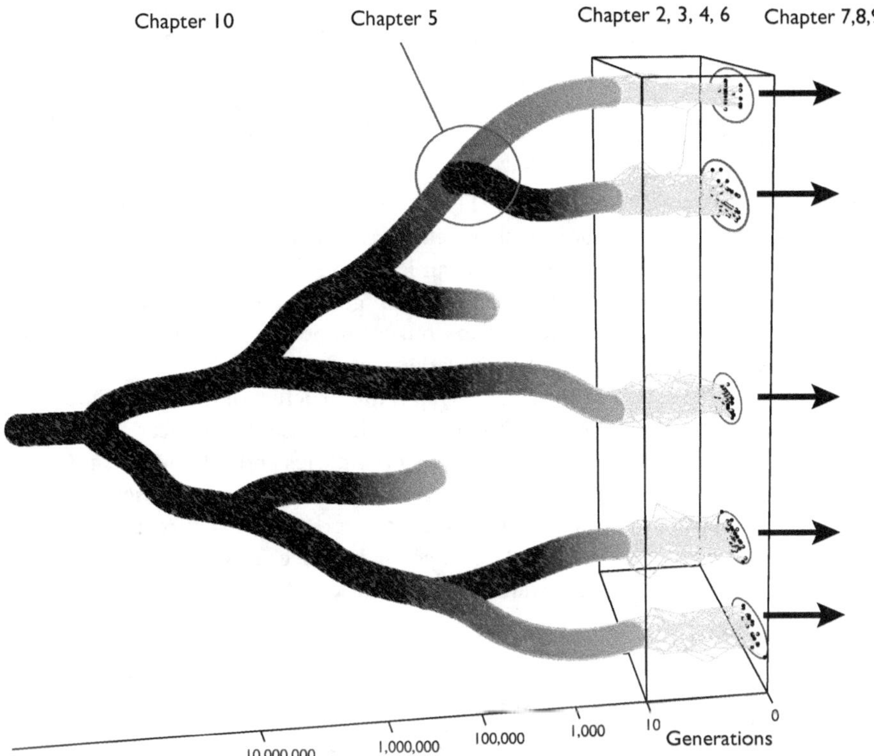

Fig. 1.1 An illustration of the evolutionary dynamics of species. Individual organisms in the present are shown as black dots on a plane representing genetic and phenotypic variation. Shared ancestry is represented by grey lines showing parent–offspring links such that vertices represent individuals during previous generations. Individuals are grouped in genetic clusters that exchange genes but with limited gene exchange between them. These are hypothetical species units denoted by ellipses. Zooming out to longer timescales, tubes representing these species units originate by speciation and are lost by extinction. Species units are important to understand not only the origin of diversity patterns in the wider clade but also how the clade will respond to future changes (represented by arrows).

mechanisms of gene transfer. The degree of ecological similarity or strength of reproductive barriers between species—which depend on the history and forces behind their initial divergence and how long they have been in contact or isolated from each other—will affect how sets of species evolve in changing conditions. The nature of species and their interactions is therefore a vital component for predicting contemporary and future evolution.

As an extra layer of complexity, contemporary evolution in turn feeds back to affect the likelihood of both speciation—that is, whether divergence occurs and the resulting new species survive long enough to leave a new 'tube' visible on the species tree—and extinction. Full specification of the dynamics of speciation, evolution of those species, and extinction requires these feedbacks to be included, which is challenging because of the wide span of timescales.

These processes together shape the large-scale dynamics of branching and expansion across the whole tree of life. Zooming out to a distance at which species themselves appear as points, whole groups of species rise and fall via chance, new innovations, or as conditions change (Fig. 1.2). We normally view this through the murky glass of the fossil record or phylogenetic reconstruction, but from our idealized viewpoint we can see the detailed dynamics of species origination and extinction as branching patterns on the tree. The pattern of growth and decline in numbers of species is not independent between groups, but can depend on ecological interactions between species making up those major clades.

This describes the scope of this book. What is the structure of diversity? Does it fall into species units and how do we delimit them? What causes diversification into multiple species? How does the nature of species boundaries and interactions influence ongoing evolution in diverse assemblages? And how do all these processes shape biodiversity? The focus is intentionally broad. Much work on the nature and origins of species has focused in detail on genetic and ecological mechanisms behind the evolution of reproductive isolation. My aim is to step back and present an overview of the

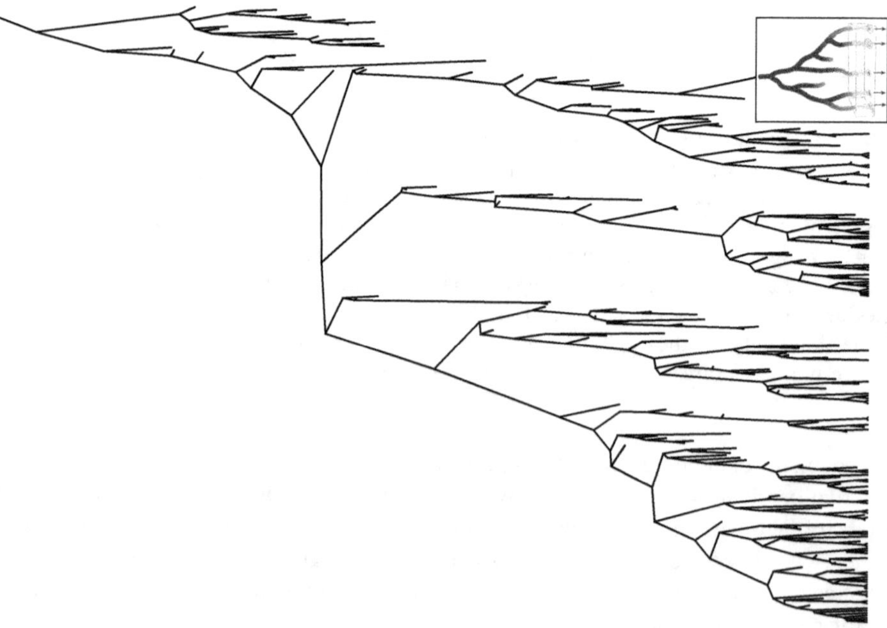

Fig. 1.2 Proliferation of species lineages over long timescales. Zooming out on the clade represented in Fig. 1.1, indicated by the rectangular inset, reveals the growth and decline of species richness through differential levels of speciation and extinction across clades.

evolutionary biology of species, incorporating their nature, origins, proliferation, and consequences. I have attempted to go back to basics. What is the pattern of nature that we are trying to explain? What are the potential processes that explain this pattern? And what are the consequences for other evolutionary phenomena?

1.3 Structure of the book

The book chapters follow the order described in section 1.2. Chapters 2 to 4 look at what species are and how to delimit them. Chapter 2 first considers the forces that cause lineages to diversify into multiple distinct and independently evolving groups and presents definitions of key concepts. I discuss whether forces of diversification act to generate discrete units, that is, species, rather than alternative diversity patterns such as a continuum of forms. The chapter aims to develop concrete theory that makes testable predictions for distinguishing alternative models of the structure of diversity.

The next two chapters test these ideas by considering empirical evidence for the existence of species versus alternative hypotheses for the structure of diversity. Focusing first on evidence for genetic and phenotypic clustering, chapter 3 outlines the theory and practice of species delimitation. Is there statistical evidence that discrete species are real and constitute the major unit of diversity in many taxa? Chapter 4 then describes methods for delimiting species based on the action of reproductive isolation and divergent selection. Prospects for using whole-genome data for interrogating diversity patterns and processes across whole clades—only recently feasible for microbes, and not yet easy for eukaryotes at such scales, but rapidly approaching—are discussed.

Chapter 5 shifts the focus to consider what causes a single species to split into multiple descendants, namely the process of speciation. Understanding speciation requires knowledge of when and why scenarios arise that promote splitting of a previously cohesive species, as well as genetic mechanisms operating once such conditions arise. I review evidence for causes of speciation from analysis of speciation patterns across clades. Dispersal and gene flow are identified as key parameters explaining speciation rates in different organisms. I ask whether speciation rates and patterns depend mainly on ecological opportunity or on intrinsic genetic properties of organisms.

Chapter 6 concludes the discussion of the nature and origins of species with an indepth look at the consequences of sex, recombination, and alternative lifestyles for species and speciation. Many authors argue that species are only found in sexual organisms and define species by reference to recombination. Others have hesitated over the reality of species in microbes, because they do not reproduce sexually (although recombination by other means is common). I evaluate the theory and evidence for the importance of recombination in generating diversity patterns, by comparing sexual and asexual clades, and microbes with alternative modes of reproduction.

The next three chapters investigate the consequences of species for contemporary evolution. Chapter 7 explores the effects of different types of species boundaries on how organisms evolve in new environments. Many studies of contemporary evolution assume that evolution can be predicted from understanding selection and genetics on a species-by-species basis. I describe examples where this assumption does not apply. For example, will a gene for antibiotic resistance spread across species boundaries? That depends in part on genetic barriers to exchange and in part on the ecological consequences of transferring a trait that affects competitive interactions.

Continuing this line of reasoning, chapter 8 explores the effects of ecological interactions among species on evolution. Species diversity evolves because lineages diversify to use distinct resources and habitats. The standing diversity of traits and resource use will therefore have a great impact on how each species evolves when faced with a change in the environment. I present simple models showing how interactions affect evolution and discuss results from experiments evolving communities of co-occurring microbial species. Evolutionary dynamics are greatly altered by the presence of multiple co-occurring species in ways that will depend on the forces behind the origin and coexistence of those species.

Spurred on by theories and results in the previous chapters, chapter 9 outlines challenges and possible solutions for predicting evolutionary dynamics of whole communities in the wild. The advantages of adopting a synthetic approach are illustrated through discussion of real-world cases, including managing gut bacteria for human health and ecosystem responses to climate change. Research questions and broad approaches are outlined to extend current work to whole communities of microbes and longer-lived eukaryotes. I argue that understanding and prediction of evolutionary dynamics will only be possible by considering whole systems of interacting species. A key challenge is to track evolution over intermediate timescales of around 100–10,000 generations that are too long to follow observationally for long-lived organisms, but too short to be resolved by fossil and phylogenetic approaches.

Chapter 10 expands to consider how the above processes influence species diversity over long timescales and broad spatial scales. Diversity patterns result in part from speciation and in part from dispersal, evolution, and extinction. Classical studies looked for traits that speed up rates of diversification. Although these studies found interesting patterns, ecological opportunity and limits seem to be more important in shaping diversity than fast diversification per se. The effect of a given trait depends on the environment. I conclude the chapter by discussing how patterns of selection and isolation shape higher-level diversity patterns. The same processes that shape genetic variation within species also shape diversity patterns above the level of species, but playing out over longer timescales across sets of interacting species.

Chapter 11 aims to synthesize these theories of species origins and consequences. An approach that incorporates the nature of species, the forces behind their origin and coexistence, and their genetic and ecological interactions is essential to tackle these questions. We are now in a position to embrace the complexity of the diversity of life.

2

What are species?

2.1 Introduction

Species represent a fundamental unit of biological diversity—so much so that we often take them for granted. Many ideas behind our understanding of species developed in a former age. Those ideas were partly shaped by the pioneers' specializations on particular organisms such as insects or birds (Dobzhansky, 1941; Mayr, 1963). Prokaryotes were initially left behind in the development of species concepts and are only recently being incorporated into the same evolutionary framework (Cohan, 2001; Barraclough et al., 2012; Shapiro and Polz, 2014). Alpha taxonomists working in museums and herbaria performed the practical business of delimiting species—according to formal rules for selecting characters and nomenclature, but not always in close contact with evolutionary concepts of the units being delimited or following conventional procedures of biological and statistical inference.

The time is ripe for a critical re-evaluation of our understanding of species. A technological revolution has opened up vast sources of new data (Ellegren, 2014; Seehausen et al., 2014) and increased capacity to model diversity patterns and evolution (Morlon, 2014). These data are allowing long-standing theories to be tested rigorously for the first time and stimulating new ideas. Yet many theories remain the same—perhaps some of those theories make simplifications that were necessary in data-deficient times but no longer. This chapter looks at ideas underpinning species, but ideas that can be formalized and tested with empirical data.

2.2 Definitions

I define a species as an independently evolving group of organisms that is genetically and phenotypically distinct from other such groups. Derived from an evolutionary species concept (Simpson, 1961), this broad definition encapsulates both the pattern of species—genetic clusters of variation separated by gaps—and the causes of that pattern (Hey, 2001). There are two sides to the coin with respect to the causes: cohesive processes limit variability within species, while independent evolution leads to divergence between species. More precise definitions will apply to particular organisms, but the broad definition provides the flexibility to compare patterns of diversification among different organisms. Note that the definition refers to shared evolutionary

The Evolutionary Biology of Species. Timothy G. Barraclough, Oxford University Press (2019).
© Timothy G. Barraclough 2019. DOI: 10.1093/oso/9780198749745.001.0001

fate as well as shared evolutionary history (Templeton, 1989): the way that a wider clade will evolve in response to changing environments depends on the pattern of independent evolution among the different sets of organisms it contains.

Independent evolution in turn is defined by the following condition: new mutations can spread within one set of organisms either by drift or by selection, but cannot generally spread to replace copies of the same gene in another set (Fisher, 1930). Many processes can lead to cohesion and interdependence of the evolution of sets of organisms, including interbreeding and selection pressures in a shared ecological niche. Indeed, ultimately the fate of all organisms on the planet is interdependent because of finite resources: in the extreme, if one primary producer attains a future abundance equal to the entire primary production on earth, other primary producers will not be able to do so. Independence is therefore a quantity that can vary in degree among different parts of the genome and different sets of organisms. A primary goal of the opening chapters is to ask whether processes leading to independence do tend to generate discrete, qualitative units of the kind envisaged as species, or whether a more quantitative description of patterns of independent evolution is required.

How do we distinguish between species and other units of diversity, such as populations? I define separate populations within a species as partially isolated from one another, usually geographically, but with ongoing gene flow maintaining cohesion across the species. They typically diverge in allele frequencies but do not accumulate fixed genetic differences (Hey, 1991). Some groupings referred to as populations might be completely isolated species by the definitions used here, but for insufficient time to accumulate diagnosable signatures that indicate independent evolution. In addition, some forces of cohesion and independent versus correlated evolution can occur at levels above species (see chapters 6 and 10). Species therefore refer to the most resolved unit of diversity demonstrating independent evolution from other sets of individuals. Of course, there are many caveats and complications to the definition—these will be addressed throughout the book. Box 2.1 provides a glossary of key terms.

Box 2.1 Glossary of key terms used in the book.

Species. An independently evolving group of organisms that is genetically and phenotypically distinct from other such groups. Derived from Simpson (1961) and Hey (2001).

Independent evolution. Defined by the condition that a new mutation can spread within a species, but it does not spread to replace copies of that gene in other species (Fisher 1930).

Cohesion. The reduction of variability within species due to processes such as gene flow, interbreeding, and shared selection pressures, which create evolutionary interdependence among a set of organisms (Templeton, 1989).

Population. Groups of individuals within a species that are partially isolated from one another, usually by geographical isolation, but with ongoing gene flow maintaining cohesion across the species.

Effective population size. The number of individuals in an idealized population under random genetic drift that would have the same levels of genetic variation and other metrics as the observed population, thus taking into account factors such as the sex ratio and variation in number of offspring per individual.

Reproductive isolation. The lack of interbreeding between two populations or species of sexual organisms (Dobzhansky, 1941; Coyne and Orr, 2004). Reproductive isolating barriers can either be pre-zygotic (e.g. mate preferences, pollen tube growth through the stigma) or post-zygotic (e.g. hybrid inviability, sterility, or ecological maladaptation).

Recombination. In population genetics, recombination refers to the bringing together of DNA from different organisms into a single organism. In sexual eukaryotes, it occurs by segregation and crossing-over. In bacteria, it occurs by a variety of mechanisms, including homologous recombination and the transfer of plasmids or viruses.

Geographical isolation. Populations or species are in different geographical regions with restricted dispersal or gene flow between them (Mayr, 1963). Geographical isolation prevents interbreeding between populations, but is typically excluded from reproductive isolation sensu stricto, which refers to cases when organisms come into contact.

Divergent selection. Selection for different trait values in different populations or species (Schluter, 2001).

Ecological speciation. Speciation driven by adaptation to different environments or ecological niches (Schluter, 2001; Nosil, 2012).

Ecological drift. Changes in the relative abundance of two or more ecologically equivalent species through random variation in birth and death rates (Hubbell, 2001).

Alpha taxonomy. The business of assigning species limits and names, in contrast to beta taxonomy that allocates species to higher groupings such as genera or families.

2.3 The shape of biological diversity: the conventional species model

Consider a sample of individuals taken from any large clade of organisms. Let's say all of the bird skin collection in the Natural History Museum in London: 750,000 specimens representing 95 per cent of bird species. Assume that species memberships of every specimen are well known, and that species were defined conceptually based on a biological species concept of reproductive isolation (Box 2.1) and delimited using diagnostic morphology, song, and other traits. Now we compile all conceivable measurements of these individuals, including morphological traits, ecological traits such as diet, and full genome sequences. We could imagine having behavioural data as well, although these are harder to collect for so many specimens. What is the expected structure of variation in the measured traits?

Among species, variation is expected to be hierarchical and nested according to a branching phylogenetic tree (Fig. 2.1A; Darwin, 1859; Hennig, 1966; Mindell, 2013). Species do not exchange genes and hence the same phylogenetic tree structures variation of all traits, including morphology, the presence and absence of genes, and genetic divergence. Exceptions occur for some traits due to convergent evolution of species adapting to similar niches.

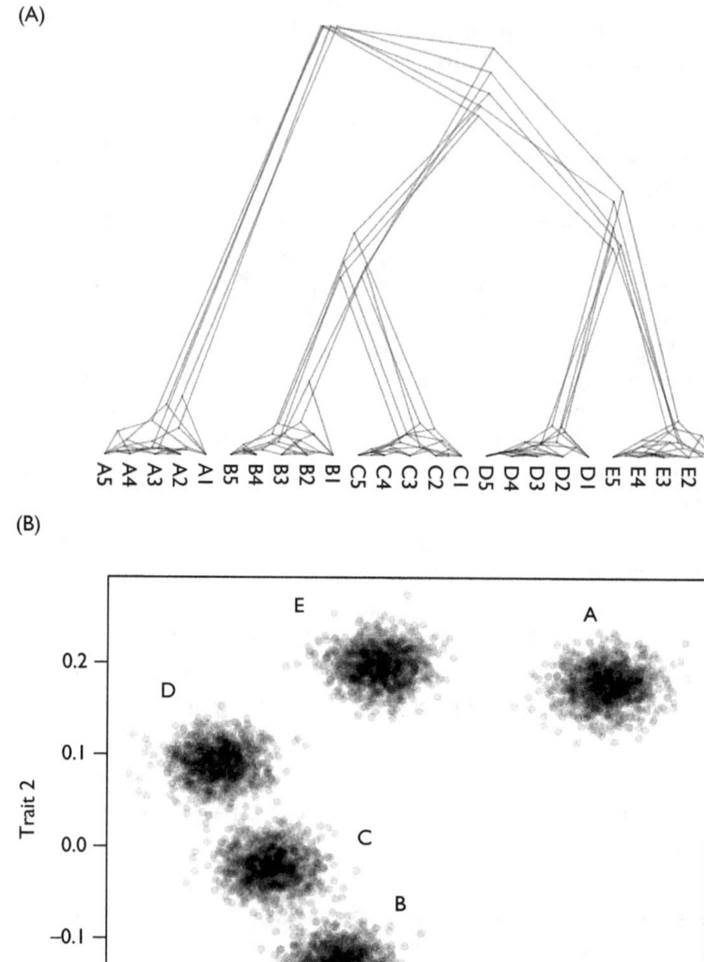

Fig. 2.1 Idealized pattern of variation expected in a hypothetical clade of five species, A to E, according to a simple, conventional species model. (A) Genetic variation across five individuals per species at multiple loci is mostly hierarchical and congruent between species, whereas each locus displays a different ancestry within species. Variation within species is lower than between species, apparent as genetic clusters with short branches compared to divergence between species. (B) Variation in sexually monomorphic phenotypic traits across 1000 individuals per species is also hierarchically distributed among species: more closely related species tend to display more similar traits. Phenotypic variation is unimodal within species, and exhibits distinct clusters for each species. Gene trees within a species tree and trait data were simulated using Phybase (Liu and Yu, 2010) and ape (Paradis et al., 2004) packages in R.

A different pattern of variation is expected within species. Random mating and recombination causes each gene to have a different pattern of ancestry (i.e. genealogy; Fig. 2.1A). Quantitative traits are therefore expected to have a unimodal distribution because recombination shuffles gene combinations across loci contributing to the trait (Fig. 2.1B; Lande, 1979). Multiple traits might have correlated variation if there are genetic or mechanistic links between them. Some traits might have multimodal distributions: sexually dimorphic traits are bimodal and there could be multiple phenotypic morphs (perhaps not in birds but, for example, in social insect castes). There might also be further structure or clines associated with geographical sub-species or populations. Nonetheless, trait variation should be non-hierarchical as long as there is sufficient gene flow to maintain evolutionary interdependence across the whole species. Hence, there will be a transition in the structure of variation at the species boundary.

The species boundary is also apparent from the amount of variation. Divergence between species accumulates over time: for example, linearly with time at silent coding sites (Zuckerkandl and Pauling, 1962) and with the square root of time for neutral, continuous traits (Felsenstein, 1985). The divergence of traits under selection depends on how selection pressures and responses to them vary among species—some traits could be under uniform selection in multiple species and remain constant, whereas others could diverge rapidly as species adapt to different environments.

In contrast, variation within species is limited by the cohesive processes of drift and purifying or directional selection to low levels of variation among individuals—even if the species has been isolated for a long period of time, variation does not increase indefinitely, but reaches an average expected level (albeit with wide variance in that expectation) because of turnover of individuals within populations (Hudson, 1991; Rosenberg and Nordborg, 2002).

If species have evolved independently from one another for long enough, the result is genetic clusters of individuals separated by gaps from other such clusters (Barraclough et al., 2003; Pons et al., 2006; Fontaneto et al., 2007; Fujisawa and Barraclough, 2013). This pattern is the consequence of evolutionary cohesion within species and evolutionary independence between them. It is expected to arise within the order of $8Ne$ generations for one diploid, neutral locus, where Ne is the effective population size (Hudson and Coyne, 2002). It will arise more quickly, and therefore be most apparent for traits under divergent selection between species, such as song or plumage involved in reproductive isolation or ecological traits such as beak shape involved in adaptation to distinct niches.

The transition from between- to within-species branching will be blurred of course. Within some recognized species, there is sub-specific variation such as geographical trends in trait values and partially isolated sub-populations, and some recently diverged species are distinguishable only from a change in their song and identical for other traits. However, many other species will be clearly distinct and unambiguously separated from one another for millions of years.

Species represent the fundamental unit for shaping these predictions. The birds might be classified into genera, families, and other higher taxa that provide predictive information about particular trait values (Holt and Jonsson, 2014). But these are labels layered onto the hierarchical variation above species—they do not correspond to theoretical units that alter diversity pattern fundamentally. In contrast, species define a conceptual transition: from the macroevolutionary processes of phylogeny to the microevolutionary processes of population genetics.

The exercise of real interest here is the reverse of this one. Instead of assuming that species are present and outlining expected diversity patterns, we want to test for the existence of species and assign individuals to species based on observed patterns. A phenotypic survey like this across 58 sympatric and parapatric species-pairs of birds found that temperate species were reliably separated in accordance with existing alpha taxonomy. However, the use of criteria based on quantitative trait data led to major changes in species limits among birds from poorly known tropical regions (Tobias et al., 2010). Even in birds, perhaps the best-known clade, we are still in the process of refining our understanding of diversity units.

More generally, given a set of genetic and phenotypic trait data, what is the most likely true model of diversity? Does it comprise independently evolving and discrete species or some other model of diversification? What alternative models of diversity might there be? One possibility is that gene exchange also occurs between species so that trait distributions become partially shuffled above the species level as well. Would shuffled sets of divergent traits generally be able to work as a functional whole? Alternatively, perhaps selection pressures are strong enough and acting independently on different traits so that each lineage evolves to an optimum for each separate trait independently of its optima for other traits. By corollary, could functional integration and patterning of selection pressures be sufficient to give the appearance of hierarchical variation? How much of hierarchical trait variation is due to evolutionary isolation per se versus the pattern of variation in selection pressures across organisms (and the way organisms are able to respond to them)?

More practically, being able to observe hierarchical variation assumes that internode intervals are relatively long. If all the diversity originated rapidly (for example, an initial adaptive radiation) with subsequent change in each lineage, the pattern of trait diversity would not depend on the timing of isolation events (which would be contemporaneous) but on the pattern of selection pressures and responses occurring in each lineage over time. Finally, variation within species could depart from an idealized random mating population for various reasons. For example, clonal reproduction (again unlikely in birds) leads to a hierarchical pattern of variation right to the tips and the closest related individuals because all genes share the same pattern of ancestry due to a lack of recombination (Barraclough et al., 2003). A hierarchical pattern of mating based on geographical location could also generate hierarchical evolutionary isolation right to the tips. Think of your own alternative model of diversity. Has anyone tested the conventional 'species' model of diversity against these alternatives?

2.4 Multiple properties of species

Section 2.3 considered a scenario assuming that a single coherent unit called species exists. But there are many properties of species and potential mechanisms for evolutionary independence that can impact on predicted patterns in subtly different ways. In order to consider alternative diversity models, we must first consider the following components, before describing the processes that influence each component:

(1) *Species are phenotypically and genetically distinct and internally coherent.* There is some variation within species, both within and among geographical sub-populations. However, evolutionary interdependence leads to coherence within species that creates a pattern of discrete clusters with gaps. In theory at least, as outlined in section 2.3, this is not simply the result of a hierarchical continuum, but apparent as a discrete shift in the pattern of variation. This is a fundamental description of the pattern of diversity corresponding to species—for shorthand I refer to it as the pattern of species—but not itself a mechanism to explain their existence.

(2) *In sexual organisms, species are reproductively isolated.* Evolutionary independence between sexual species requires restriction to gene flow: full random mating and recombination rapidly erodes species differences (Fig. 2.2). For species living in the same place, evolutionary independence requires behavioural or genetic barriers to the production or survival of hybrid offspring. Species living in different areas often have automatically restricted gene flow because they do not meet. These concepts are not restricted to sexual eukaryotes; equivalent definitions apply in prokaryotes based on arenas for recombination. For example, mechanisms of homologous recombination or plasmid transfer might only function within a particular clade and not between clades (see chapters 6 and 7).

(3) *Species are ecological units.* The main unit for ecological models of diversity is the species. There can be ecological variation within species, such as ecotypes of plants (Hufford and Mazer, 2003), adaptive polymorphism (Skulason and Smith, 1995), and ecological dimorphism between sexes (Slatkin, 1984). Also, related species (for example, within a genus) can be ecologically similar and found in different areas. Yet, most (but not all—see section 2.5) ecological models consider multiple co-occurring species within an area to be ecologically distinct from one another. This is partly because evolutionary independence allows species to accrue ecological differences. It also reflects the role of divergent selection in driving population divergence and speciation (Sobel et al., 2010). The criterion of ecological coherence versus distinctiveness has particular prominence in definitions of species in clonal prokaryotes and eukaryotes, developed further in section 2.6).

(4) *Species exist in a geographic setting that is a key parameter for their divergence.* Geographical isolation promotes divergence of species (Allmon, 1992), by preventing homogenizing forces from operating even in the absence of reproductive isolating mechanisms. Geographical and non-geographical barriers to gene flow can lead to similar patterns of diversity, as represented by patterns of trait variation among individuals. However, species patterns could be strengthened in geographical contact, through selection against hybridization (known as reinforcement) or character displacement and niche divergence.

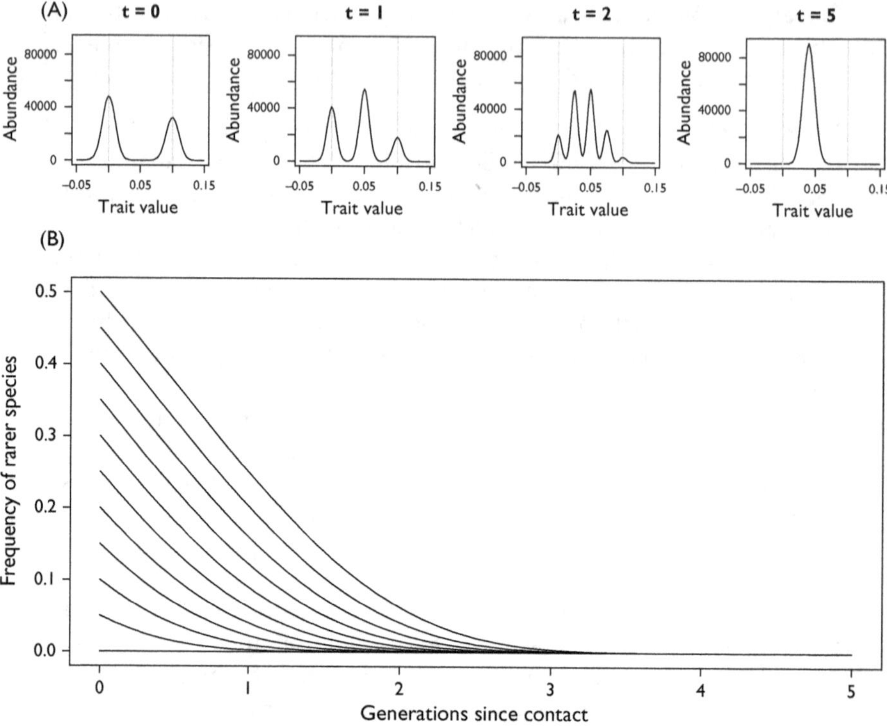

Fig. 2.2 Diversity is lost rapidly with random interbreeding and no divergent selection. Two initially genetically distinct and isolated populations that formed in geographical isolation establish contact at time $t = 0$ but lack reproductive isolation. (A) Assuming random mating with discrete generations and no selection, the bimodal distribution of a quantitative trait at the start is eroded to a unimodal distribution after five generations via a series of intermediate stages, with multiple modes reflecting populations of F1 and F2 individuals, and so on. Vertical grey lines show mean trait value of initial populations. (B) Frequency of pure-bred individuals of rarer species relative to the total number of individuals of both species declines deterministically to less than 1 in 1000 after four generations irrespective of starting frequency, p. Lines are shown for different starting frequencies in intervals of 0.05. Frequency is p^2 after one generation, p^4 after two, and therefore $(p^2)^t$ after t generations.

Together, these properties of species define and reflect evolutionary dependence and independence. A new mutation arising in one species can spread and replace alleles at that locus in its own species, but not in other species (Fisher, 1930). That mutation could be neutral, in which case the limit on its spread is due to drift and barriers to recombination, or it could be positively selected, in which case the limit could also reflect a different selective optimum in different species. This broad definition applies to asexual organisms as well: in them, evolution occurs by spread of whole genotypes and extinction of other genotypes. The asexual clade could still be structured, however, such that a genotype is able to spread and replace the descendants of only a subset of the clade (Cohan, 2001;

Barraclough et al., 2003). This could arise through geographical isolation or ecological divergence—the same mutations are not beneficial in all sub-clades.

Reproductive isolation has become the primary criterion for defining species in sexual organisms. It is so fundamental that the process of speciation has become virtually synonymous with the evolution of reproductive isolation. There are good reasons for this view (Coyne and Orr, 2004). However, there has been confusion when considering organisms that do not reproduce sexually. What happens if we step back momentarily from the definition of species by reproductive isolation and instead focus on the pattern of species defined in property (1) above—that is, phenotypic and genetic cohesiveness versus distinctiveness (Barraclough, 2010)? What processes are important for generating this pattern of diversity? Observing a pattern of species requires more than conditions that favour divergence of populations—it also requires that the balance of formation and persistence of species promotes a standing-level of diversity (Barraclough and Herniou, 2003).

2.5 Separating out the effects of alternative mechanisms

Having considered multiple aspects of species and parameters affecting their formation, now imagine a series of models where these different properties and parameters of species operate separately. What would the structure of diversity be? The following considers the effects of each mechanism alone before building up models with multiple mechanisms. Simulations of the population model described in Box 2.2 are used to illustrate alternatives (Fig. 2.3).

2.5.1 No diversifying processes

Assume that within a wider clade, all individuals belong to a single interacting population with no internal structure or units of diversification at all. The pattern of diversity for measured traits would depend on the level of recombination. A strictly asexual population would display hierarchical variation indicative of a single coalescent process with chance gaps (Higgs and Derrida, 1992; chapter 3). This reflects turnover of individuals through drift and selection (only uniform selection can operate in the present scenario, but it would speed up turnover of individuals). A sexual population would display a hierarchical pattern of variation for each gene, but variation would average across genes to generate a unimodal distribution for multilocus traits (Barraclough et al., 2003). The total amount of variation in the clade would depend on the processes affecting turnover: random birth and death of individuals and the strength of selection operating on the population. This scenario is hard to imagine for a real clade of any size, as its members would necessarily encounter different conditions and find themselves in different geographical areas. However, there might be an effectively similar scenario where conditions for isolation and divergent selection come and go too quickly to generate consistently isolated and discrete species, and instead allele frequencies constantly fluctuate in different parts of the wider population. For example, diversity patterns in some bacteria and viruses could reflect ongoing dynamics of local epidemic radiation and decline from a single, interacting pool.

Box 2.2 A simple model of diversification.

I use a simplified but flexible model to illustrate diversification for multiple reproductive modes and under different conditions (section 2.5, Fig. 2.3). The model is based on the Wright–Fisher model, which assumes discrete generations, finite population size N, and random sampling to produce subsequent generations (Ewens, 1979). The simplest version has two continuous phenotypic traits: a reproductive trait that determines which individuals can mate and an ecological trait that determines survival in a given environment. Each trait is coded by m additive loci, and mutations occur at rate μ with normally distributed effect sizes on the trait value (i.e. an infinite alleles model). There is additional normally distributed, environmental phenotypic variation. The following sequence of events occurs at each generation.

(1) **Reproduction**. Each female produces f offspring. In sexual populations, mates for each female are chosen at random, with replacement either separately for each offspring (i.e. polygamy) or a single mate for each female (i.e. monogamy). Sexuals can be selfing or non-selfing hermaphrodites, or dioecious. Mate choice occurs either at random or more readily between individuals with more similar reproductive trait values: the probability of two individuals mating declines as a Gaussian function of the difference in trait values. The trait could be thought of as body size or flowering time, for example. Offspring genotypes are determined by Mendelian inheritance with mutation. Alternative modes of reproduction suitable for prokaryotes are considered in chapter 6.

(2) **Dispersal**. The model can be subdivided into two or more geographical regions. Each offspring has a probability d of entering a pool of dispersers. The dispersers then colonize geographic regions at random to fill spaces left by other dispersers.

(3) **Survival**. From the pool of Nf offspring, N survive to form the next generation. This occurs either at random (i.e. pure drift) or with probability defined as a declining Gaussian function as the ecological phenotypic trait of an individual is further from the optimum phenotype in that environment.

Diversification arises through multiple processes that can be manipulated separately. Divergent selection is specified by multiple phenotypic optima, either in different regions or in the same place. Reproductive isolation results from divergence in reproductive traits, which can be correlated with the phenotypic trait under selection or uncorrelated, or there can be no reproductive isolation if individuals mate at random. Reproductively isolated sets of species can be ecologically equivalent—in which case, survival is determined by a combined carrying capacity irrespective of species—or ecologically distinct with separate limits to the population size of each species. Code to simulate the model in the R statistical programming language (R Core Team, 2018) is provided with the web materials for the book.

2.5.2 Just reproductive isolation

Now imagine the same model but there are organismal traits that determine who mates with whom. Initially, consider that no diversifying forces are at play on those traits, simply variation that accumulates within the initially panmictic population (Fig. 2.3A).

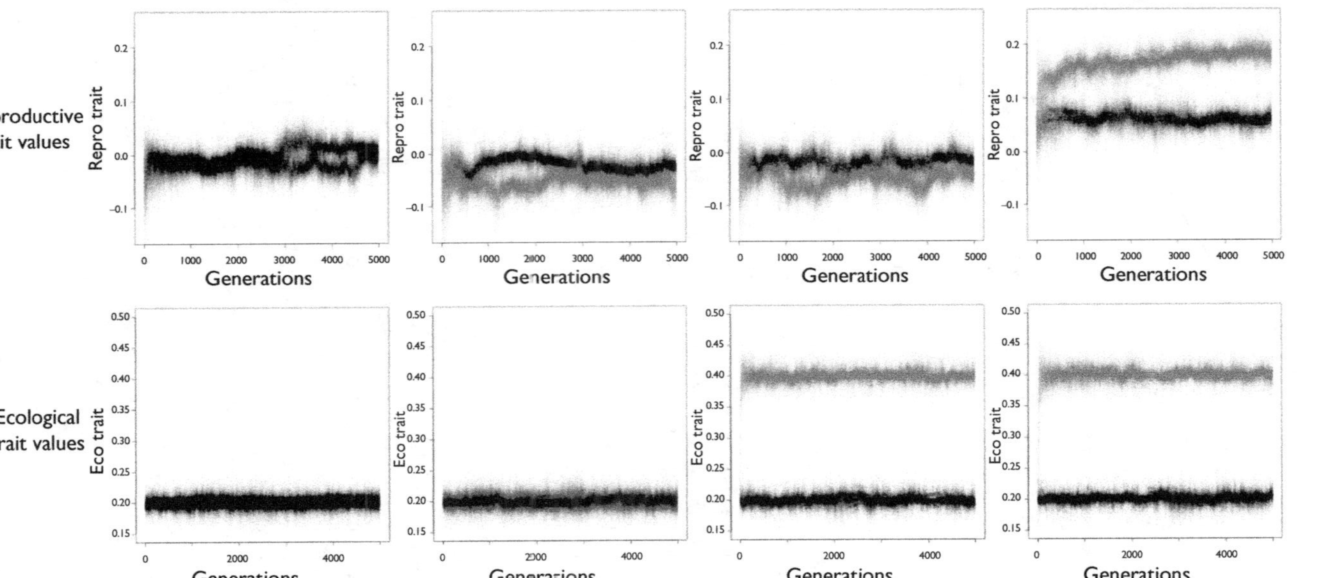

Fig. 2.3 Patterns of diversification in a reproductive (top row) and ecological (bottom row) trait in four scenarios building up diversifying mechanisms simulated using the model described in box 2.2. Density of shading indicates the frequency of individuals with that phenotype. (A) Assortative mating based on similarity of a continuous phenotypic trait in a single geographical region with uniform selection on the ecological trait. Reproductively isolated groups sometimes emerge but are short-lived and lost through ecological drift within relatively few generations. (B) Individuals are now split into two geographical regions (coloured black and grey, respectively) with no dispersal at time zero. The optimum phenotype for the ecological trait is the same in both regions (0.2), that is, under uniform selection, and so ecological trait distributions overlap. Reproductive trait values diverge slowly because of the advantage of having a reproductive trait value close to the population mean. (C) Each region now has a different optimum phenotype for the ecological trait (0.2 and 0.4, respectively), and consequently the ecological trait diverges rapidly but the reproductive trait remains relatively uniform between populations. (D) Ecological and reproductive traits are genetically correlated (50 percent of their 10 loci in common) and so now reproductive traits diverge rapidly. Total number of individuals was 1000, with 500 in each region, and simulations ran for 5000 generations. Simulation code and other parameter values are available from the website accompanying the book.

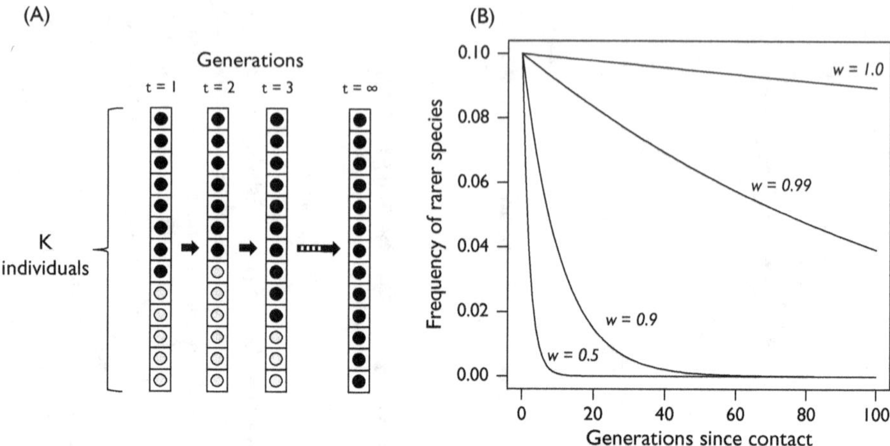

Fig. 2.4 Loss of diversity through ecological replacement is slower than by random inter-breeding. Two geographically isolated populations come into contact and are reproductively isolated but use the same ecological resources. (A) The sum of the number of individuals of both species, $N_1 + N_2$, is therefore jointly limited at a fixed carrying capacity, K, and offspring that survive to the next generation are chosen at random irrespective of which species they belong to (black versus light grey circles). (B) If the two species are equivalent competitors, and relative fitness of the rarer species $w = 1.0$, then average frequency of the rarer species declines slowly (over the order of K generations but with wider variation among trials). Diversity is lost more rapidly with asymmetric competition. For example, when the rarer species at the start has relative fitness $w = 0.5$, it goes extinct on average within 10 generations. Results are shown for $K = 1000$ and $N_1 = 100$, $N_2 = 900$ at the start.

Higgs and Derrida (1992) and Fraser et al. (2007) modelled a scenario where sufficient trait divergence evolves by chance to create independently evolving units. In bacteria, it might arise from the exponential decline in the probability of homologous recombination as sequences diverge. In sexual organisms, such a process might be facilitated by facultative selfers or asexuals that can reproduce from low numbers even without the presence of compatible mates initially, and by large mutation events such as chromosomal and ploidy changes that generate instantaneous reproductive isolation from progenitors. Other potential drivers could be the origin of arbitrary mate preferences (even non-genetic) that create assortatively mating sub-populations (Gavrilets, 2004).

Once such reproductively isolated units arise, in principle they can diverge neutrally over time and accumulate genetic variation. However, there are two difficulties. First, if initially there is mating across the entire group, individuals that acquire assortative mating within a sub-group would be at a disadvantage as their number of potential mates is reduced. Second, if there were no ecological differences among species, ecological drift will still occur (box 2.1)—as modelled in ecological neutral theories like Hubbell's unified theory of biodiversity—and this will place an upper limit on the amount of genetic and phenotypic variation that can arise. The standing level of richness would be a balance between rates of formation and rates of stochastic extinction. Indeed, by this mechanism alone, a pattern of distinct genetic and phenotypic clusters would not arise for any traits except those involved in mate choice (Fig. 2.3A; Barraclough et al., 2008)—reproductive isolation alone is not sufficient to explain the existence of species.

2.5.3 Just geographical structure

All organisms have a physical location, and hence geographical location is an inevitable feature of the real world. In this scenario, the only potential source of diversification is geographical isolation. Geographical location might change through Brownian motion dispersal from a point of birth or include rare long-range dispersal events. Organisms only interact (both reproductively and ecologically) with individuals in their local area. If dispersal rates are sufficiently low (e.g. less than one migrant per generation (Slatkin, 1985)) that groups of individuals become isolated by their location, but not so low as to prevent expansion and colonization of new areas (Pigot et al., 2010), evolutionarily isolated and divergent species will arise in each location. This would seem particularly likely with discrete geographical areas, such as islands, isolated lakes, or patches of favourable habitat, rather than a continuum of isolation by distance. Whenever organisms come into contact again they interbreed at random and interact neutrally through ecological drift in this scenario. The chance of observing multiple divergent species within a single area would therefore depend on the rate of re-entering into sympatry, the rate of loss of genetic divergence through interbreeding (Kleindorfer et al., 2014), and the rate of extinction through ecological replacement.

Random mating (Fig. 2.2) causes much faster erosion of diversity than pure ecological drift (Fig. 2.4), which explains the importance of reproductive isolation in sexual organisms. Ecological replacement would be more rapid, however, if one population uses shared resources more effectively than the other, that is, has higher relative fitness. Also, loss of genetic divergence through interbreeding is slower with lower outcrossing rates (falling to zero with no outcrossing). There might be scenarios therefore where ecological replacement becomes more important than interbreeding in limiting divergence. Speciation reversal in whitefish due to eutrophication involved loss of both reproductive isolation and divergent selection maintaining coexistence (Vonlanthen et al., 2012).

2.5.4 Reproductive isolation and geographic structure

Combining the two scenarios described in sections 2.5.2 and 2.5.3, reproductive isolation can now arise in geographical isolation, because of either changes in mating preferences by drift or the origin of post-zygotic isolation due to genetic incompatibilities. This might reduce the chance of erosion of species differences by recombination in secondary contact, although the rate of divergence in reproductive traits will be slow without selection for different traits in different regions (Fig. 2.3B; Rice and Hostert, 1993; Gavrilets, 2004). Furthermore, this scenario still assumes that there is no divergent selection or ecological distinctiveness between species, and therefore ecological drift and replacement would still limit the standing diversity of species present in a local region.

2.5.5 *Just ecological heterogeneity*

In this scenario, different resources are available in one place without mechanisms for reproductive isolation or geographical structure, corresponding to classical niche axis models of Roughgarden (1976) or the ecotype model for bacteria by Cohan (2001). An asexual population is expected to differentiate into an infinite series of clones adapted to each location on the niche axis. This model does not necessarily generate discreteness per se, unless the underlying resources have a discrete distribution. Alternatively, there might be constraints on the precision of resource use by organisms or stochasticity, so that there are finite limits to specialization that generate discreteness even with a continuous resource distribution. In a sexual population, recombination restores multilocus traits in the population to a unimodal distribution after random mating, and hence adaptation to the most abundant resource or a broad generalist phenotype is expected. Polymorphism at a single locus leading to adaptation to different niches could arise through balancing/frequency-dependent selection or by ecological dimorphism between the sexes (Slatkin, 1984), but not multilocus divergence of autosomes if there is fully random mating.

2.5.6 *Ecological divergence and reproductive isolation*

This scenario represents classical models of sympatric speciation (reviewed in Gavrilets, 2003, 2004). For a sexual population in a heterogeneous environment, ecological divergence can promote reproductive isolation either as a pleiotropic by-product (for example, adaptation to distinct hosts leading to indirect effects on mating encounters) or by direct selection for assortative mating that permits ecological divergence (reinforcement). Both mechanisms create units that are both reproductively isolated and ecologically divergent. The degree of differentiation depends on the balance between selection and gene flow: in theory, differentiation can be preserved even with appreciable gene flow as long as selection is strong enough.

2.5.7 *Ecological divergence and geographical structure*

In this scenario, ecologically distinct populations can evolve in different geographical areas, without the need to acquire specific mechanisms of reproductive isolation. If habitats are arranged patchily in local areas, this might even allow coexistence within a geographical region through ecological isolation, that is, selection against hybrids. What features of the distribution of environmental heterogeneity in relation to geographical structure affect the pattern of diversity that evolved? Steep transitions might more strongly favour divergence into discrete entities than smoother gradients. Many models consider simplified environments of one dimension (including mine in Box 2.2 and Fig. 2.3)—but what happens with multiple traits and environmental axes? The distribution of different dimensions could be correlated, nested, or contradictory, and might vary in their pattern of temporal variation—all factors that would affect the outcome.

2.5.8 *Reproductive isolation, ecological divergence, and geographical structure*

If all three mechanisms operate simultaneously (as they likely do in sexual clades), how do they interact to shape the pattern of diversity? As argued in section 2.5.6, there are reasons to expect an association between reproductive isolation and ecological divergence: ecological divergence based on multiple traits is only possible between reproductively isolated units, and reproductive isolation might only evolve readily when driven by divergent selection (Fig. 2.3C,D). Geographical structure should promote the origins of both, although both reproductive isolation and ecological divergence might be strengthened in secondary contact by reinforcement and character displacement, respectively. The pattern of distinct species might be most apparent within local communities and become less clear at larger scales encompassing more closely related taxa (Bergsten et al., 2012). This expectation might be altered, however, in more complex environments. For example, if organisms are adapting to ongoing environmental change, the strength of selection for reproductive isolation versus gene flow would depend on the relative strengths of diversifying versus unifying selection—in bacteria, it has been posited that the genome is split between genes under selection for low gene flow between species and those that benefit from gene transfer between species (Wiedenbeck and Cohan, 2011). To what extent are species an incidental outcome of the existence of diversifying forces versus optimized to enhance adaptation? I return to this issue in chapter 7.

2.6 Which mechanisms are most important?

A preliminary evaluation of the relative importance of each process and plausibility of the different models can be made by considering putative examples of particular cases (Table 2.1).

Are there cases of reproductively isolated but ecologically equivalent species that co-occur in the same place? The most likely scenario is when reproductive isolation arises so rapidly that it drives a standing level of diversity despite ecological drift. Cichlid fish in the African Great Lakes come to mind: many of the most closely related species are nearly identical except for male breeding colours and female preferences. Rapid shifts in mating characteristics might be explained in part by the genetic architecture of their association with sex chromosomes (Seehausen et al., 1999), and the diversity is prone to rapid collapse if conditions no longer favour visual mating cues (Seehausen et al., 1997). Male colour and visual pigments do reflect ecological adaptation to local light regimes at different depths, however, and hence there is still an environmental component to their adaptive divergence (Keller et al., 2013).

Are there cases of reproductively compatible but ecologically distinct forms co-occurring in the same place? Mallet (2008) presents a list of cases of ecologically distinct, co-occurring races belonging to the same species with intermediate levels of gene flow, such as head and body lice of humans (Li et al., 2010). They are termed races rather than species because gene flow is above 1 per cent. Partial reproductive isolation in

Table 2.1 Putative examples of particular scenarios of diversification involving some but not all of the mechanisms or features typically associated with evolution of a species pattern

Scenario	Putative example	Reference
Reproductively isolated, co-occurring species, with little ecological divergence	Closely related cichlid fish in Lake Victoria? (but depth is important)	Keller et al. 2013
Ecologically divergent races, no reproductive isolation? (See chapter 6)	Coexisting forms of *Sulfolobus islandicus* in a hot spring	Cadillo-Quiroz et al. 2012
Diversified clade with ongoing hybridization among species	Sympatric species of ducks in the genus *Anas*	Kraus et al. 2012
Geographically isolated, but weak reproductive and ecological divergence	Salamanders in the genus *Plethodon* in the Appalachian Mountains	Wiens et al. 2006
Long-lived lineage that occupies large geographical area but has not diversified	Planktonic foraminiferan *Hastigerina pelagica* contains two globally distributed sub-clades	Weiner et al. 2012
Diversified clade of asexual organisms	Bdelloid rotifers display genetic clusters and divergent selection on morphology	Fontaneto et al. 2007

these examples arises due to fine-scale spatial distributions, habitat selection, or culturally based assortative mating. One case where it has been proposed that ecological selection limits gene flow between two co-occurring populations without other barriers to gene flow is two co-occurring groups of the hot spring archeon *Sulfolobus islandicus* (Cadillo-Quiroz et al., 2012). The two forms differ in metabolic gene complements and growth phenotypes, yet share compatible molecular mechanisms for gene transfer, and the estimated recombination rate does not correlate with genetic divergence (as would be expected if homologous recombination was constrained by divergence). The existence of large genome regions (continents) of divergence between them is compatible with some form of restriction to gene flow and recombination, however (see chapter 6). I am not aware of cases of sympatric divergence in the face of full random mating in eukaryotes—indeed, all models of sympatric divergence require a mechanism for restricted gene flow based on emerging trait differences or non-random encounters. There are more and more examples, however, of divergence proceeding in the face of partial gene flow (Peccoud et al., 2009).

Are there cases of whole diversified clades of sexual organisms with incomplete reproductive barriers, in which geographical or ecological isolation is sufficient to prevent interbreeding and gene flow? Mallet (2008) and Harrison and Larson (2014) present evidence for high levels of hybridization in some clades and porous species boundaries. Ducks are a classic example, because they hybridize readily in sympatry but sustain high diversity by geographical and ecological divergence (Fig. 2.5). Phylogenomics confirms high historical levels of gene flow among sympatric species (Kraus et al., 2012).

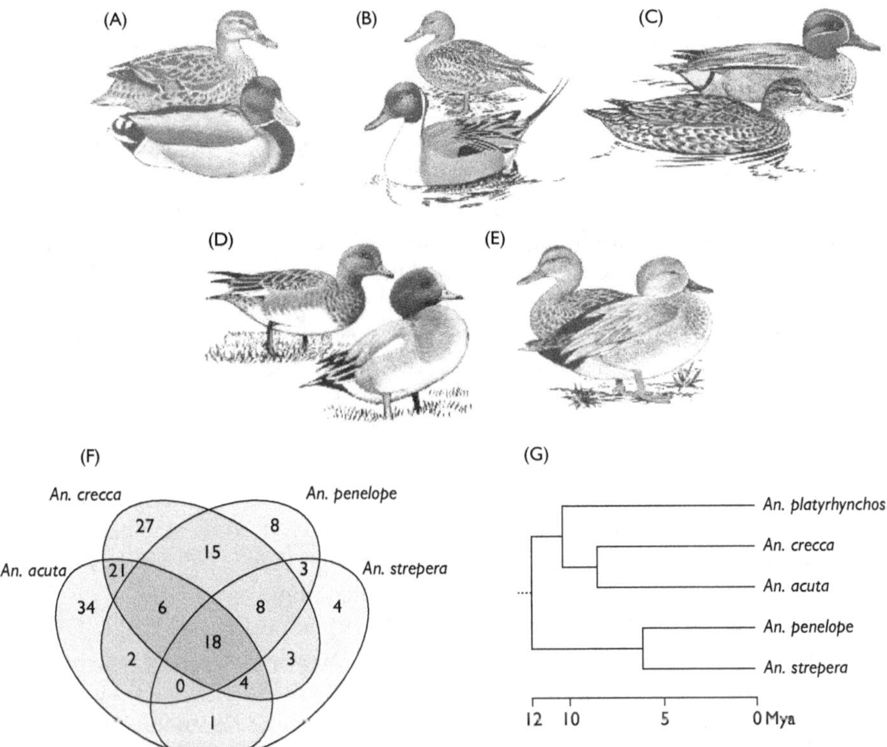

Fig. 2.5 Ducks provide a classic example of an ecologically and morphologically diverse clade of species that nonetheless hybridize readily. A survey of 364 single nucleotide polymorphisms (SNPs) from (A) the mallard, *Anas platyrhynchos* in the genomes of four other species from the genus *Anas*; (B) pintail, *A. acuta*; (C) teal, *A. crecca*; (D) European wigeon, *A. penelope*; and (E) gadwall, *A. strepera* revealed high levels of shared SNPs. SNP sharing is above levels expected from incomplete lineage sorting (F) based on divergence times (G) and effective population sizes of each species. (Reprinted from Kraus et al. (2012) under a creative commons license: drawings originally from the artwork stocks of the Wildfowl and Wetland Trust, Slimbridge, UK.)

Information also comes from case studies of related species that hybridize and exchange genes across hybrid zones (Martin et al., 2013; Stoelting et al., 2013). Many hybridizing species pairs exhibit genomic islands of high divergence relative to background levels (Ellegren et al., 2012; Soria-Carrasco et al., 2014). Genomic islands could indicate speciation driven by strong divergent selection in the face of gene flow. However, other processes can lead to the same genomic pattern of weakly differentiated genome regions, aside from speciation with gene flow (Nachman and Payseur, 2012; Renaut et al., 2013; Cruickshank and Hahn, 2014). The question in need of systematic evaluation is how do levels of gene flow in nature map to patterns of genetic and phenotypic divergence across entire clades? Surveys of genetic case studies may bring a biased view because genetic crosses are only feasible between species pairs that hybridize.

Are there cases of geographically diversified clades with weak ecological divergence or reproductive isolation? North American salamanders in the genus *Plethodon* are a classic example of so-called non-adaptive radiation. Species occupy distinct geographical ranges but are ecologically and morphologically uniform (Kozak and Wiens, 2006). Although the number of cases of hybridization events decreases with time since divergence, even some species separated for more than 8 million years are able to hybridize (Wiens et al., 2006). Yet, where species do co-occur in sympatry, there is character displacement in head morphology associated with dietary specialization (Adams, 2010). Rates of morphological divergence and the origin of reproductive isolation are low in this clade, perhaps associated with uniform ecological conditions between the geographically fragmented mountain habitats that they live in (Kozak and Wiens, 2006). In contrast, geographical isolation has been found to predict genetic divergence better than ecological divergence in *Anolis* lizards (Wang et al., 2013).

Are there cases of long-lived lineages found over large geographical areas and varied ecological conditions that have not diversified into separate species? Many marine species have cosmopolitan distributions. Some of these are split into cryptic species indicative of localized gene flow (Clarke et al., 2015). Others do indeed seem to have massive geographical ranges (Di Giuseppe et al., 2015), although they can differentiate across other dimensions, such as depth. For example, a planktonic foraminiferan displayed two cryptic species separated by depth but both with almost global ranges (Weiner et al., 2012). On land, the barn owl is one of the most widely distributed vertebrates—New World and Old World forms have been split into separate species, but there is still low genetic variation within each region (Nijman and Aliabadian, 2013). Many mosses have cosmopolitan distributions that appear to reflect true global dispersal (McDaniel and Shaw, 2005) as well as slow rates of morphological evolution. Are these patterns due to widespread gene flow constraining diversification or because these taxa occupy similar ecological conditions that are found repeatedly worldwide, which they can exploit thanks to their long-distance dispersal abilities?

Are there cases of diversified clades of asexual organisms? The difficulty is in finding clonal organisms that have survived for long enough to undergo speciation. Most cases are recent, such as in the flowering plant genera *Alchemilla* and *Taraxacum*, and display an amorphous continuum of genetic and phenotypic variation, rather than divergence into discrete clusters (Sepp and Paal, 1998). Bdelloid rotifers are a putative case of ancient asexuals that do display discrete genetic clusters and divergent selection on morphology (Birky et al., 2005; Fontaneto et al., 2007). Prokaryotes display discrete genetic clusters even in clades with low recombination rates (Acinas et al., 2004; Barraclough et al., 2009). This topic is developed further in chapter 6.

I expand further on testing the contribution of different processes in chapters 3 to 6, but this quick evaluation indicates that both reproductive isolation and ecological divergence are important in sexual organisms; although neither is sufficient alone for creating the pattern of species, ecological divergence seems to play the driving role and reproductive isolation the facilitating role (Sobel et al., 2010); geographical isolation alone is a weak force for generating the pattern of species, although it is important for facilitating initial divergence; geographical contact can strengthen patterns of differentiation; and a pattern of species can evolve in organisms with alternative life-styles and is not restricted to sexual eukaryotes.

2.7 Alternative models for diversity

Having summarized the main mechanisms generating diversity and the possible separate effects of those, I now return to the question posed at the start—what alternative models of diversity might there be in addition to the conventional species model?

2.7.1 *No evolutionary independence*

With no evolutionary independence (Fig. 2.6A), all the organisms within a clade comprise a single interacting metapopulation. The strength and direction of diversifying forces is shifting continually and no consistent emergence of evolutionary independence arises. Total variation would depend on the size of the metapopulations and dynamics of fluctuating diversifying forces.

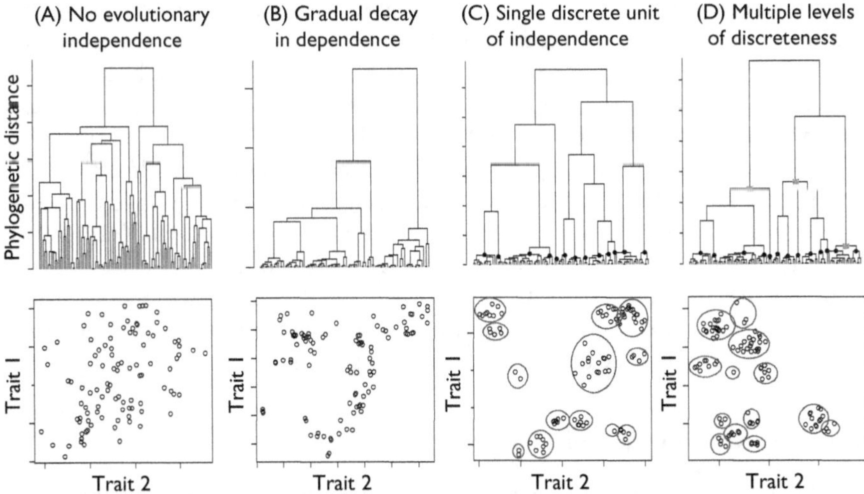

Fig. 2.6 Hypothetical gene trees (top row) and bivariate plots of two continuous phenotypic traits (bottom row) under four alternative models of structure of diversity. (A) No evolutionary independence, all individuals interact through interbreeding and shared selection pressures. (B) Gradual decline in evolutionary dependence between more distantly related organisms, as envisaged for the exponential decline in homologous recombination rates in bacteria or decline in interfertility between yeast strains. (C) Units of evolutionary independence are at a single level, namely the species, indicated by black circles on tree and ellipses on bivariate plot. (D) There are multiple hierarchical levels of evolutionary independence, additionally represented by squares on tree and shaded ellipses on bivariate plot. (Figure expanded and adapted from Barraclough and Humphreys (2015) with permission.)

2.7.2 Gradual decline in evolutionary dependence

There might be a gradual decline in the probability of interbreeding and ecological distinctiveness (Fig. 2.6B). Combined with non-random sub-sampling of individuals, this might give the appearance of discrete units, or a particular level of similarity might be used as an arbitrary and convenient cut-off. Such a model has been proposed in bacteria: the probability of homologous recombination declines exponentially with increasing genetic divergence (section 6.5). Other mechanisms of gene exchange, such as transfer of plasmids, can operate at wider levels but probably also decline in frequency between more distantly related cells. In this scenario, there is no transition or unit of particular significance that emerges. It has also been proposed for eukaryotes (Mallet, 2008; Peccoud et al., 2009).

2.7.3 Discrete and single-level species

A single level of discrete and independently evolving entities is present (Fig. 2.6C). The various diversifying forces act upon organisms to generate coherent and distinct groups, albeit with fuzzy boundaries because of the ongoing formation of those units. A key question is then why? Why should these multiple forces combine to generate simple discrete units rather than a more complex pattern? Does discreteness result from discreteness of the environment or from organismal responses? For example, reproductive isolation might be the key transition that frees up ecological divergence and marks a transition from microevolutionary dynamics to macroevolutionary dynamics. The Higgs and Derrida (1992) model has reproductive isolation originating from gradual genetic divergence, but once it happens there is self-organization and a shift to alternative dynamics and accumulation of greater levels of trait divergence than previously occurred in the population. Gourbiere and Mallet (2010) modelled shapes of the species boundary under alternative models for the accumulation of reproductive isolation—a snowball effect where incompatibilities evolve at an accelerating rate over time, a linear relationship, and a decelerating effect. They found some evidence for all three models in different organisms.

2.7.4 Discrete but with multiple levels

Diversifying and cohesive forces act on organisms to generate independently evolving groupings, but some processes operate at different levels or on different parts of the genome (Fig. 2.6D; Barraclough, 2010; Humphreys and Barraclough, 2014). For example, in bacteria, genetic clusters are clearly apparent for core genomes, which in part are associated with recombination. However, accessory genome or other mobile elements such as plasmids can be transferred among separate genetic clusters. The level of sharing of those elements has not been widely investigated but is probably not unrestricted and instead limited to particular sets of taxa (Smillie et al., 2011). Different arenas of recombination could therefore be present—narrow ones for housekeeping genes and wider ones for accessory genome (Wiedenbeck and Cohan, 2011; Barraclough et al., 2012).

In principle, a similar scenario could arise in eukaryotes: islands of genomic divergence between ecologically divergent species, maintained by ecological selection against heterozygotes at those loci, but allowing gene flow at other regions between species. For example, a study of orchids found low post-zygotic isolation within genera but high isolation between them (Scopece et al., 2010). Distinct species are still present in those genera, but there is the potential for gene flow at non-differentially adapted loci across the whole genus. Similarly, plastid and mitochondrial genomes are known to cross species boundaries relatively easily—diversifying forces on those genes could act at a broader level than forces operating on nuclear genes. Would organelle genomes cross species boundaries so readily if there were strong divergent selection acting on them? A key difference between prokaryotes and eukaryotes, however, is that the dominant means of recombination in prokaryotes involve uptake of DNA by one cell from a donor, whereas in sexuals it involves random shuffling of genes from two parents. This makes it harder to exchange some genes but not others in sexuals—unless via a series of backcrosses. There will be costs and benefits of being able to control gene exchange more. Multiple levels need not result from reproductive isolation and gene flow.

The examples above also considered the case of ecological drift of reproductively isolated species. If ecological drift occurred within wider sets of species, it could introduce some evolutionary interdependence above the level of reproductively isolated species, such that ecological trait variation and phylogenetic diversity is structured by ecologically defined higher taxa (Humphreys and Barraclough, 2014). This idea is explored more in chapter 10.

Beyond these examples, there have been few actual models of alternative diversity scenarios and even fewer empirical tests.

2.8 Other representations of diversity

Species are a convenient and efficient unit for representing diversity patterns. But now that more data are available for representing diversity, such as phylogenetic data, traits, and whole-genome sequences, might there be other more information-rich ways to conceptualize and portray diversity patterns and the processes generating those patterns? Ideally, we would document the pattern of phenotypic and genetic variation across a clade; the probability of recombination occurring between each set of individuals; the probability that selection on a given gene or trait is the same between each set of individuals; or a composite measure of the probability that a copy of gene A in individual 1 could replace the descendants of copies of gene A in individual 2 at a future generation, which could vary across the genome.

Networks are a popular representation for interacting sets of individuals, with the benefit of a depth of existing theory for considering their properties (Proulx et al., 2005). Would it be better to develop network visualizations of diversity instead of classifying into discrete units? For example, we could portray location as a summary of a multidimensional trait, connecting lines as the strength of reproductive isolation,

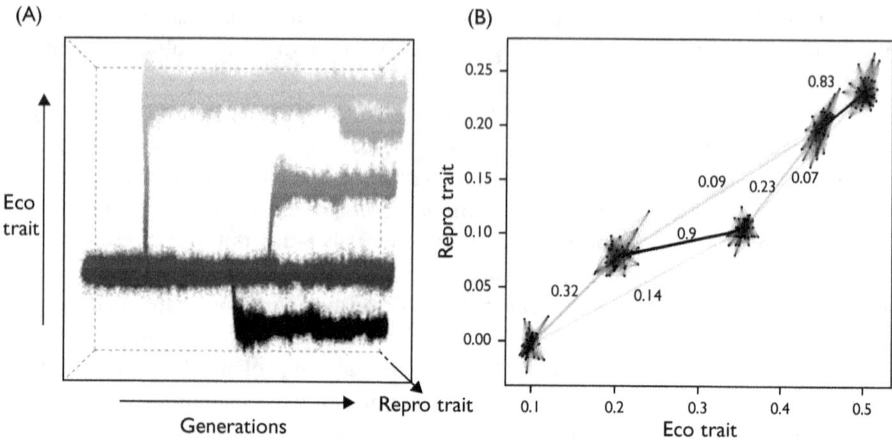

Fig. 2.7 Example of visualizing diversity using a network of evolutionary interdependence. (A) Simulation of a diversified clade of five species adapted to different ecological optima in separate geographical regions that became subdivided successively over time (total K = 1000, 2500 generations, model in box 2.2). The reproductive trait evolves as a correlated response as in Fig. 2.3D. (B) Pattern of phenotypic diversity at end of simulation, with links mapped on representing mean probability of successful interbreeding between individuals in separate phenotypic clusters should they come into contact (line shading: white, 0 per cent, incompatible; black, 1.0 per cent, fully compatible). Links within clusters are shown with 5 per cent opacity to facilitate visualization: all within-cluster pairs are at least 98.9 per cent compatible.

and colour to show ecological type (summarizing multidimensional differences; Fig. 2.7). A network can be used to model how the whole system evolves in a changing environment, and to predict how genes spread through the network (Lieberman et al., 2005). Then alternative patterns for the network can be formulated—clear discrete units, versus gradual change, versus multiple hierarchies with core and accessory genomes—to see how they affect subsequent dynamics in changing environments.

What other conceptual and visual representations could we use? Tree representations are useful for between-species patterns (in clades with minimal lateral transfer) or single loci, but they emphasize the history of divergence more than the ongoing processes maintaining discrete units. Adaptive landscapes have been widely used to discuss speciation but have limitations. Chiefly, they are difficult to measure, especially across multiple species in a clade across which fitness might be hard to measure in a comparable way. Also, formulation of a static landscape in three dimensions (fitness and two genotype axes) might not adequately capture the dynamics that would occur in changing environments or across multiple dimensions (Gavrilets, 2004)—for example, incorporating reproductive isolation among different genotypes as well. We need new conceptualizations that go beyond a static and historical map of diversity for a dynamic picture that aids prediction of future evolutionary responses. Testing ideas raised in this chapter in detail will require new approaches to portray and envisage diversity patterns and processes.

2.9 Conclusions

Species might indeed be the best unit for describing diversity patterns, and even if not, they provide a convenient and efficient approach. However, we can now test the 'species model' against hypotheses of alternative models of the structure of diversity. As well as providing better insights into how diversity originates, finding departures from strict assumptions of a single discrete unit of diversity could affect how organisms evolve in changing environments. If so, it might be better to use a more complex model of diversity that specifies the pattern of action of multiple diversifying forces across a clade. Key questions are: To what extent do patterns of diversity reflect discreteness in the environment versus discreteness in how organisms respond to continuous environments? And do the multiple processes causing diversification tend to result in discrete units?

3

The evidence for species—phenotypic and genetic clustering

3.1 Introduction

Chapter 2 outlined a range of scenarios for how diversifying forces influence patterns of variation within clades. I now turn to how to detect evolutionary species and test alternative hypotheses for the structure of diversity (i.e. test the ideas in chapter 2). There has been a growth in evolutionary methods for species delimitation, driven by the increasing availability of molecular data (Sites and Marshall, 2003; Pons et al., 2006; Knowles and Carstens, 2007; Yang and Rannala, 2010; Fujita et al., 2012). These methods focus on finding evidence of genetic isolation based on population genetic theory. However, the nature of units being detected is not always stated explicitly.

As argued in chapter 2, different diversifying forces affect patterns of variation in different ways. It is important therefore to consider these alternatives when designing methods of species delimitation—does the method detect reproductive isolation, divergent selection, or some other process? For example, it is normal to assume a lack of gene flow as a criterion for designing tests. However, not all speciation models require zero gene flow at neutral loci (see chapter 2). If speciation is driven by divergent selection, what criteria should be used then to define species units across clades? Also, for a given criterion, such as a lack of gene flow, what level of gene flow is biologically relevant for delimiting species versus populations within a species? Would it be better in a data-rich age to estimate the network of gene flow across a clade rather than using particular criteria to label sets of individuals as belonging to different species or not? Most excitingly, the use of several signatures offers the potential to reconstruct how multiple diversifying forces acted on a set of organisms. We can then test whether simple species units exist or whether a more complex pattern is required to describe diversity. Questions raised in chapter 2 that have been traditionally the realm of philosophical or theoretical debate can be addressed by formal empirical and statistical methods.

Chapters 3 and 4 describe empirical tests for the existence of species from surveys of phenotypic and genetic variation. Multiple signatures are discussed in order of the types of processes they address and increasing data demands. I emphasize concepts behind species delimitation rather than methodological details. The present chapter focuses on the most commonly available data at present, namely phenotypic and single-locus data. Does life fall into distinct phenotypic and genetic clusters indicative of

The Evolutionary Biology of Species. Timothy G. Barraclough, Oxford University Press (2019).
© Timothy G. Barraclough 2019. DOI: 10.1093/oso/9780198749745.001.0001

independently evolving species? Chapter 4 then explores alternative causes of independent evolution and outlines how whole-genome information might be used to infer diversity models across whole clades—only recently feasible for microbes and not yet easy for eukaryotes at such broad taxonomic scales.

3.2 Phenotypic clusters

The traditional way to delimit species is to describe the morphology of a representative set of specimens for a clade, and then to choose diagnostic characters that define species. The emphasis since the rise of cladistics has been on discrete characters, such as the presence or absence of a particular structure (Wiens and Servedio, 2000) or, more recently, diagnostic nucleotides in a DNA marker (Jörger and Schrödl, 2013). This approach is simple and useful for both phylogenetic reconstruction and identification, but it does not test for the presence of independently evolving species. For that we need measures of variation within and between putative species and then statistical analyses to compare observed patterns to the predictions of different models of species limits.

Continuous traits are more useful than discrete characters for testing for the presence of independently evolving species. If there is gene flow and uniform selection across a clade of focus, then we expect a unimodal, multivariate normal distribution

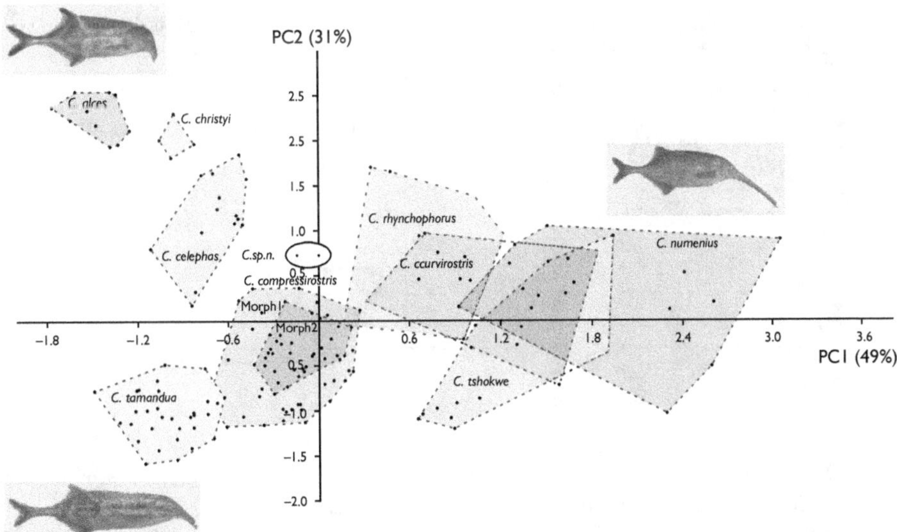

Fig. 3.1 Morphometric variation among named and unnamed taxa within the genus *Campylomormyrus* of weakly electric fish from the Congo Basin. Morphological variation was summarized as the first two axes of a principal component (PC) analysis on 20 morphological landmarks and 100 sliding semi-landmarks. The status of morphospecies was interrogated with molecular markers in the source paper. (Reprinted with minor edits from Lamanna et al. (2016) with permission.)

in quantitative traits coded by multiple genes (see chapter 2). If, instead, there are multiple reproductively isolated species that experience different selection pressures, then we expect separate morphological clusters of individuals belonging to each species. A morphometric survey of variation across a focal clade will therefore reveal whether discrete, independently evolving species are present or not. For example, fish in the genus *Campylomormyrus* live in the Congo Basin and are of interest because of their communication and mate choice via weak electric signals, which might play a role in reproductive isolation. Morphometric analysis reveals morphologically discrete species taxa that often differ in electrical signals (Fig. 3.1; Lamanna et al., 2016).

Good examples of this approach come from the fossil record, which is reliant on morphological traits for species delimitation. However, detecting discrete clusters within a continuous distribution is statistically challenging—many methods require the user to define groups a priori or to set the number of groups that are then fitted optimally from the variation. Instead, in order to test evolutionary hypotheses of whether discrete clusters are present or not, we need to estimate the number of clusters from the data and compare models representing alternative delimitations. One approach is to decompose a multimodal distribution into an optimized number of overlapping normal distributions, that is, the expected distribution for polygenic traits within a single panmictic population. Ezard et al. (2010) developed this approach to delimit morphological species of fossil planktonic foraminifera (Fig. 3.2).

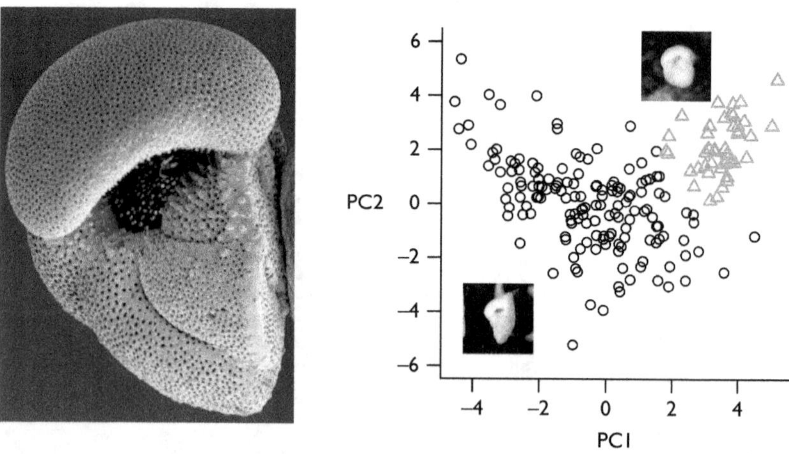

Fig. 3.2 Morphometric variation in extinct planktonic foraminifera *Turborotalia* from the late Eocene. Morphological variation was measured from micrographs (left) as a defined set of linear traits, areas, and angles, which were then summarized by principal components (PC) analysis (right). The number of morphological clusters (i.e. species) was estimated using Gaussian mixture models: the model providing the best description of data, in this case two species (black circles versus grey triangles), was selected using the Bayesian information criterion. At an earlier time period, only a single morphological species could be detected, allowing timing of divergence between the two species to be inferred (data not shown). (Reprinted with minor edits from Ezard et al. (2010) under creative commons license.)

Large morphological datasets of this kind are relatively rare for extant taxa and mainly come from the heyday of numerical taxonomy (a branch of systematics later eclipsed by cladistics). Rieseberg et al. (2006) used these data to investigate the reality of species groupings in plants and animals by comparison with overlapping datasets on levels of reproductive isolation. Over 80 per cent of cases from 218 plant and animal genera revealed phenotypic clusters, and 75 per cent of those clusters in plants matched reproductively isolated lineages based on crossing studies (Fig. 3.3).

The key advantage of phenotypic data is that ecologically and reproductively relevant traits can be sampled (Leache et al., 2009). Comparisons of multiple traits can reveal which traits are particularly strongly differentiated among closely related species versus those under uniform selection pressures across a wider clade (Fontaneto et al., 2007; Solis-Lemus et al., 2015). Another question of interest is whether the phenotypic distribution tends to have clear gaps between clusters or whether they are partially overlapping. Different patterns might be expected depending on whether there is ongoing gene flow among species (Harrison and Larson, 2014) and on whether selection pressures tend to promote distinct forms or more continuous variation. More comparisons like those of Rieseberg et al. (2006) would be fruitful to identify biotic traits or conditions that promote particular patterns of differentiation; for example, more or less discrete distributions, differentiation more or less evenly

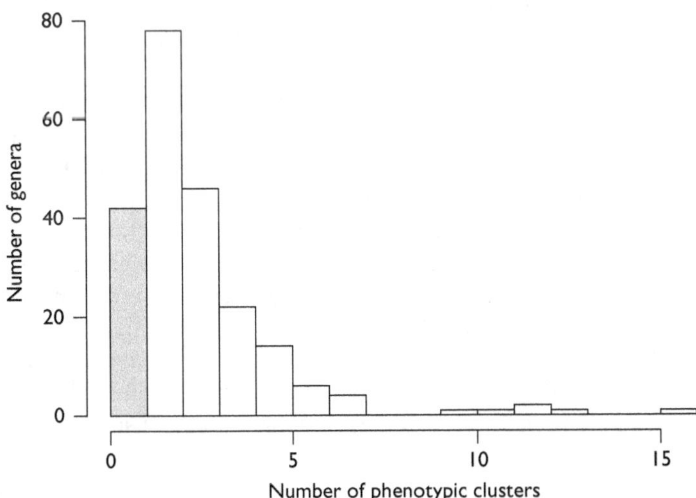

Fig. 3.3 Prevalence of discrete phenotypic clusters across 164 plant and 54 animal genera from meta-analysis by Rieseberg et al. (2006). Shaded bar indicates number of genera comprising just a single cluster in multivariate analysis of morphological characters. In most cases, phenotypic clusters corresponded to reproductively isolated units as judged by overlapping crossing data. Although these units did not always correspond to taxa recognized as species by taxonomists, the data support the conclusion that reproductively independent and phenotypically discrete units are prevalent in both animals and plants. (New figure using the original data.)

distributed across multiple traits, and reproductive and ecological traits showing stronger or weaker differentiation.

A key disadvantage with phenotypic data is that variation reflects plasticity as well as genetic divergence, making it difficult to test alternative evolutionary models formally. Integrating phenotypic and molecular data might ameliorate these limitations, a topic I return to later. Another disadvantage with phenotypic data is that it is hard to measure comparable data across large clades. Matching up homologous traits and measuring lots of specimens in the same orientation is more challenging than sequencing DNA and finding orthologous genes. Methods that automate image capture and measurement in a production line equivalent to high-throughput DNA sequencing offer potential for a major breakthrough (Hudson et al., 2015). High-throughput metabolic phenotyping is feasible for culturable microbes, and potentially by single-celled measurements for unculturable microbes as well.

3.3 Genetic clusters: single locus

The most abundant source of genetic data for delimiting species comprises DNA sequences of a single marker locus often referred to as DNA barcodes (Hebert et al., 2003). Mitochondrial DNA markers such as cytochrome oxidase I in animals, plastid markers such as *matK* in plants (Lahaye et al., 2008), the internal transcribed space (ITS) region in fungi (Schoch et al., 2012), and the 16S ribosomal RNA gene in prokaryotes (Acinas et al., 2004) have revolutionized systematics. These markers are easily sequenced and comparable across broad taxonomic scales and have opened up the systematics of groups like bacteria that were opaque to traditional taxonomy. For evolutionary studies, they can be used to reconstruct relationships among species and estimate genetic variation within species across unprecedented taxonomic scales. DNA barcodes therefore provide a wealth of data for genetic species delimitation—called DNA taxonomy (Tautz et al., 2003)—as well as for their original purpose in animals and plants as a tool for identification against a reference database with existing species names. Of course, relying on a single locus is severely limited compared to using multiple unlinked markers (Knowles and Carstens, 2007; Yang and Rannala, 2010; Dupuis et al., 2012)—more about this later—but no dataset for multiple markers is currently available with the same taxonomic breadth and depth. So how well can we delimit species and test evolutionary models using single loci alone?

3.3.1 Theory

Consider the expected pattern of variation at a single locus sampled across a clade comprising several independently evolving species (Fig. 3.4; Barraclough et al., 2003; Fujita et al., 2012; Fujisawa and Barraclough, 2013). Assume that the sample includes multiple individuals of each species. In practice, we would not know species identities a priori, but we could sample intensively from a wide range of morphotypes, geographical areas, and habitats in order to ensure an adequate sample. A key quantity that affects the power to discriminate independently evolving species from barcode data is the amount of genetic variation within species relative to genetic divergence between species (Meyer and Paulay, 2005; Puillandre et al., 2012).

As outlined in chapter 2, the level of genetic variation within a species depends on the mutation rate and the rate of coalescence due to drift and selection (see Appendix 3.1 for a user's guide to coalescence). In the simplest case of constant population size, random mating, and no selection, the expected average pairwise sequence divergence $\theta = 2N\mu p$ substitutions per site, where N is the population size, μ is the mutation rate in substitutions per site per generation, and p is the effective ploidy level of the marker (Hudson, 1991; Hudson and Coyne, 2002; Rosenberg and Nordborg, 2002). For a diploid, biparentally inherited marker such as a typical nuclear gene, $p = 2$. For a haploid marker such as bacterial 16S, $p = 1$. For a haploid and uniparentally inherited marker such as mitochondrial or plastid DNA, $p = 0.5$ because only half the population (females) can pass copies to the next generation, assuming an equal sex ratio. The expected genetic distance to the most recent common ancestor within the population, shown by black circle nodes in Fig. 3.4, is also θ substitutions per site. The variance in θ is exponentially distributed, so across species we expect marked variation even if population sizes are equal—many species will coalesce to a common ancestor within less than θ units of genetic distance, but a few will take much longer. In reality, population sizes are likely to vary among species, for example according to a log-normal distribution (Green and Plotkin, 2007), which will introduce further variation among species in θ. Finally, estimates of θ also depend weakly on the number of individuals that are sampled, i, such that $\theta_{obs} = \theta(1-1/i)$. Hence, for a sample of 10 sequences, θ is typically underestimated by 10 per cent on average.

Other factors will affect genetic variation beyond the simple model, including breeding system, historical changes in population size, and purifying or directional selection on the gene or on genes it is linked to (Hein et al., 2005). Most of these reduce population variation compared to the neutral expectation. Often, some or all of these factors are lumped together into the term effective population size (Ne), which in its broadest use is the effective number of copies of the gene in the population. This is helpful when θ is a nuisance parameter that affects other tests. However, if we want to understand factors affecting differentiation in a given marker, it is better to think about multiple factors separately. More work is needed to test correlates of within-species genetic variation across clades (Bazin et al., 2006; Fujisawa et al., 2015)—do parameters highlighted by neutral coalescent theory predict variation in nature? For all the complications, however, genetic variation within species as expressed by θ is typically less than a few per cent for standard barcode markers in most organisms.

The second component of differentiation, genetic divergence between species, K, depends upon mutation rate and the time since species became isolated. Immediately after an event causing isolation of two species, genetic divergence will be obscured by variation within each species. Only gradually over time will signatures of isolation emerge. The first signature of restricted gene flow after the initial isolation is a change in allele frequency; that is, some variants of the gene inherited from the common

(A)

(B)

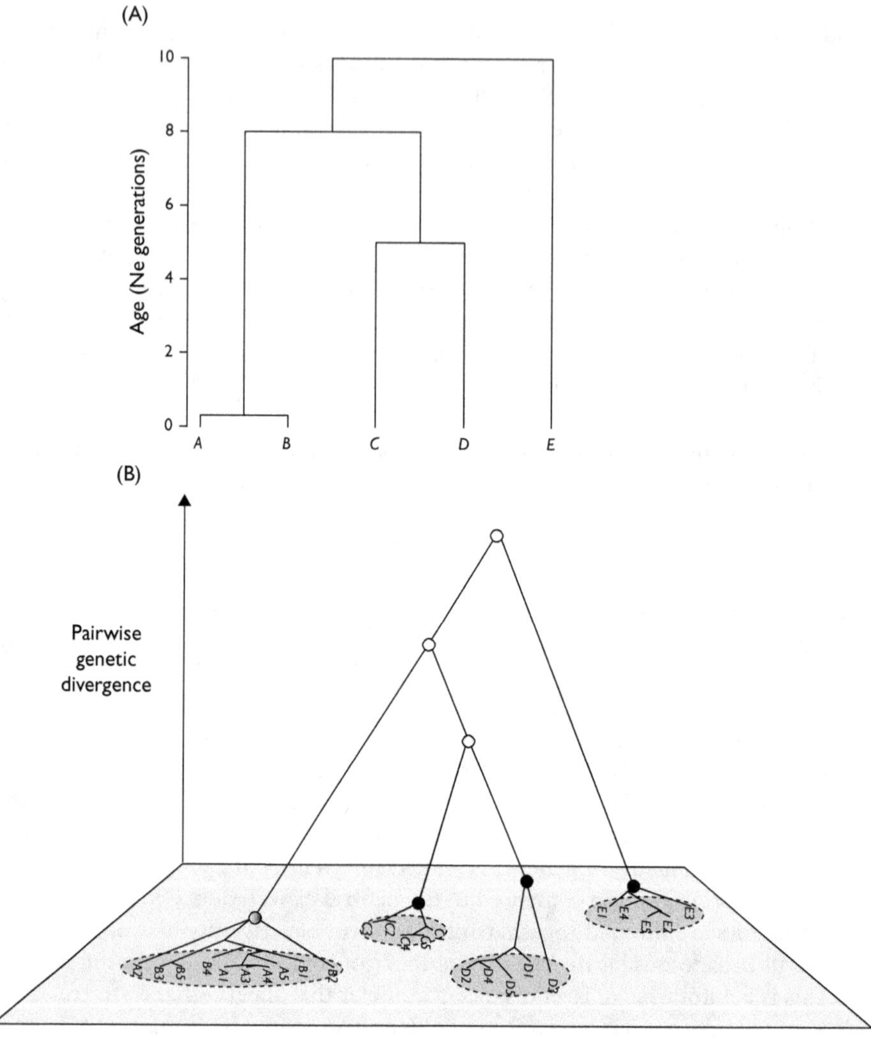

(C)

Branching type	Event	True age	Reconstructed age
Speciation	(A, B, C, D) versus E	10	10.3
(between species)	(A, B) versus (C, D)	8	8.1
	A versus B	0.3	Not recovered
	C versus D	5	5.4
Coalescence	TMRCA species A	Exponential, mean = 1	1.2
(within species)	TMRCA species B	Exponential, mean = 1	1.1
	TMRCA species C	Exponential, mean = 1	0.2
	TMRCA species D	Exponential, mean = 1	2.1
	TMRCA species E	Exponential, mean = 1	0.4

ancestral species will be more frequent in one species than the other. This is not a strong signature of independent evolution because it can also arise in partially isolated populations still connected by appreciable gene flow (Morjan and Rieseberg, 2004). The first strong signature of isolation is the presence of one or more fixed differences: mutations that are fixed in one species and absent in the other. Under neutrality, this criterion should evolve, on average, within around $0.5N$ generations for a uniparental, haploid marker such as mitochondrial DNA or within N generations for a haploid gene such as bacterial 16S (Fig. 3.5; Hey 1991). This signature is not expected to arise by chance between two random samples in a single population (Hey, 1991) and was used in early molecular methods for species delimitation such as population aggregation analyses (Davis and Nixon, 1992).

Once divergence time exceeds the expected level of variation within species, a signature of monophyly of one or both species arises (Fig. 3.5; Hudson and Coyne, 2002; Rosenberg, 2003). Under neutrality, reciprocal monophyly of both species is expected after approximately N generations for uniparental and haploid mitochondrial or plastid DNA or $2N$ for bacterial 16S (Hudson and Coyne, 2002; Rosenberg and Nordborg, 2002). Eventually, as divergence time further exceeds coalescence time within species, the final signature of genetic clusters emerges: apparent as groups of closely related individuals separated by long stem branches from other such groups (Fig. 3.5D; Barraclough et al., 2003). Stronger clustering than expected under the null model of a single species is apparent after around $4N$ generations of isolation for mitochondrial or plastid DNA or $8N$ for bacterial 16S (Fig. 3.5, middle row, explained in Box 3.1). This is the first signature that is detectable from single-locus data alone, such as 16S sequencing of a litre of seawater (Acinas et al., 2004). Using fixed differences or monophyly to delimit species requires a prior hypothesis for putative species, such as morphospecies or geographical populations, which is lacking in such cases. Because my goal is to test whether species exist rather than check existing species hypotheses, I will focus mainly on this latter signal of clustering.

Fig. 3.4 Pattern of genetic clustering for a single-locus marker. (A) An arbitrary species tree was designed for five fully isolated species (A–E). Species A and B diverged very recently, 0.3 Ne generations ago. (B) A gene tree was simulated onto the species tree assuming neutral coalescence and that effective population size is equal for all species. A single random realization of the gene tree is shown with vertical axis of age and grey shading to help visualize clustering in genospace at the present. White circles indicate speciation events recovered in the gene tree. Black dots indicate the most recent common ancestral (MRCA) node within each species. Grey circle indicates MRCA of species A and B, which are not monophyletic because their divergence time is more recent than realized times to MRCA in both species. (C) True and reconstructed ages for nodes signifying speciation events and MRCA within each species. Reconstructed ages for speciation events are older than true ages because of sorting of polymorphism in marker gene present at time of split (Edwards and Beerli, 2000). Expected time to MRCA (TMRCA) within species has a mean of 1 but is exponentially distributed: most species have TMRCA less than 1 but a few have much older TMRCA (e.g. species D). Trees were generated with Phybase package (Liu and Yu, 2010) in R and redrawn in three dimensions manually.

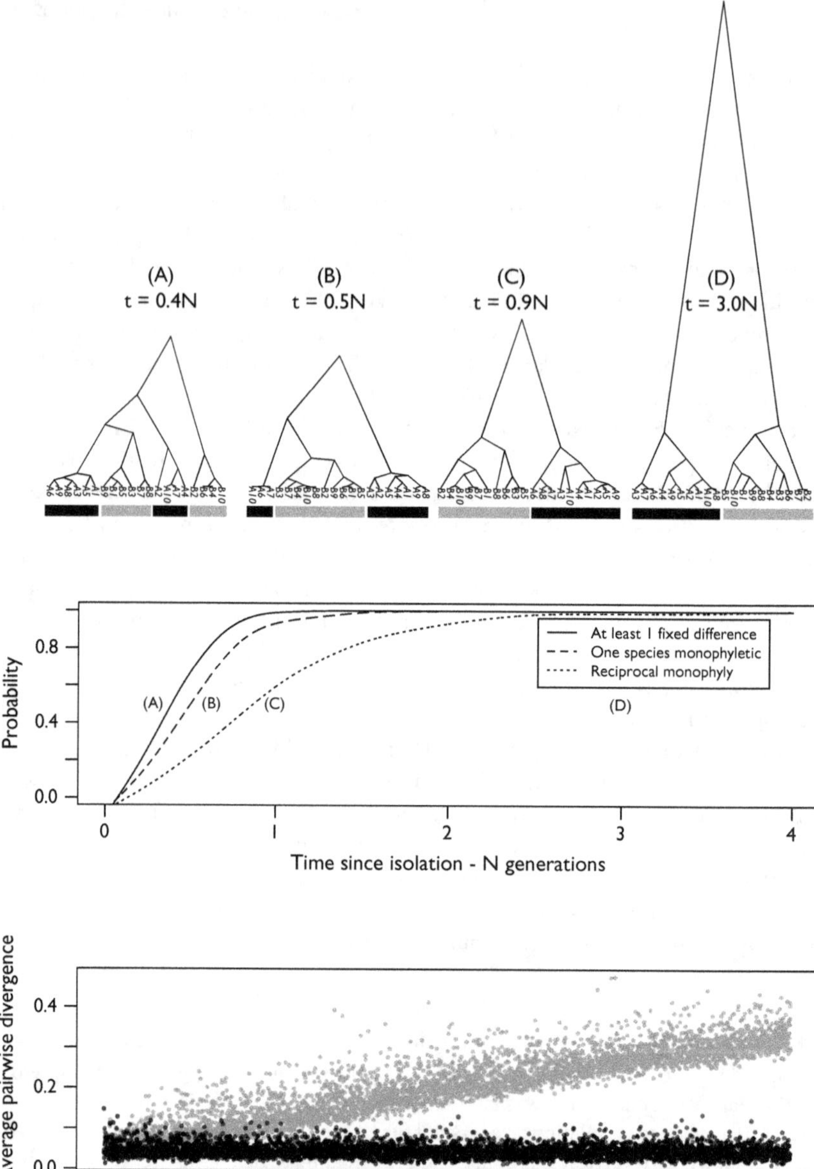

Fig. 3.5 Simulation results for genetic variation in a haploid, uniparentally inherited marker such as mtDNA during divergence of two species. (Top row) Example gene trees at different times since isolation: labelled A, B, C, D. (Middle row) Probability of observing each of three signatures of isolation over time. (Bottom row) Levels of average pairwise divergence between species (grey) and within species (black). Genealogies were simulated with the sim.coaltree.sp function of Phybase (Liu and Yu, 2010) and sequences with the seqSim function of Phangorn in R (Schliep, 2011). Code for recreating the plots is provided in the website accompanying the book.

Box 3.1 Delimiting species using genetic clustering—the two-species case.

Suppose we wish to compare a hypothesis of two species to a null hypothesis of one species under the standard neutral coalescent model. What divergence time is required to be confident of correctly delimiting two species based on a signature of distinct clusters? The answer provides an instructive example of using coalescent theory to guide species delimitation.

Birky et al. (2005, 2010) proposed a 4× rule for delimiting species; namely that we can infer two species with 95 per cent confidence when the ratio of the average pairwise divergence across the ancestral node to the average pairwise divergence within each sub-clade is > 4 (= K/θ in the two-species model, here called the Birky index; Fig. 3.6A). This was based on theory for the probability of observing reciprocal monophyly under the two-species hypothesis after a given divergence time T (Hudson and Coyne, 2002). It assumes, however, that two samples were taken from two hypothetical species and found to be reciprocally monophyletic. In a discovery setting, instead we only have a single sample of all thef individuals and no prior hypothesis of species membership. What is an appropriate threshold of the Birky index to accept the two-species model now?

Simulation shows that it is common for the two stem branches descending from the most recent common ancestor of a sample to be relatively long even in the one-species model. (On average, the time to coalescence of these two lineages is half the time to coalescence for the entire sample.) The upper 95 per cent confidence limit of the Birky index under the one-species hypothesis (dashed line in Fig. 3.6B) indicates a 10× rule to be more appropriate: only then can we reject the null hypothesis at the 5 per cent level. A 4× rule would yield a 43 per cent chance of rejecting the null model when it is true.

When can we be confident of rejecting the null hypothesis? Simulations show that after around 4N generations for haploid maternal markers there is a 50 per cent chance of correctly rejecting the null model using the 10× rule, rising to 95 per cent after 8N generations

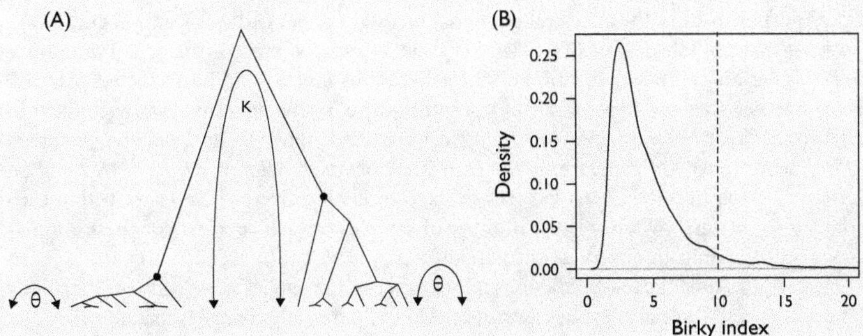

Fig. 3.6 (A) Birky index is the pairwise divergence across the root node divided by average pairwise divergence within each sub-clade (indicated by black filled circles), which equals K/θ in a two-species model. (B) Probability density of Birky index under the null model that all individuals belong to a single species, based on 10,000 simulations of a neutral coalescent sampling of 20 individuals. Dashed line shows the 95 percentile.

Box 3.1 *Continued*

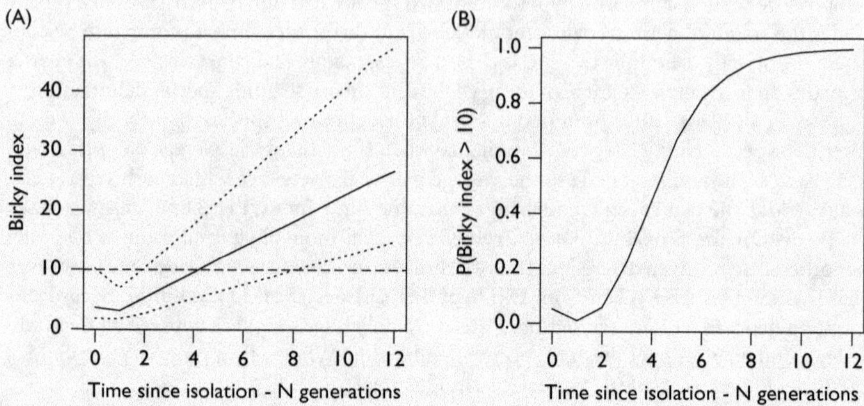

Fig. 3.7 (A) Median and 95 percent confidence limits of Birky index for genealogies simulated for two species isolated for varying amounts of time, from a sample of 10 individuals of each species. (B) Probability that Birky index > 10 across the simulated genealogies.

(Fig. 3.7). The power to delimit species from single-locus clustering is therefore relatively low for two species, although it increases when extended across a clade of multiple species (Box 3.2).

Why is $4N$ generations the expected time until $K/\theta = 10$? We might expect it to be $5N$ generations: pairwise divergence is twice the divergence time, hence a divergence time of $5N$ would yield a Birky index of 10 for a mitochondrial marker with $\theta = N\mu$. The estimate of K from the genealogy overestimates the true divergence time, however, due to sampling of polymorphism from the ancestral common ancestor species (Edwards and Beerli, 2000). The reconstructed estimate of $K = (2T\mu + \theta_{anc})$ on average, where T is the actual time since species isolation expressed in numbers of generations and θ_{anc} is the expected pairwise divergence between the two randomly chosen lineages in the ancestor that were fixed in each species in turn after isolation (assuming reciprocal monophyly). The simulations assumed $\theta_{anc} = \theta$; hence after $4N$ generations the expected observed value of $K = 2 \times 4N\mu + \theta$ on average $= 9\theta$ for an mtDNA marker. The average observed pairwise divergence within each species, $\theta_{obs} = \theta(1-1/i)$, where i is the number of samples of each species. For these simulations, $i = 10$, yielding an observed Birky index, $K/\theta_{obs} = 9\theta/0.9\theta = 10$.

One useful feature of these predictions is that the criterion applies irrespective of the ploidy of the locus—a 10× rule also applies for bacterial 16S, and the only change is that now $\theta = 2N\mu$; hence it will take twice as many N generations to reach the same stages. Another simplifying feature is that mutation rate should affect K and θ linearly, and therefore it cancels from the calculation of the Birky index. This assumption might be violated, however, if multiple substitutions lead to underestimation of K relative to θ, which would reduce the power of delimitation.

To conclude, the power to delimit two species based on genetic clustering depends on present-day θ in both species, time since isolation T, θ of the common ancestor, and potentially the mutation rate of the marker, under a 'simple' neutral model of population divergence.

Within a wider clade there will be species at various levels of divergence from their nearest related species. Consequently, some will be fully differentiated as genetic clusters, whereas others will have diverged too recently to leave a detectable signature in single-locus variation (e.g. species A and B in Fig. 3.4). The power to detect independently evolving species depends on the typical mean and variance of θ across species relative to the distribution of divergence times between species. The latter depends on the dynamics of diversification, namely speciation rate, extinction rate, and any diversity-dependent effects (for example, if speciation rates slow down as the number of species in a clade increases (Fujisawa and Barraclough, 2013)). Hence, power does not depend solely on population variation. For example, a clade might have a large effective population size (e.g. widely dispersing and well-mixed marine bacteria populations) such that the expected time to reciprocal monophyly for 16S is many millions of generations. Yet, if the net diversification rate is slow enough that related species typically diverged tens of millions of years ago, well-differentiated genetic clusters would still be detectable.

The above predictions assume strict isolation, that is, zero gene flow between diverging species. A final consideration for applying these predictions to species delimitation is how gene flow affects the probability of detecting each signature. How complete must isolation be in order to detect species? Might we incorrectly delimit species when there are in fact appreciable levels of gene flow? Classical theory by Slatkin (1991) showed that if gene flow is below 1 migrant per generation ($N_e m < 1$, where m is the migration rate per individual), then separate populations diverge at neutral loci as predicted above without encumbrance (Fig. 3.8). If gene flow is greater than 1 migrant per generation, then the whole meta-population is effectively panmictic and no genetic clusters should be apparent. There is therefore a relatively strict criterion for defining independent evolution at neutral loci—gene flow less than 1 effective migrant per generation—and the above signatures reflect it robustly (Fig. 3.8).

3.3.2 Evidence

Pons et al. (2006) used these predictions to delimit species in tiger beetles of the genus *Rivacindela*. Members of the genus are found living on numerous salt lakes in the arid interior of Australia (Fig. 3.9A,B). A few species had been described by traditional taxonomy prior to the study, but beetles from many lakes remained undescribed, providing an interesting test-case for DNA taxonomy. Up to six individuals per lake were sampled, with an additional six per morphotype if obvious morphological differences were observed at each site. Each individual was sequenced for three mitochondrial genes, which were concatenated to represent a single locus because the genes are fully linked.

Different methods were used to delimit species based on each signal of divergence outlined in section 3.3.1. A new method called the Generalized Mixed Yule Coalescent (GMYC) model was developed to test for significant genetic clustering based on optimizing a threshold for the switch from between-species to within-species branching

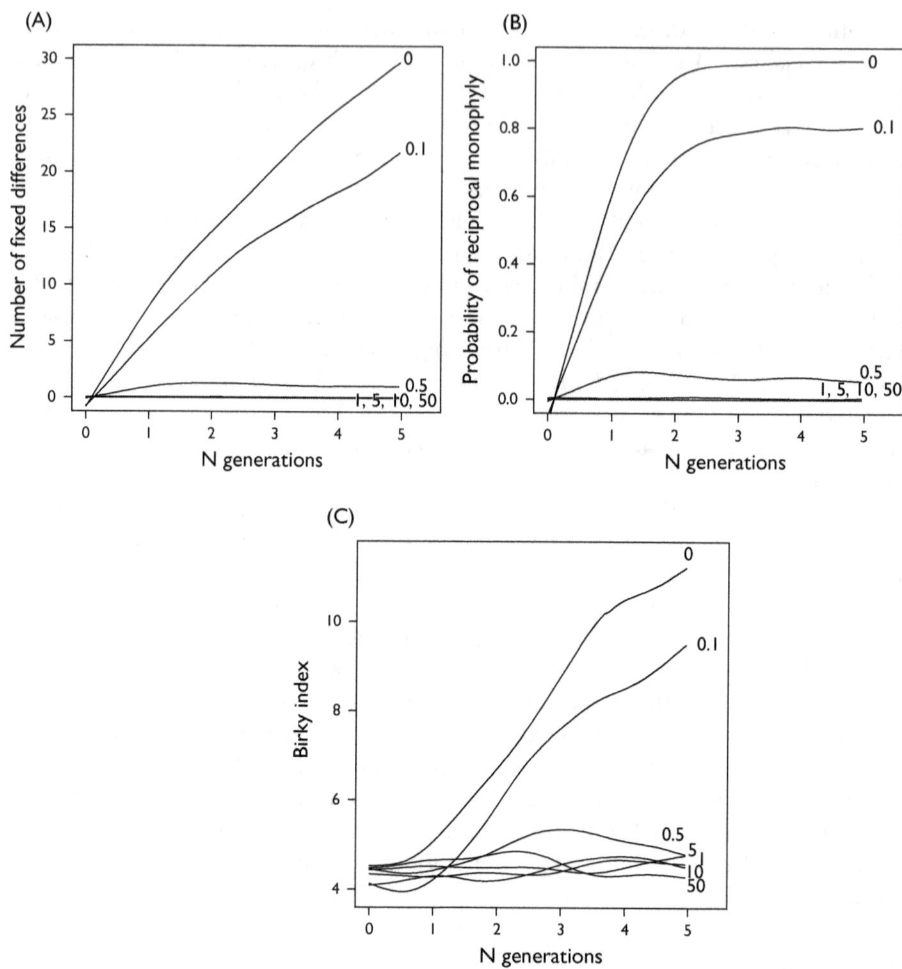

Fig. 3.8 Simulation results for (A) number of fixed differences, (B) probability of reciprocal monophyly, and (C) Birky index indicative of distinct genetic clusters (defined in Box 3.1) under varying levels of gene flow between two populations. Detectable signatures of independent evolution are obtained when gene flow is less than 1 migrant per generation. Forward simulations of the model outlined in Box 1.2 were used (code available with the accompanying materials).

(Box 3.2, Fig. 3.9). Despite expecting signals to differ in their power to detect divergent species, the results were remarkably consistent in this case and the estimated number of species ranged from 46 to 48. This was because individuals on a lake and within a morphotype often comprised distinct genetic clusters that were highly divergent from sister species on nearby lakes. These species diverged as populations became isolated on salt lakes formed as Australia dried out since the Miocene. Average pairwise differences within species were much lower (0.5 per cent) than

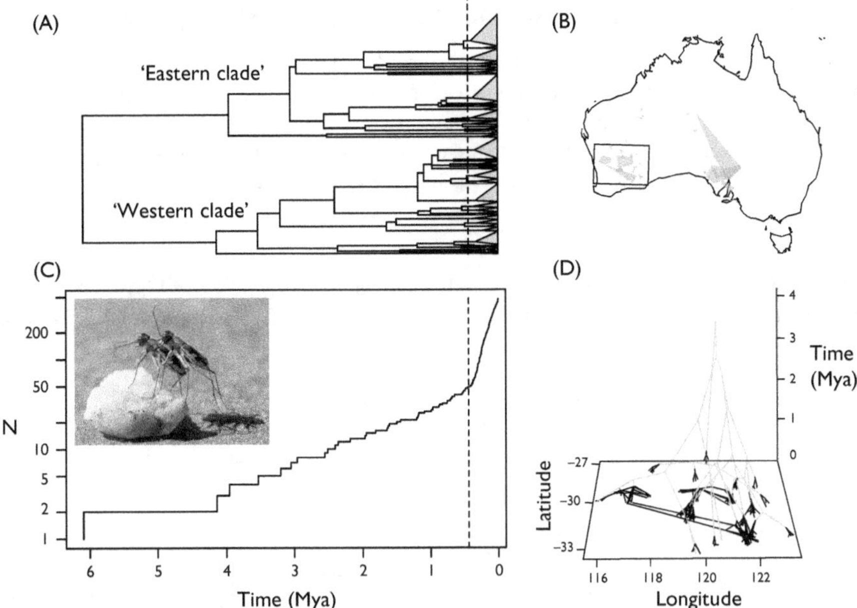

Fig. 3.9 Delimiting species of tiger beetles in the genus *Rivacindela* using mitochondrial single-locus data redrawn with data from Pons et al. (2006). (A) Phylogenetic relationships of 43 species delimited using the Generalized Mixed Yule Coalescent (GMYC) model showing branching within species collapsed to a grey triangle, with depth equal to TMRCA of each species. (B) Distributions of each species within Australia. Most species occupy single salt lakes, here visible as tiny flecks of grey. Rectangle shows range of 'western clade'. (C) Log lineages-through-time plot showing the upturn in apparent rate of branching towards the present, indicative of transition to within-species branching. A pair of courting *Rivacindela salicursoria* from Lake Lefroy are shown (photograph by Sophien Kamoun, with permission). (D) Genealogy of sampled individuals in the 'western clade' with tips plotted by latitude and longitude (with jitter to separate individuals from same locality). Within-species branching is shown in black and is frequently restricted to a single salt lake. Three-dimensional plotting is adapted from the phylomorphospace3d function in the phytools package (Revell, 2012).

the average divergence between sister species (2.2 per cent), hence the strong pattern of clustering.

The tiger beetles represent a primarily allopatric case in which population-focused methods looking for fixed differences and reciprocal monophyly between lake populations could also be used. What about patterns of genetic variation in a co-occurring assemblage? Acinas et al. (2004) sequenced bacterial 16S in a litre of marine bacteria. They used a sliding threshold of per cent divergence to delimit species to show that there was not just a uniform hierarchical pattern of sequence variation, but there was a particular transition that defined distinct genetic clusters. No additional information about geographical populations or traits was required to detect this signature. Contrary to many earlier assertions that species are not meaningful units in bacteria, this result

Box 3.2 The Generalized Mixed Yule Coalescent (GMYC) approach.

Pons et al. (2006) and Fontaneto et al. (2007) introduced a likelihood method for delimiting species from patterns of significant genetic clustering, later revised and described in detail in Fujisawa and Barraclough (2013).

Under the null model that all the sequences in a sample belong to a single species, the likelihood of branching intervals (x_i in Fig. 3.10) assuming a neutral coalescent is:

$$L_{(x_i)} = be^{-bx_i}$$

where $b = \lambda n_i (n_i - 1)$, n_i is the number of lineages during interval i, and λ is the coalescent rate (which equals $1/(2Np)$ for a neutral coalescent, where p is effective ploidy of the marker).

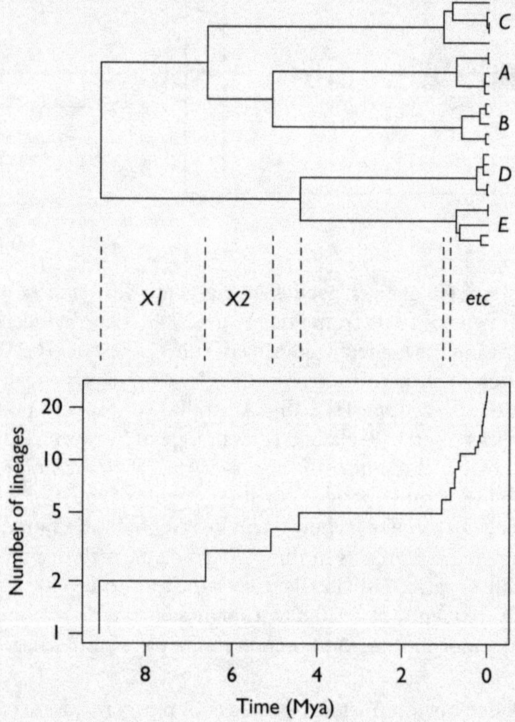

Fig. 3.10 (Top) Genealogy for a set of five well-differentiated species with branching intervals *x1*, *x2*, etc. Within-species branches are shown in bold. (Bottom) Log of the number of lineages over time reveals an upturn at the shift from between-species branches to within-species branches.

Under the alternative model that the sample comprises multiple isolated species, the likelihood of branching intervals depends on an alternative expression for b, namely:

$$b = \lambda_{spec} n_{i,spec} + \sum_{j=1,k} (\lambda_j n_{i,j} (n_{i,j} - 1))$$

where branching intervals deeper in the tree reflect a Yule process with constant birth rate λ_{spec} and those nearer the tips (bold in Fig. 3.10) reflect coalescence occurring independently in different species. The λ parameters therefore represent net diversification rate for between-species branches and coalescent rate for within-species branches. Additional parameters, p_{spec} and p_j, were added for each branching process to allow smooth changes in branching rates from the root to tips, hence 'generalizing' the Yule and coalescent models (Pons et al., 2006).

The number of possible species delimitations is larger than the number of possible trees. To simplify the problem of finding the best delimitation, the original version of GMYC optimizes the location of a single threshold to split branches into within- and between-species branching events. The transition threshold is visible as a change in slope on a log-lineages through time plot (Fig. 3.10). Assuming the same coalescent rate in each species for frugality, the alternative model has four parameters (two λ and two p; Fujisawa and Barraclough, 2013) and the null model has two parameters (one λ and one p). A log likelihood ratio test is therefore used to test the null hypothesis. Later versions were developed to allow multiple thresholds in different parts of the tree (Monaghan et al., 2009) and using the Akaike Information Criterion to select a 95 per cent confidence set of delimitations (Fujisawa and Barraclough, 2013).

The method has been evaluated by simulations and empirically (Fujisawa and Barraclough, 2013; Tang et al., 2014a). It performs best when there are multiple species within a clade and average sampling of around five individuals per species or more. Obtaining a robust ultra-metric tree is a vital step. A derived Poisson Tree Process (PTP) method has been developed that uses branch lengths directly, removing the need to reconstruct ultrametric trees by assuming a simple mutation model (Zhang et al., 2013).

showed that a signature consistent with independently evolving groups is also present in bacteria—and maintained without any geographical barriers. The pattern has since been confirmed in a wide range of bacteria (Barraclough et al., 2009). A key point is that these data are not consistent with an even branching process of relationships coalescing back to the common ancestor of all life—there is a shift in process that occurs.

How general is this pattern? Do any large or old clades not display a signature of genetic clustering in single-locus data? The GMYC test for genetic clustering has now been applied to a wide range of clades including animals (Monaghan et al., 2009), plants (Lahaye et al., 2008), fungi (Millanes et al., 2014), heterokonts (Vieira et al., 2014), single-celled eukaryotes, bacteria (Barraclough et al., 2009), and viruses (Herniou et al., 2015). The typical pattern is of genetic clustering and, where existing hypotheses of species are available, the level of clustering is approximately at the species level, or slightly below and indicative of multiple cryptic species within taxonomic species. Often these are geographically isolated populations that appear morphologically similar but the genetic clustering is consistent with strong isolation between them. Compiling metrics of the degree of genetic clustering across clades, we could seek environmental conditions or traits that favour particularly strong or weak levels of clustering. Interestingly, coalescence times within species tend to be more uniform than expected under a neutral coalescent model. This could be partly an artefact that species with much older coalescence times tend not to be detected but confused with interspecific divergences. Alternatively, it would suggest that there are additional cohesive processes acting on barcode markers within populations, such as purifying selection or demographic processes (Bazin et al., 2006).

Widespread evidence for distinct genetic clusters is consistent therefore with the existence of discrete, independently evolving groups, otherwise known as species. Perhaps this is the most comprehensive statistical evidence we have that life has diversified into species, rather than constituting a hierarchy within which taxonomic species are just arbitrary labels. But are there alternative explanations for or problems with this interpretation?

3.3.3 Partial isolation and sampling

One obvious issue is sampling. Intermediate genetic forms might exist but be absent from our sample. For example, sampling from one geographical region might miss closely related populations found elsewhere. Indeed, the degree of differentiation as measured by relative levels of divergence between and within taxonomically defined species—the barcoding gap—does tend to decline as sampling encompasses broader geographical scales (Bergsten et al., 2012). This occurs both because more closely related taxonomic species are included in the sample and because the variation within each species increases as more samples are taken from across its range—consistent with partially restricted gene flow across species ranges.

It has been argued that incomplete sampling of a species comprising subdivided populations could lead to the appearance of discrete genetic clusters (Lohse, 2009). With low but appreciable levels of gene flow, connected populations in a chain are expected to share similar genetic variation, and if we sampled all of them there should be no signature of isolation across the whole meta-population. But if we only sampled from extreme populations at the end of the chain, is it possible to infer separate genetic clusters by mistake?

Zhang et al. (2011) investigated this question using linear stepping-stone models of discrete demes with migration between neighbouring demes, building on the Slatkin (1991) model. With complete sampling of neighbouring demes, their method of species delimitation correctly identified demes as separate species when there were < 0.1 migrants per generation; as part of a single species with > 10 migrants per generation; and was indecisive with 1 migrant per generation, which represents the traditional boundary for neutral differentiation. What happens when only the end demes in a linear chain were sampled? The authors showed that the effective migration rate between them is approximately m/i, where i is the number of steps between adjacent demes exchanging m migrants per generation. Consequently, there is an increased chance of delimiting separate species, but the effects are small and only apply for a narrow range of migration rates where the delimitation method was already indecisive.

The risk of incorrect delimitation also depends on the scale of sampling. Many single-locus studies sample across large clades. Incomplete sampling of populations might introduce additional structure within species, but it should be less distinct than divergence between separate species that have been diverging for millions of years (Papadopoulou et al., 2008, 2009). Clades with intensive sampling still display significant genetic clustering (Meyer and Paulay, 2005; Kerr et al., 2007), so genetic clustering does not seem to result from patchy sampling alone.

3.3.4 *Independent limitation causes clustering, not reproductive isolation*

Assuming that genetic clusters are not artefacts of sampling, let's return to the processes that cause genetic clustering in single-locus data. In chapter 2, I outlined several processes that contribute to diversification. Which ones cause clustering at a single locus? The key point is that genetic clusters at a single locus do not provide direct evidence for reproductive isolation. Somewhat startling at first for those (like me) raised on the biological species concept, it is obvious when you think about it— *coxI* sequences in a clade of animals provide no direct information on recombination between individuals, since mtDNA does not recombine (except arguably in rare circumstances). Instead, a significant pattern of genetic clustering in (maternally inherited) mtDNA tells us, strictly, that females in one cluster are limited independently from females in another cluster, such that drift or selection on the marker gene occurs independently in each cluster. This sounds strange and unfamiliar, but two thought experiments help to clarify.

Consider first a set of reproductively isolated species that do not interbreed but that are ecologically equivalent as defined in the neutral theory of biodiversity (Hubbell, 2001). That is, the probability that offspring survive and recruit into the population is equal across species and depends on the density of individuals belonging to all species, rather than just the density of their own species. With a sufficient speciation rate to counteract loss through ecological drift, this model can generate a standing diversity of species within a region (see chapter 10). What is the expected pattern of variation in mtDNA? The answer is a single coalescent for the whole meta-community without any genetic clustering (Fig. 3.11; Barraclough, 2010). Each female has the same probability of leaving offspring in the next generation, irrespective of species, but the total number of individuals is limited. Species are arbitrary labels with no mechanistic impact on variation in any single marker locus. If you sampled multiple nuclear markers you could detect a clear signal for reproductive isolation (more in chapter 4) but not genetic clustering for mitochondrial DNA. Note that this is not simply because of insufficient time for the signal to arise due to ecological drift—there is no force operating to cause genetic cohesion and divergence. The only single loci that might indicate species divergence in this scenario would be those coding for the traits underlying reproductive isolation.

Consider next the case of a single interbreeding and panmictic population. Every male has the same probability of mating with every female. However, females are restricted to separate areas, whereas males disperse and select females at random. For example, imagine insect populations with winged males and flightless females, separated by river channels and limited by available habitat or resources. In this scenario there is no reproductive isolation between the separate populations, because nuclear

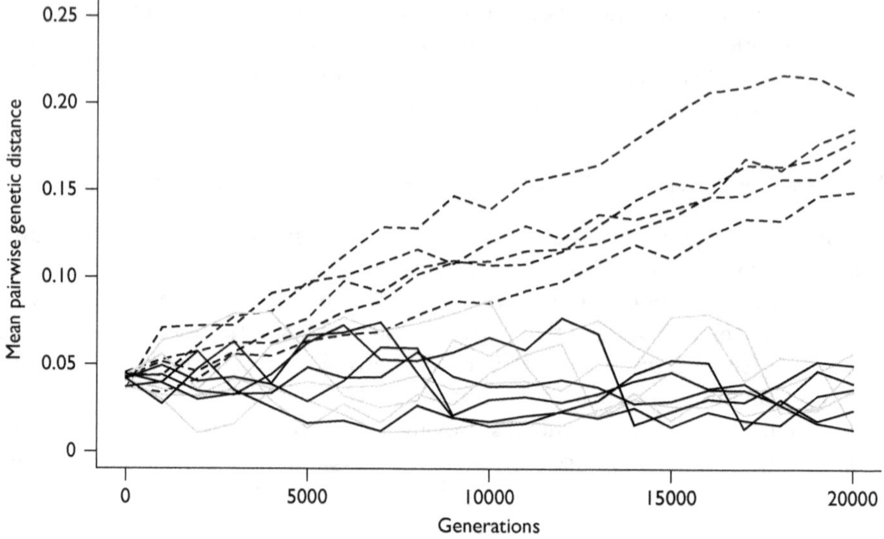

Fig. 3.11 Simulation results showing that divergence at a neutral, single-locus marker depends on independent limitation. (Solid lines) All individuals belong to a single population with 5000 individuals (five replicates shown). (Dashed lines) Individuals belong to five separate species that are evolving independently. Mothers of offspring for the next generation were chosen at random with replacement from the 1000 individuals of each species in turn. (Grey lines) A metacommunity of 5000 individuals, which is jointly limited but split into five reproductively isolated species at time zero. In this case, the 5000 offspring contributing to the next generation were chosen at random with replacement from mothers irrespective of species membership. (Redrawn with permission from Barraclough (2010).)

genes flow freely via male dispersal. Nonetheless, genetic clusters will emerge for mtDNA (or any other maternally inherited marker) if the populations persist in isolation for long enough for divergence time to exceed the coalescence time. A given mitochondrial DNA haplotype can replace another in its own population but not in the other populations. The cohesive force in this case is competition for available resources and survival of offspring, not recombination, hence the term 'independent limitation' (Barraclough et al., 2003), or equivalently 'demographic exchangeability' (Templeton, 1989).

The above scenarios are not necessarily realistic, but they illustrate the mechanism that causes genetic clustering at a single locus. In many cases, ecological limits and reproductive barriers coincide, and then mitochondrial clusters will correspond to evolutionary species. But even then, technically, the reason for clusters is that reproductively isolated species are ecologically discrete units and hence the number of females in each species is limited independently. Clusters indicate independent arenas within which drift occurs. For other single-locus data, such as 16S in prokaryotes (Acinas et al., 2004), there is a similar interpretation. If the bacteria reproduce clonally and the genome is strictly linked, then clusters indicate that there are independently limited populations: for example, adapted to different ecological niches or found in different geographical areas. If instead there is recombination, then clustering at the 16S locus shows that replication of that gene is limited separately in different clusters—most likely this represents ecological limits and drift acting on the core genome rather than just this gene. In principle, there could be different patterns for different genes (e.g. a resistance gene with high rates of horizontal transfer; Fraser et al., 2007).

Nobody would call mtDNA clusters associated with non-dispersive females separate species, so how reliable are single loci for demonstrating that life evolves into discrete and independently evolving species? There are other potential problems beyond sex-biased dispersal, in particular the observation of high rates of introgression of mtDNA and plastid DNA across species boundaries. Introgression causes distinct species to share organelle variation even when nuclear DNA is differentiated (Funk and Omland, 2003). These problems have led some authors to dismiss organelles as reliable markers of species boundaries.

While these problems will affect accuracy in particular delimitation problems, there are grounds to believe they do not confound interpretation of genetic clustering as evidence for species. First, low female dispersal rates are not sufficient to generate discrete clusters—there need to be barriers to the transfer of mtDNA haplotypes between populations for an extended period of time below the thresholds for neutral divergence (i.e. < 1 migrant per generation). Such an effect has been demonstrated in the extreme case of wingless, tunneling females of the scarab beetle genus *Pachypus* (Eberle et al., 2019), but there is so far no general evidence that mtDNA leads to greater overestimation of species clustering in organisms with male-biased dispersal, such as mammals, relative to female-biased dispersal, such as birds, despite theoretical reasons to expect it (Dávalos and Russell, 2014). Second, the conclusion is robust to introgression across species boundaries, which if anything would reduce the true signal of clustering.

3.3.5 Single locus conclusions

Single locus data offer many benefits for broad surveys of species limits, principally the depth of individual and species sampling that can be achieved. In animals and plants, organelle markers have the advantage of faster coalescence times than nuclear markers and hence greater power to discriminate species, but the disadvantage of sampling only maternal genealogies (or paternally for chloroplasts in gymnosperms and some angiosperms). Whole chloroplast or mitochondrial genomes still count as single-locus markers, because of strict linkage. Multiple unlinked markers are greatly preferable if available because they sample multiple genealogies rather than just a single gene tree, and consequently offer power to discriminate more recently diverged species. We should expend every effort to make multilocus data applicable over equivalent scales to DNA barcodes. But, at present, feasibility still greatly weighs in favour of single loci, and only single-locus data are robustly collected for environmental DNA applications using short-read sequencing technology. The benefits of multiple genomic markers are also limited if recombination rates are low—then any sufficiently variable marker will reflect the history of the whole genome (see chapter 6).

3.4 Conclusions

Single locus variation in both prokaryotes and eukaryotes indicates that life has diversified into discrete genetic clusters consistent with independently evolving species. Genetic relationships among organisms are not fractal, that is, self-similar at all levels. Instead there is a transition in branching patterns, approximately around the traditional species level, where evolutionary processes change. Other models of diversity described in chapter 2 might apply in particular clades, but discrete units visible from single loci are common. This conclusion is supported by the more limited number of broad surveys of morphological variation: animal and plant taxa contain multiple phenotypic clusters with gaps between them. What these results cannot answer, however, is what processes cause differentiation into discrete species.

Appendix 3.1 A user's guide to coalescence

Coalescent theory can be difficult to navigate. Here are a few tips that I find useful.

- Coalescence theory predicts genealogical relationships in timescales of generations or generations multiplied by the effective population size; genetic distances are obtained by overlaying a mutation model onto the genealogy.
- Coalescent theory predicts levels of variation proportional to the population size, the number of copies of a gene per individual (i.e. ploidy level), and whether the gene is bi- or uniparentally inherited.
- Predictions are often given in coalescent time units (i.e. units of θ), which can be confusing when converting between different types of markers. Many theory papers report results for diploid markers, which are rarely used as single locus markers, for example.
- Predictions of monophyly and genetic clustering should be independent of mutation rate as long as the mutation rate is high enough to reconstruct genealogy and not so high that saturation leads to an underestimate of divergence between species. The number of fixed differences does depend on mutation rate as well as genealogy.
- Haploid and uniparental markers such as mtDNA or plastid DNA have higher resolution to detect recent divergence, under neutral conditions, because their coalescent times are shorter. This is a desirable property for species delimitation. There might be other problems with focusing on such markers, however, if they tend to cross species boundaries easily (called introgression) or maternal history does not reflect overall species history.

- Neutral coalescent theory is useful to judge the likely relative power of different approaches, but their power in real cases is an empirical question that depends on empirical levels of variation and divergence, which might not match theory.
- Coalescence is a stochastic process with high variation. A single-gene tree says relatively little about the process it originated under. Sample size can be increased by sampling multiple loci, or in some cases by sampling multiple species in order to improve the inference of general properties of within-species variation.
- Pairwise genetic divergence K is twice the divergence time T multiplied by mutation rate, μ.

4

Why are there species? Arenas of recombination and selection

4.1 Introduction

Chapter 3 documented the 'pattern' of species, namely that life is packaged into genetic clusters consistent with independently evolving groups. The next question is why. What mechanisms generate evolutionary independence and maintain discrete genetic groups? Classic accounts identified two main explanations for the existence of species (Maynard Smith and Szathmary, 1995; Coyne and Orr, 1998). First, species might be a consequence of sex and reproductive isolation. Sexual recombination maintains coherence within species, while reproductive isolation promotes divergence between them. Second, species might result from divergent selection and adaptation to discrete ecological niches. Single-locus markers are practical across large taxonomic scales for documenting broad patterns of genetic clustering, but they do not provide direct information on reproductive isolation and divergent selection. I therefore continue the topic of evolutionary species delimitation by discussing direct methods for inferring units of reproductive isolation and divergent selection. After presenting approaches inferring reproductive isolation from marker loci, I discuss the potential for inferring networks of reproductive isolation and divergent selection from whole-genome data.

4.2 Arenas of recombination from multiple loci

Reproductive isolation is hard to measure directly. In principle, crosses can be attempted for a broad sample of individuals across a clade to measure the frequency of successful matings, fertilization, and hybrid viability and fertility. A meta-analysis of direct measures of reproductive isolation across animals and plants revealed high levels of consistency between reproductively isolated groups and phenotypic clusters (Rieseberg et al., 2006; Chapter 3). Even if the experiments work, however, they do not necessarily reveal the level of reproductive isolation in natural conditions or over multiple generations, and many organisms cannot be bred in controlled conditions. An alternative approach using genetic data is desirable to quantify the pattern of reproductive isolation and interbreeding within clades and to test for units of reproductive isolation.

The Evolutionary Biology of Species. Timothy G. Barraclough, Oxford University Press (2019).
© Timothy G. Barraclough 2019. DOI: 10.1093/oso/9780198749745.001.0001

One question is whether a clade contains discrete reproductively isolated groups (i.e. species), or whether a more complex pattern of gene flow is apparent. For example, there might be varying levels of gene exchange across the genome or a gradual decline in the probability of successful gene exchange between more distantly related organisms, as has been proposed for yeast and bacteria. Initially I focus on inferences of reproductive isolation from a set of arbitrary marker loci, but return later to consider the use of genome data, including information on loci of adaptive significance.

4.2.1 Theory

Consider a clade of sexual organisms that comprises several reproductively isolated species that exhibit random mating within them and no interbreeding between them. Reproductive isolation could derive either from intrinsic barriers (e.g. mating incompatibility) or from extrinsic barriers (e.g. geographical isolation); the predicted signatures of isolation will be the same. A sample is taken of dozens of unlinked nuclear genes from multiple individuals of each species. In real cases of species delimitation, we do not know species limits ahead of time, but it is assumed that the sample is deep enough to include multiple individuals of each reproductively isolated unit. Within a diploid species, each individual has two copies of each locus: one from a maternal grandparent and one from a paternal grandparent. Whether the first copy comes from the maternal grandfather or grandmother is random (rarely, it might constitute a recombined amalgamation of both due to crossing over). There is therefore a 50 per cent chance that each copy shares the same ancestor as a given copy found at a second unlinked locus. Tracking back through time, each locus therefore displays a separate random trial of the coalescent process. A strict consensus tree of gene trees across separate loci within species would yield an unresolved polytomy (Fig. 4.1A; Koufopanou et al., 1997).

Now consider the pattern of variation between two species that diverged T generations ago. If T is much longer than the expected coalescence time of each locus within each species, it is expected that alleles sampled from species A will cluster with each other separately from alleles sampled from species B, for every locus that is sampled (Fig. 4.1D). A consensus tree would resolve the divergence between the two species, but be unresolved within species (Koufopanou et al., 1997; Barraclough et al., 2003). In other words, we expect congruence between species but incongruence within species. For example, the ancestry of a sample of humans varies at different nuclear loci, but all loci recover humans as divergent from chimpanzees. This signature of genealogical concordance versus discordance has been widely used to delimit species in fungi and led to identification of cryptic species that were not distinguishable by morphology (Koufopanou et al., 1997; Taylor et al., 2000).

The problem with a strict criterion of congruence is that it has low power—reciprocal monophyly across all of several loci takes many Ne generations to evolve, much longer than reciprocal monophyly at a single locus (Fig. 4.1; Hudson and Coyne, 2002). However, more sophisticated approaches based on the probability of observing correlated topologies across a sample can detect reproductive isolation well before the signature of strict congruence emerges (Knowles and Carstens, 2007), and indeed

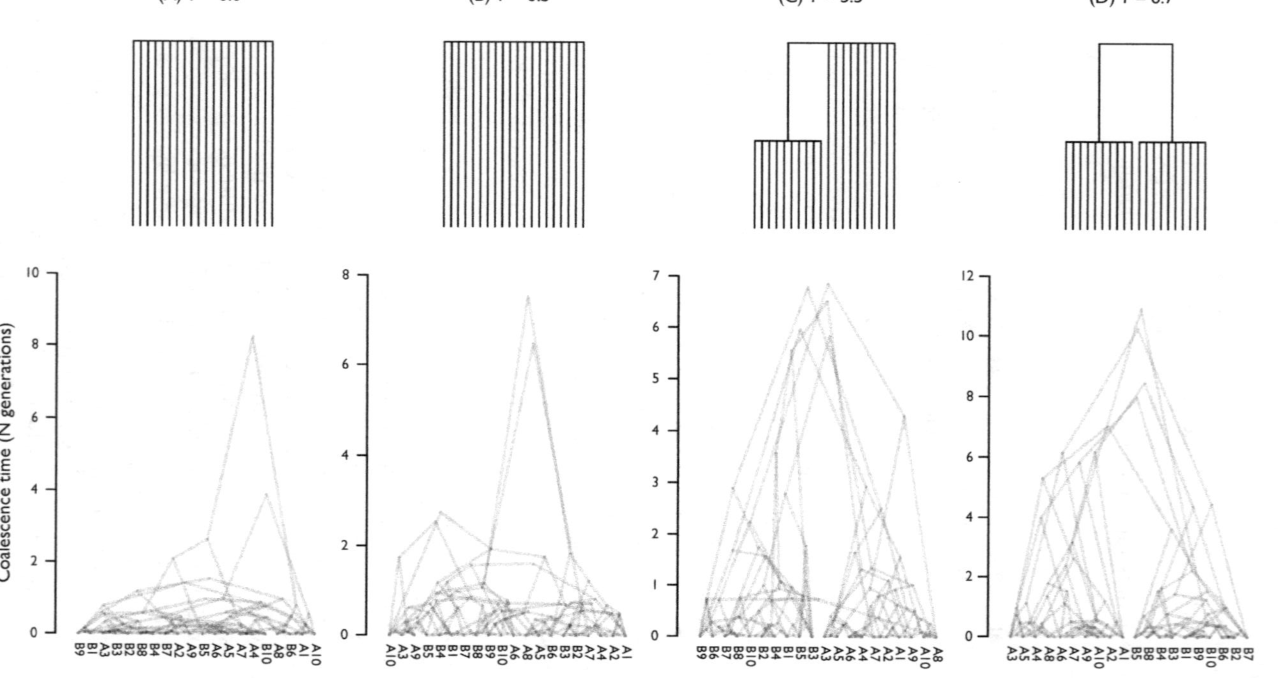

Fig. 4.1 Expected patterns of variation for multilocus data sampled from two species, A and B, under increasing divergence times (T). Genealogies of five loci were simulated under a neutral multispecies coalescent. The top row shows a strict consensus of the five gene trees in each case; the bottom row shows separate gene trees superimposed. Coalescence times are in units of effective population size (N_e) times number of generations. (A) Immediately at the time of divergence, no signature of divergence is present and the genealogy at each locus represents a separate random trial of a neutral coalescent. (B) After $0.5N_e$ generations, correlations start to emerge among loci that allow statistical methods to detect significant isolation between individuals belonging to A and B (Knowles and Carstens, 2007). (C) After $5.5N_e$ generations, there is a 50 per cent chance of observing monophyly of one of the species (Hudson and Coyne, 2002). (D) After $6.7N_e$ generations, there is a 50 per cent chance of observing reciprocal monophyly of both species, rising to 95 per cent chance after $11.8N_e$ generations. Simulations were run in the sim.coaltree.sp function of Phybase (Liu and Yu, 2010) and visualized using the densiTree function of Phangorn (Schliep, 2011) in R.

even before a single locus is expected to be reciprocally monophyletic. As a result, multilocus approaches offer greater power than single-locus approaches to detect isolation as long as enough variable, unlinked loci can be sampled.

There has been a recent growth in methods to delimit species using multiple locus data (Fujita et al., 2012). The most thorough approaches use a multispecies coalescent model (Knowles and Carstens, 2007; Yang and Rannala, 2010). The underlying model calculates the likelihood of observing a set of sampled gene trees given a particular species tree (defined by the topology and times of speciation events and by the membership of sampled individuals within each species, e.g. Fig. 3.4). Parameters for the model are the effective population sizes defining variation within each species (see chapter 3) and divergence times for each speciation event, T. Inferring the correct species tree and delimitation involves finding the species tree with the highest likelihood of yielding the observed data given the model (or in a Bayesian framework, finding species trees with the highest posterior probabilities).

Finding the best species tree and assignment is simple in principle but hard in practice (Carstens et al., 2013). One problem is that the number of possible species trees and delimitations is too large to search exhaustively. The intractably large number of possible tree topologies is familiar to students of systematics. The number of possible assignments of K individuals into i species, where the correct number of species is unknown and hence can vary in principle between 1 and K, is also intractably large. The number of possible species trees comprising both topology and the assignment of individuals to species combines both aspects and hence is even larger— because for each way of assigning individuals to species (above $i = 2$), there are multiple possible species tree topologies. For example, for a sample of three individuals there are three possible rooted trees, but seven possible species trees incorporating different species assignments (Fig. 4.2).

Searching 'species assignment and tree space' is therefore computationally more challenging than searching 'tree space'. To make the problem harder still, changing the species delimitation changes the dimensionality of the model, by altering the number of parameters (the set of θ and T). This contrasts with typical phylogenetic tree searches where varying topology does not change the number of parameters in the substitution model. Added features beyond standard Bayesian approaches are therefore required, such as a reversible-jump Markov chain Monte Carlo (MCMC) to evaluate models of different size (Zhang et al., 2011), which further complicates the calculations.

As a result of these challenges, methods using the multispecies coalescent to delimit species are computationally intensive. Implementation often relies on limiting the scale of the problem. For example, a common approach is to use a fixed guide tree, which removes the need to consider alternative species tree topologies. The species tree topology is assumed to match the guide tree, and at each node the method tests whether sister clades belong to the same species or different species. Guide trees are typically obtained from a phylogenetic tree of concatenated sequences, or a consensus of the separate gene trees at each locus. While these approaches generally give a reasonable estimate of the species tree topology (they are statistically consistent with the multilocus coalescent (Mirarab et al., 2014)), there are scenarios in which the correct

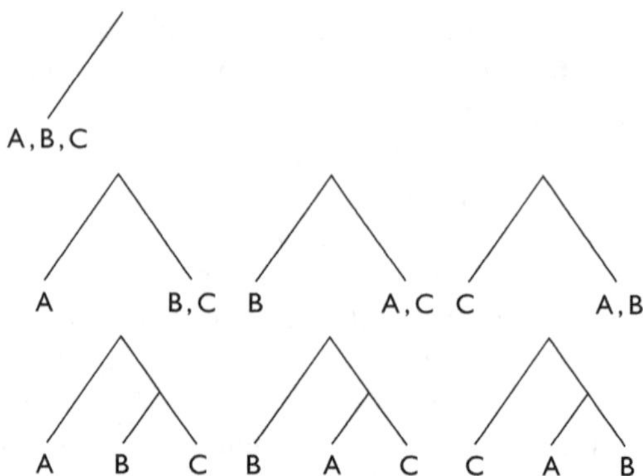

Fig. 4.2 The large number of possible assignments of individuals to species and species trees. A rooted tree of three individuals has three possible topologies: (A,(B,C)), (B,(A,C)), or (C,(A,B)). But there are seven possible species trees connecting three individuals allowing for all different assignments to species: all individuals might belong to the same species (top row), the individuals might belong to two species (middle row), or they might all be separate species but with three different possible topologies of the species tree (bottom row).

species tree does not match the most common topology across loci (Degnan and Rosenberg, 2006). Another simplification is to consider only a restricted set of alternative hypotheses of interest rather than all possible assignments—that is, to use the model for species validation, rather than for *de novo* delimitation.

My primary focus, however, is testing the reality of species units using *de novo* delimitation, rather than validating taxonomic species hypotheses. Less computationally intensive methods are needed in order to reconstruct the pattern of reproductive isolation across a wider clade at scales equivalent to those covered by single-locus data in chapter 3. One option is to use heuristic approximations or 'tricks' to simplify likelihood calculations and/or exploration of tree space, making larger-scale analyses feasible. For example, one method uses an approximation of the model together with a reformulation of the model priors of the Bayesian analysis to remove the need for reversible jump MCMC and simplify calculations (Jones et al., 2015). Other such methods will no doubt emerge to open up multispecies coalescent approaches, in the same way that computational tricks and simplifications opened up model-based phylogenetic reconstruction to scales that were unfeasible when these methods first appeared.

An alternative to the full multispecies coalescent approach is to use different signatures to infer reproductive isolation. For instance, with a large enough sample of heterozygote individuals, the network of associations of alleles within diploid individuals can be used to infer fields of recombination (Doyle, 1995; Hausdorf and Hennig, 2010; Dellicour and Flot, 2015): alleles within the same species are sometimes found in the same individual (assuming random mating), but alleles in different species are not (assuming sufficient divergence time that no ancestral polymorphism is retained in each species). Another simplified method decomposes genealogies into triplets of individuals and uses coalescent predictions for the topology of triplets (Box 4.1). Such methods will

Box 4.1 Delimiting arenas of reproductive isolation using the topology of rooted triplets.

Fujisawa et al. (2016) introduced a simplified method for delimiting species based on signatures of reproductive isolation that scaled to larger datasets than full multispecies coalescent methods available at the time. The method starts with a guide tree for the entire sample, which could simply be the phylogenetic tree derived from concatenation of separate loci into a single matrix. It then extracts all the rooted triplets consistent with the guide tree.

For each triplet (for example, of alleles sampled from individuals A, B, and C) there are three possible rooted trees (Fig. 4.3).

The method counts how often each topology is observed across the set of multiple loci that have been sampled. If the three individuals belong to a single species with random mating, it is expected that on average one-third of loci should display each topology. The probability distribution of the number of counts is therefore simply an equal rate trinomial distribution:

$$P_W\left(n_1, n_2 n_3\right) = \frac{N}{n_1! n_2! n_3!} \left(\frac{1}{3}\right)^N$$

where n_i is the number of loci supporting topology i out of N loci in total. In contrast, if, for example, individual A belongs to a different species than B and C, there will be a skewed distribution of topologies, with more loci supporting topology 1. The probability depends on the length of the internal branch defining the species containing B and C:

$$P_B\left(n_1, n_2, n_3 | v = 1, \lambda\right) = \frac{N}{n_1! n_2! n_3!} \left(1 - \frac{2}{3} e^{-\lambda}\right)^{n_1} \left(\frac{1}{3} e^{-\lambda}\right)^{n_2 + n_3}$$

where $v = 1$ indicates that topology 1 is the correct species tree and λ is the length of the internal branch leading to B and C measured by coalescent time units on the species tree. If the branch is short, there is still a high probability of alternative topologies due to incomplete lineage sorting, whereas if it is long, all of the loci are likely to show topology 1. These expressions can be used in a Bayesian framework to calculate the posterior probability that each node in the guide tree is within a species or a branching event between species—based on prior uncertainty of what the correct species tree is. An indication of the simpler approach is that the posterior probability can be calculated analytically in this model: Markov chain Monte Carlo methods are not required, in contrast to full multispecies coalescent approaches.

Finally, a dynamic programming algorithm was devised to step through the guide tree efficiently and reduce the number of calculations needed to implement the model to delimit species across the whole sample. The method, called tr2, provides a statistical framework for classical methods of delimitation based on genealogical concordance versus discordance, without requiring strict concordance to define species memberships. It works conservatively but robustly on simulated data and compared to empirical analyses using other methods.

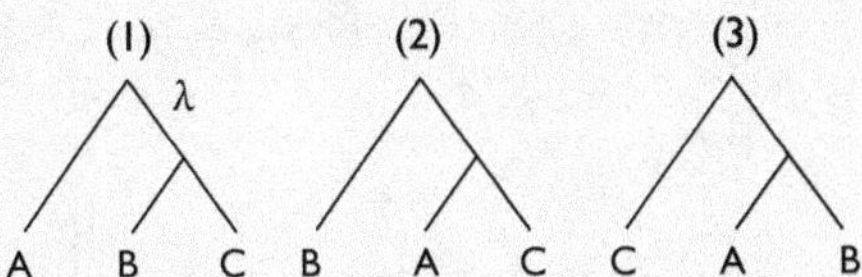

Fig. 4.3 The three possible rooted trees.

be less powerful than the full multispecies coalescent, since they discard some useful information, but they open up the possibility for applications at larger taxonomic scales.

All of these methods at present seek to optimize delimitation of reproductively isolated groups—that is, discrete sets of individuals with random mating within sets but none between them. Their success depends critically on the amount of gene flow. In theory, neutral divergence between populations occurs when the number of migrants per generation, M, is below a threshold of approximately 1 (Wright, 1931; Slatkin, 1985). Above that threshold, populations display some genetic differentiation but divergence does not increase linearly with time. Ideally, methods for detecting reproductive isolation from a set of arbitrary nuclear markers should therefore delimit species when $M < 1$ but not when $M > 1$, with a narrow band of indecision around this transition (see section 3.3.1).

In practice, simulations show that the above methods can fail to correctly delimit species if there is additional structure within species (i.e. partially isolated populations) or low levels of gene flow between species (Box 4.2). For example, a multispecies coalescent analysis of 50 simulated loci delimited separate species in a case of two populations that diverged recently and retained a level of migration ($M = 10$) above the threshold that would permit phylogenetic divergence of those populations (Jackson et al., 2017). This problem can be fixed by requiring a threshold divergence time to detect species, but it indicates that the most powerful methods can be oversensitive to partial gene flow unless it is explicitly modelled.

New methods are being developed to allow to incorporate population structure within species or allowing for speciation with gene flow (Dalquen et al., 2017; Jackson et al., 2017). At present, these are being formulated with the assumption that a correct delimitation of individuals into discrete species units exists, albeit with added complications of population structure and/or some level of gene flow between species (Box 4.2). An alternative approach, and one that would allow us to test rather than assume the existence of discrete reproductively isolated species, would be to infer the network of gene flow connecting sets of individuals in a sample, and to test that network against alternative models of how reproductive isolation is distributed across a clade (see section 4.4).

Another question for future work is how to combine organelle and nuclear genes. This is desirable because of the greater variability and faster coalescence times of organelle markers, but there is a problem if mtDNA or plastid genes show a different history to nuclear markers, because of their different mode of inheritance. Current methods allow for the different ploidy level of those markers (i.e. by optimizing a separate θ value for different types of loci), but not for different histories among markers. The multispecies coalescent could be used to test for significantly different histories between nuclear versus organelle ancestries.

4.2.2 Evidence

Datasets of multiple loci for multiple individuals across large clades are still relatively rare for animals and plants. Most studies focus on species complexes of closely related taxa and often include taxa that have not been recognized as separate species

Box 4.2 What units do multispecies coalescent methods delimit?

The multispecies coalescent model was formulated originally for the scenario of populations isolated by zero gene flow (i.e. migration rate) since a point of separation T generations ago. It is the most powerful method for detecting such units and for rejecting the alternative model that they are simply repeated samples from the same randomly mating population. The complication arises because there can be intermediate levels of gene flow (Fig. 4.4).

Theory states that two populations should diverge phylogenetically at neutral markers, as if evolving independently, as long as the effective number of migrants per generation between them, M, is < 1, and certainly when it is < 0.1. In contrast, divergence between populations is constrained if M is > 1, and certainly when > 10. There is therefore a relatively narrow transition zone between evolving independently and maintaining cohesion between the populations.

Zhang et al. (2011) evaluated the performance of multispecies coalescent models in a stepping-stone model of partial isolation with symmetrical migration between a linear series of populations. With a sample of just a few loci, the method BPP (Bayesian Phylogenetics and Phylogeography) was relatively robust to this scenario: there was a tendency to over-splitting when intermediated populations were not sampled, but only over a narrow range of migration rates that in any case define a zone of indecision between splitting and lumping (around $M = 1$).

Subsequent models showed, however, that oversplitting can occur in some scenarios. With more loci or asymmetric migration between populations, the methods can identify significant isolation between populations even when gene flow would not permit long-term divergence (Jackson et al. 2017). Barley et al. (2018) introduced methods for assessing the model fit of the multispecies coalescent to help diagnose cases where assumptions are violated. Dalquen et al. (2017) developed likelihood models for scenarios with partial gene flow that could be used to devise delimitation methods that do not assume strict presence or absence of gene flow, but allow for additional structure of partially isolated populations within species.

In using these revised methods, a decision still needs to be made about the criterion for delimiting isolated units. For example, it is now widely acknowledged that speciation sometimes occurs in the presence of gene flow, so should we allow for detection of separate units in cases with appreciable levels of gene flow between them, using these methods? My view is no, as this is confounding different aspects of diversification. The aim of these methods is to detect barriers to gene flow. This is the signature that interbreeding sets of individuals are present that are separated by reproductive isolation (*sensu lato*, including geographic barriers to interbreeding as well as pre-zygotic or post-zygotic incompatibility). Additional methods and data are needed to diagnose speciation with gene flow (see section 4.4).

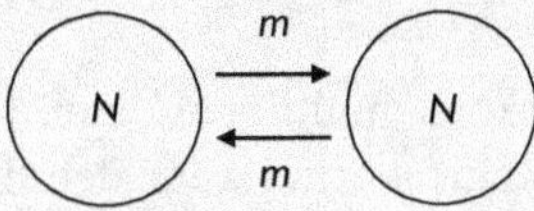

Fig. 4.4 Simple island model of two populations, both of effective population size N connected by migration of a fraction m individuals per generation, where $M = Nm$ (number of migrants per generation).

Box 4.2 *Continued*

Based on the observation that multispecies coalescent methods often delimit units below conventional species, Sukumaran and Knowles (2017) argued for the existence of two types of entities: units of genetic isolation that are mostly transient and do not go on to evolve as separate species, and a higher-level entity recognized as true species. The criterion for defining those higher entities was not fully explicit: their models assumed that a unit called species exists separately from a unit called populations. If genetic evidence indicates that there is appreciable gene flow between some entities delimited by multispecies coalescent methods, I would agree that these should be regarded as populations rather than species. If instead the methods reveal units below conventional species that are fully isolated (i.e. with $M \ll 1$), however, then these do represent independently evolving units of diversity. Whether the conditions for independence and survival will persist is difficult to ascertain, but under present conditions they evolve separately because of barriers to gene flow. Practical taxonomy for classification might sensibly choose not to call them separate species, but studies aiming to test for evolutionary units of diversity based on barriers to gene flow should acknowledge them as fully independently evolving units.

taxonomically (e.g. Fig. 4.5). Generating reliable nuclear markers that work across broad taxonomic scales is harder than collecting organelle data. This is changing rapidly with the rise of genomic techniques like restriction site-associated DNA (RAD) sequencing or exon capture that allows hundreds of single nucleotide polymorphism (SNP) markers to be generated for non-model organisms (e.g. Paun et al., 2016; Moritz et al., 2018). Even so, many existing studies test prior hypotheses of putative species rather than performing unguided searches.

To paraphrase the findings, multilocus methods generally do delimit separate units consistent with reduced gene flow. Whether these demonstrate discrete, reproductively isolated units according to a definition of gene flow below the threshold needed for neutral divergence of the sampled markers (i.e. $M \sim 1$) is less clear (Sukumaran and Knowles, 2017). For example, Chan et al. (2017) delimited species of frogs using BPP and other methods, but further population genetic methods demonstrated appreciable levels of gene flow in some cases (Fig. 4.6). They lumped such cases into connected metapopulation lineages, which they identified as species. Together with simulation results described in section 4.2.1, this indicates that the methods might be oversplitting and detecting genetic isolation between partially isolated populations with ongoing gene flow. Nonetheless, clear cases of strict isolation are also detected in most studies.

In contrast to animals and plants, multilocus delimitation has been used routinely in fungi for decades. Morphological criteria for defining fungal species are notoriously difficult (e.g. many sexual and asexual stages of the same species, referred to as teleomorphs and anamorphs, were named separately), yet good nuclear markers are relatively easy to find. The standard approach uses patterns of concordance versus discordance among unlinked loci to infer species limits, called the genealogical concordance phylogenetic species concept (Taylor et al., 2000). Many fungal studies are

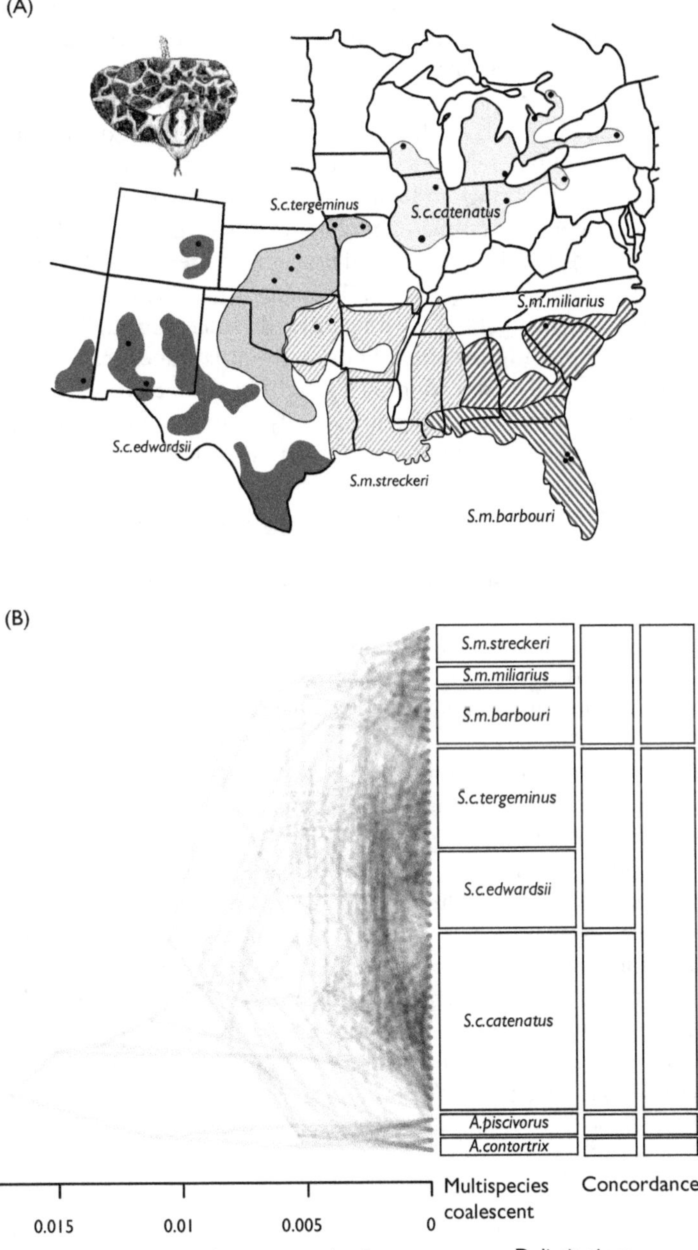

Fig. 4.5 Delimiting species of rattlesnakes in the genus *Sistrurus* using 18 nuclear loci (Kubatko et al., 2011). (A) The two taxonomically described species, *S. catenatus* and *S. miliarius*, both contained several sub-species that differ in morphology and colouration and have allopatric

at broader taxonomic scales than typical for animal and plant studies. Evidence for reproductively isolated groups within fungal clades is therefore common (Fig. 4.7A). Because genealogical concordance is more conservative than multispecies coalescent methods, these data provide robust evidence for strict reproductive isolation shaping variation in the sampled markers. These results alone, however, do not rule out more complicated patterning of reproductive isolation, such as networks of isolation of varying degrees (e.g. declining gradually with genetic distance). A further advantage of fungi is the abundance of estimates of pre- and post-zygotic reproductive isolation from crossing data, which have been used to validate inferences of reproductive isolation from molecular data (Fig. 4.7A). These data provide evidence for declining levels of reproductive compatibility with genetic distance (Fig. 4.7B).

One issue with applying these methods to fungi is the effect of recombination rates on signatures of diversification. Fungi vary in the frequency of sexual versus asexual reproduction and, even when meiosis occurs, selfing mechanisms often reduce the effective outcrossing rate. The multilocus methods outlined above assume that species are outcrossing sexuals. In a strictly asexual population, genealogies would be identical across all loci and the consensus would be fully resolved right to the level of individuals. Clearly the methods are redundant in this case. But might intermediate levels of assortment due to cyclical asexuality or high levels of inbreeding limit the power of the approach? Common intuition is that even a rare level of sexual reproduction and recombination is sufficient to generate incongruence within species that is distinguishable from congruence between species. Future methods could jointly infer recombination rates and rates of gene flow between sets of individuals.

The other major group with abundant multilocus data is bacteria and, to a lesser extent, archaea. Initially, the focus was on multilocus sequence typing (MLST) of a set of seven so-called house-keeping genes that perform basic cellular functions (Urwin and Maiden, 2003). Different but overlapping genes were used for different taxa in order to maximize variability. Often the initial design had a narrow taxonomic focus on particular pathogens, such as *Staphylococcus aureus*, and was used primarily to discriminate different clinical isolates (i.e. genotypes or strains) that might differ in disease properties or reveal the history of epidemic spread. In other words, the focus was on genetic variation within a single interacting lineage or species. However, several databases were expanded to incorporate multiple related lineages.

Fraser et al. (2007) used MLST data to investigate the nature of bacterial species. They found evidence of four multilocus clusters separated by gaps in genotype space

Fig. 4.5 Continued

distributions, and were included as putative species units. (Map reproduced with permission from Kubatko et al. (2011).) (B) Overlaid gene trees for all 18 loci including two outgroup species in the genus *Agkistrodon*. Boxes on the right indicate groupings by alternative delimitation methods: multispecies coalescent using *BEAST or BPP delimit all six in group taxa as separate species, whereas patterns of concordance of rooted triplets using the tr2 method (Box 4.1) yielded two less well-resolved delimitations. (Drawing of rattlesnake by C.E. Barraclough.)

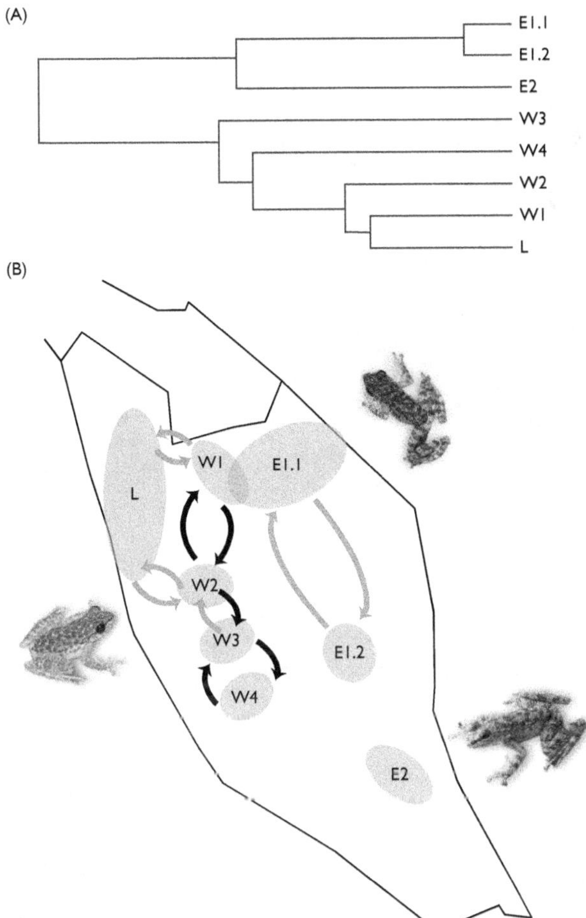

Fig. 4.6 Delimitation of species of Malaysian torrent frogs in the *Amolops larutensis* species complex by Chan et al. (2017). (A) Mitochondrial gene tree of eight populations. All eight populations were reciprocally monophyletic in mtDNA, and delimited as separate based on > 17,000 unlinked SNP loci analysed by multilocus methods BFD* and iBPP (which included morphological trait data as well as SNP data—see section 4.3). (B) Just three independently evolving species were inferred based on estimates of gene flow between populations using Fastsimcoal: west (including the population labelled L), east 1, and east 2. Grey arrows indicate that estimated contemporary number of migrants per generation (*M*) is between 0.1 and 1.0; black arrows indicate *M* > 1.0. One hybrid was sampled between eastern (E) and western (W) clades, but other individuals in the zone of overlap between W1 and E1.1 were pure breeding for west versus east genotypes, respectively. (New diagram based on data and using illustrations from Chan et al. (2017) with permission.)

in the *Streptococcus mitis* species complex, which corresponded to named taxa based on phenotypic tests (Fig. 4.8). The rate of recombination per nucleotide in these bacteria was estimated at three times the mutation rate, which was shown by modelling to be high enough to require barriers to gene flow for multilocus divergence to occur.

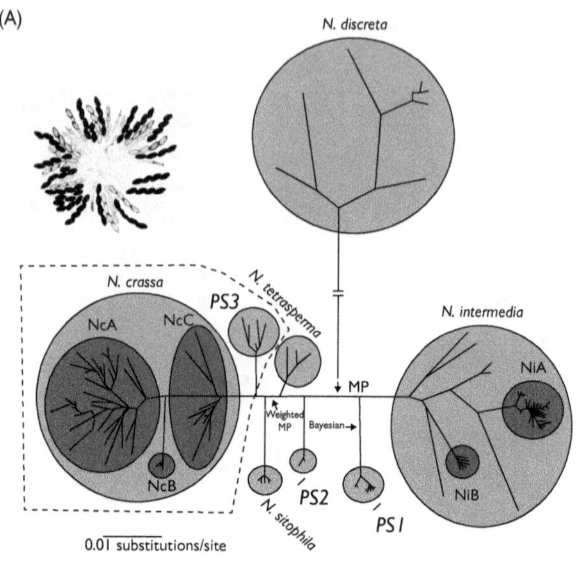

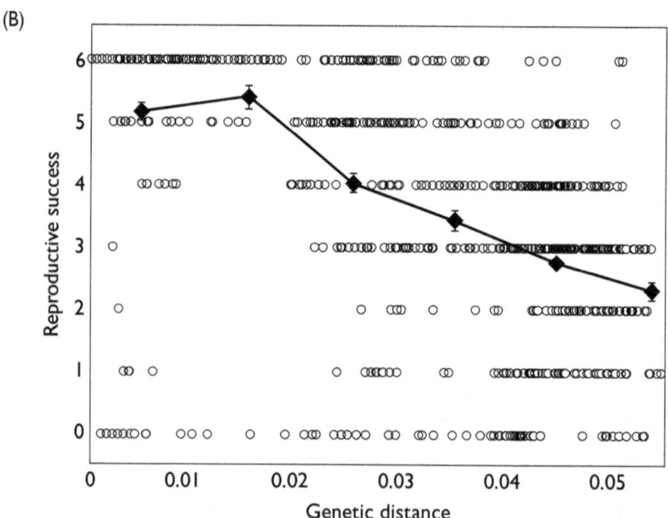

Fig. 4.7 Evidence of reproductively isolated groups of individuals with the fungal genus *Neurospora*. (A) Unrooted tree summarizing relationships of a global sample of individuals based on combined analysis of four anonymous nuclear loci. Genealogical concordance was applied to identify eight species indicated by light-grey shaded ellipses. Additional sub-groups within some of those clades were also recovered congruently (darker grey ellipses), but were deemed sub-groups because they would leave other species paraphyletic. PS1 to PS3 indicate phylogenetic species 1 to 3. Mating experiments were used to interrogate five of the eight species under the biological species concept: *N. intermedia*, *PS1*, and *PS2* displayed full mating potential within but not between species,

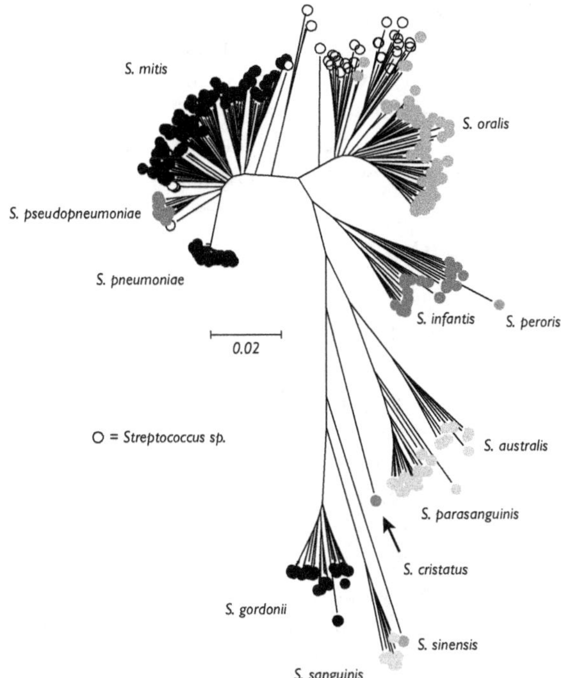

Fig. 4.8 Genetic clustering of strains of the *Streptococcus mitis* group of bacteria surveyed with 7 house-keeping genes involved in core cellular functions. Multiple, distinct clusters are apparent on the concatenated gene tree, many of them corresponding to named species. The four species identifed as clusters by Fraser et al. (2007) were *S. pneumoniae, S. pseudopneumoniae, S. mitis,* and *S. oralis.* Reprinted with minor modification under the creative commons license from Bishop et al. (2009), who sampled more taxa and an extra gene compared to Fraser et al. (2007).

Recombination occurred between loci within species clusters, whereas most genes were not exchanged between clusters. Interestingly, one gene was transferred between clusters: *ddl*, which is associated with penicillin resistance. This supports the idea of a mosaic model of bacterial species with limited gene exchange at core house-keeping loci, but wider transfer of genes under strong and variable selection pressures.

Similar patterns have been found in other bacterial clades with appreciable levels of recombination. For example, Fujisawa et al. (2016) detected reproductively isolated

Fig. 4.7 Continued

whereas PS3 was found to be fully compatible with some *N. crassa* strains (dashed box indicates inferred biological species boundary in that case). (B) Reproductive success of crosses declines with genetic distance between strains. Success was scored on an ordinal scale: 0–1, sterile, no reproductive structures; 2, some structures but no ascospores; 3–4, < 15 per cent black ascospores produced (see inset in A), indicates mature products of successful mating; 5, 15–50 per cent black ascospores; 6, > 50 per cent black ascospores. Black line and diamonds show means within bins of genetic distance, with standard errors. (Reprinted from Dettman et al. (2003a, 2003b) with permission.)

groups in the *Bacillus cereus* complex of environmental and insect-associated bacteria using the method in Box 4.1, although one clade displayed fully congruent relationships to the tips, indicative of low levels of recombination in that clade. A sample of ecologically relevant genes involved in virulence demonstrated incongruence relative to the clustering for house-keeping genes. I discuss causes of divergence in bacteria and the effects of recombination in chapter 6.

4.2.3 Multiple loci conclusion

Multiple loci are becoming ever more tractable for broad surveys and it will soon be feasible to estimate the pattern of interbreeding versus recombination across clades and genome regions. We already know roughly what that pattern will look like. Units with negligible gene flow are generally present because we do observe phylogenetic divergence across multiple markers: the alternative would be an extremely long stem branch leading to a single interbreeding metapopulation with limited variation. In fungi and prokaryotes, there is evidence for significantly discrete units for core genes, but with wider arenas for exchange for some genes, namely those under strong and variable selection. We also know that genes can transfer between otherwise phylogenetically distinct species via hybridization or introgression (Mallet et al., 2016). I discuss the significance of gene flow between species further in chapter 7. Finally, we know that partially isolated populations that retain enough gene flow to prevent long-term phylogenetic divergence at neutral markers are common within species (Morjan and Rieseberg, 2004).

There remains work to be done, however, connecting statistical models and multilocus data to quantify these patterns. Many multilocus methods still require tweaking around the edges to infer sensible units—such as specification of a threshold for acceptance of species status, or comparison with other data sources (such as morphology) to back up conclusions. While methods should always be used with brains switched on and using multiple data sources, this does imply that the methods currently do not quite 'do what they say on the tin', and leave open a subjective element to assessment of genetically isolated groups. The main advance for multilocus methods in the future, however, will come from connecting inferences of gene flow to inferences of selection across whole clades. While the methods described in this section are more powerful than single-locus methods, the genes used in most multilocus studies to date are still not directly involved with sexual or ecological traits targeted by drivers of species divergence. In all the above examples, marker variation is assumed to be neutral.

4.3 Units of divergent selection

The focus so far has been on evidence of genetic isolation—separate arenas for drift from single-locus markers (see chapter 3) or separate arenas for reproductive isolation from multilocus markers (see section 4.2). The final signature of independently evolving species that I consider is evidence for separate arenas for natural selection. Can we use surveys of reproductive and ecological traits, or the genes underlying those traits, to delimit units with shared selective optima that experience divergent selection from other such units? Few studies of species delimitation have tested for this signature, but potentially it is of more interest for understanding mechanisms of diversification than patterns of variation across neutral or effectively neutral markers. This is more challenging than looking for units of genetic isolation, however, because only a few key traits and underlying genes might be targeted by divergent selection. Many traits and genes are under uniform selection pressures across multiple species or diverging neutrally. I discuss the potential for using DNA sequence data and phenotypic trait data in turn.

4.3.1 Genetic data

The easiest scenario for identifying units of divergent selection from DNA sequence data would be if a single marker gene tended to experience different selection pressures among species. Some studies have looked for signatures of divergent selection on single loci such as the cytochrome oxidase I (*coxI*) gene, the widely used DNA-barcoding marker for animals that codes an enzyme used in oxidative phosphorylation. The hypothesis is that the amino acid sequence of *coxI* might change to adapt to local differences in energy demands, temperature, or oxygen levels.

One standard method to detect divergent selection (typically referred to as positive selection in the molecular evolution literature) is the McDonald–Kreitman test (McDonald and Kreitman, 1991; Booker et al., 2017), which tests for a significantly higher ratio of amino acid divergence to polymorphism (Ka/Pa) than expected based on the ratio of divergence to polymorphism (Ks/Ps) at silent sites, which are assumed to be neutral. Divergent selection drives amino acid divergence between species, while purifying selection maintains low amino acid variability within them. The test normally assumes known species or population units for comparison, but for a gene under divergent selection between species, we could imagine delimiting units that optimize a model of molecular evolution with separate rates of amino acid change within units versus between them (Fig. 4.9). For example, for a given species delimitation we calculate the likelihood of a model that estimates separately the Pa/Ps ratio on within-species branches (ω_{intra}) and the Ka/Ks ratio on between-species branches (ω_{inter}). The evidence of divergent selection is that ω_{inter} is significantly greater than ω_{intra}. We then search alternative delimitations: for example, stepping a threshold through the tree as in the GMYC method described in chapter 3. The best model would represent the optimum delimitation of units of divergent selection (Fig. 4.9).

Sadly, the dominant pattern for *coxI* is purifying selection acting across whole clades: both divergence and polymorphism of amino acid sites are low relative to silent sites (Kerr, 2011; James et al., 2016). Variation with species (either taxonomically defined or delimited using methods such as GMYC) tends to display $\omega_{inter} < \omega_{intra}$ because of inefficient selection against weakly deleterious mutations in the linked mitochondrial genome (Barraclough et al., 2007): amino acid mutations that are selected against over longer timescales and hence never become fixed are still polymorphic within species. There is evidence for adaptive substitutions in particular environments, such as in

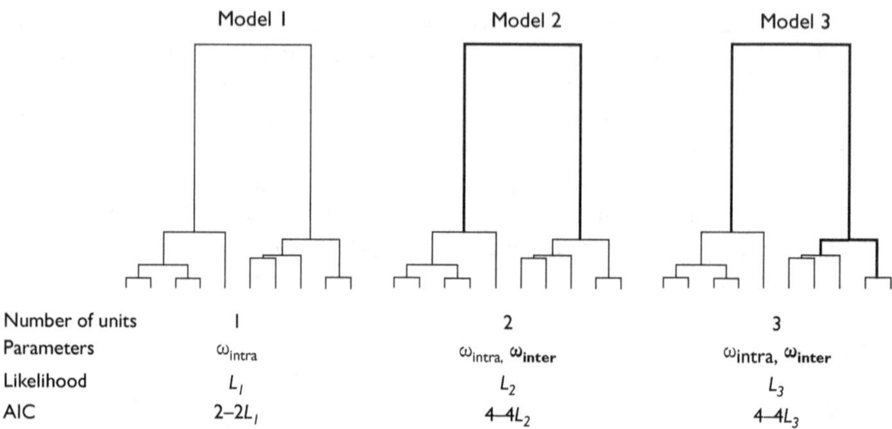

	Model I	Model 2	Model 3
Number of units	I	2	3
Parameters	ω_{intra}	$\omega_{intra}, \omega_{inter}$	$\omega_{intra}, \omega_{inter}$
Likelihood	L_I	L_2	L_3
AIC	$2-2L_I$	$4-4L_2$	$4-4L_3$

Fig. 4.9 Hypothetical method for delimiting units of divergent selection on a gene. A model of molecular evolution is fitted with a separate Ka/Ks ratio between species (ω_{inter}) and Pa/Ps within species (ω_{intra}). Model 1 assumes that the whole sample belongs to a single species experience of uniform selection on the gene. Model 2 assumes that there are two separate species that experience divergent selection between them but purifying selection within them ($\omega_{inter} > \omega_{intra}$). Thicker branches indicate between species branching. Model 3 assumes a different delimitation with three separate species experiencing divergent selection between them. A more complex model could be fitted with a separate ω_{intra} for each separate species. Optimum delimitation under the scenario of divergent selection acting on the gene is given by the lowest Akaike information criterion (AIC) score, which indicates maximum likelihood relative to number of parameters in the model. The approach could be extended to survey multiple loci in a whole-genome survey.

species adapting to parasitic lifestyles (Pentinsaari et al., 2016), but not for general changes in selective pressure across species. Oh well, it is probably too good to be true to hope that the ~ 600 base pairs of standard barcoding marker also act as a general marker for divergent selection among animal species.

Surveys of whole genomes offer better prospects to detect loci under divergent selection. Whole-genome sequences, or samples of them from transcriptome or RAD sequencing, can be surveyed to find genes that are more differentiated (i.e. higher fixation index, Fst, or Ka/Pa ratio) than expected from background levels (Romiguier et al., 2014). At present, the emphasis is on finding loci under selection between pairs of known species, or on looking for divergent sites between species delimited by other means (e.g. Fig. 4.10). However, if sampled for multiple individuals across a wider clade, these data could be used to delimit units that experience divergent selection. For example, we used the approach described above—linking GMYC-style searching of alternative delimitations with likelihood calculations in PAML (Phylogenetic Analysis by Maximum Likelihood software) of a model with separate ω_{inter} and ω_{intra}—to delimit units of divergent selection in the *Bacillus cereus* species complex of bacteria from alignments of ortholog genes (see chapter 5). The units were broadly similar to those inferred by GMYC based on patterns of genetic clustering in house-keeping

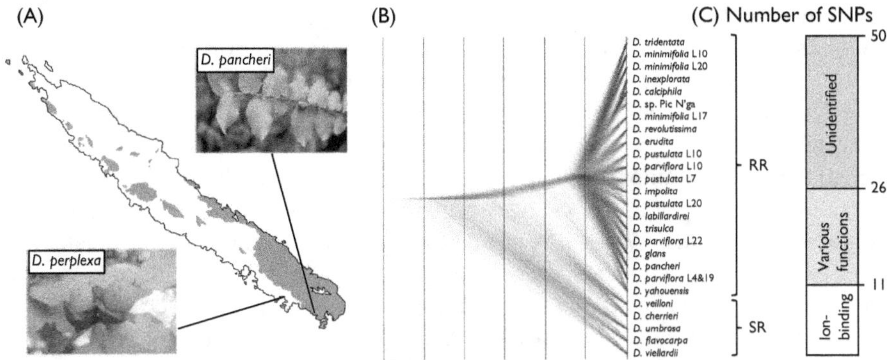

Fig. 4.10 Radiation of the tropical tree genus *Diospyros* under strong diversifying selection in New Caledonia. (A) New Caledonia is a continental fragment with complex geological history leading to a mosaic of soil types, including ultramafic soils with high heavy-metal content (shaded dark grey). A clade of 24 species of *Diospyros* has diversified largely over the last 2 million years to occupy all major terrestrial vegetation and soil types on the island. Species are ecologically and morphologically well differentiated (two examples shown) but exhibit extremely low variation in conventional molecular markers. (B) A species tree inferred from 1506 RADseq SNPs using SNAPP (a Bayesian multispecies coalescent model). Species units were assumed to be those recognized morphologically, rather than delimited *de novo* from genome data. A STRUCTURE analysis identified only two genetically isolated units: the slowly radiating (SR) group and rapidly radiating (RR) group. (C) A set of 50 SNPs were identified that exhibited high differentiation indices (Fst) in at least two separate comparisons between species pairs on ultramafic versus volcanic soils (i.e. indicative of parallel changes). Over one-third of highly differentiated loci matched to ion binding functions, which could reflect adaptation to different heavy-metal content of the soils. (Map reprinted from Grandcolas et al. (2008) with permission, photographs from Turner et al. (2013), and species trees reprinted from Paun et al. (2016) under creative commons license.)

genes. If there are simple species units, we expect to find that congruent units apply for several loci: a single delimitation model is sufficient to explain variation in Ka/Ks ratios. Not all loci would necessarily show significant divergent selection, but those that do would reflect the same species units.

There are several challenges to applying genome screens across whole clades. First, it is likely that different loci are involved in reproductive and ecological divergence in different lineages. Comparative analyses work best when the same trait or gene is demonstrating the same process and pattern across multiple species. How can we make general statements about the effects of selection on diversification if different genes are targeted in different species? There might be sets of particular genes that are under strong diversifying selection in all species, for example, those involved in defence against pathogens (Enard et al., 2016), which are likely to evolve rapidly and to different sets of coevolving pathogens in each species, or those under sexual selection or conflict. This would greatly simplify the delimitation of units experiencing divergent selection pressures.

Another challenge relates to metrics used for divergent selection. The McDonald–Kreitman and PAML approaches are designed for detecting differences in *rates* of change, and have relatively low power except in the case of strong selection between recently diverged species. Imagine, for example, that adaptation to a new environment involves changes in a relatively small number of amino acids—say three in a protein that is 300 amino acids long. A few changes can have a large effect on protein function, and many adaptive changes in proteins involve changes in relatively few amino acid positions. When divergence times are short and the accumulation of silent substitutions between species is low, the excess of amino acid divergence is detectable against the background expected rate under purifying selection in a single environment (defined as ω_{intra} above). As the number of silent substitutions accumulates, however, and the expected number of amino acid substitutions also accumulates, these three changes are harder to detect against background expectations based on polymorphism, unless ω_{intra} is so low that the level of amino acid polymorphism within species is virtually zero.

Only if the selection pressures in the two species entail continued amino acid change (for example, resistance genes in an evolutionary arms race with an antagonistic parasite) or a shift in a sizeable number of amino acids (e.g. MHC loci) do we predict a significantly higher sustained rate of Ka/Ks between species than within species. If selection involves progress to a new optimum (for example, during adaptation to a new niche), its effects are only detectable for a relatively short time until they merge into background divergence. Ideally, for present purposes, we want to identify separate adaptive optima rather than evidence of high rates of change. We want to know whether sets of individuals have the same or different selected optima for a set of loci. Optima might vary in a hierarchical way: some genes have different optima between species, whereas others have shared optima among closely related species but different optima between higher clades (e.g. genera). Unfortunately, detecting separate optima requires additional evidence for the functional consequences of observed variation: that is, its effects on protein function or on individual fitness in different environments.

Another conceptual problem with these approaches is that they assume that genes act independently and that patterns of divergent selection can be inferred by surveying statistical patterns across a set of loci. But does selection on the genome really work like this? Selection acts on genes via phenotypes, and the measured unit of phenotypes is a trait. Multiple traits in turn are not independent if overlapping sets of genes encode them. Interactions between multiple genome regions could have large effects on phenotype without greatly changing coding sequence at any locus. What effect does a more realistic genome to phenome map have on the power to detect patterns of selection from DNA sequence data (Messer et al., 2016)?

4.3.2 Phenotypic traits

The easiest way to collect data on adaptive traits is to measure phenotypic traits associated with mating preferences or ecological niche use. Integrating these data with information from genetic markers could refine interpretation of patterns compared to analysis of morphology alone (see section 3.2). For example, consider the evolution of a phenotypic trait between two fully isolated species under alternative models of the nature of selection acting on the trait. If the trait is neutral and additive, within a panmictic sexual species we predict a normal distribution with expected variance $2N_e\sigma^2_m$ (Lynch and Hill, 1986), where σ^2_m is the rate of accumulation of mutational variance per generation (i.e. the input of variance per generation due to mutation). Between two species, variance increases linearly with the divergence time T with the same mutational parameter: the expected variance is $2T\sigma^2_m$ (Lande, 1976; Lynch, 1990). The expected ratio of interspecific to intraspecific variance is therefore T/N_e for a neutral additive trait. In addition, we can calculate the expected ratio of interspecific divergence to average phenotypic pairwise distance within species as a phenotypic version of the Birky index used in chapter 3 for genetic distances (Fig. 4.11).

If the trait is under divergent selection instead, with a different phenotypic optimum in each species, $\phi 1$ and $\phi 2$, phenotypic variance within species decreases, due to stabilizing selection on the trait. In turn, the variance between species increases as long as $(\phi 1 - \phi 2) > \mathrm{sqrt}(4T\sigma^2_m/\pi)$: that is, the difference between the two optima exceeds the expected divergence under neutrality (Fig. 4.11). Phenotypic differentiation would therefore be higher than for neutral expectations (Lande, 1976). In contrast, a trait under uniform selection in each species would have a lower value of phenotypic

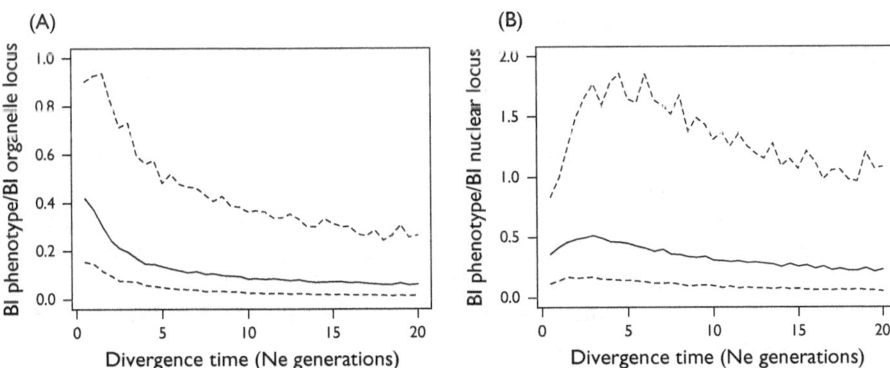

Fig. 4.11 Simulated values of the Birky index (BI) for a neutral phenotypic trait relative to (A) a neutral organelle locus (e.g. mitochondrial or plastid) and (B) a neutral nuclear locus. Genealogies were simulated assuming a neutral multispecies coalescent model with the sim.coaltree.sp function of Phybase (Liu and Yu, 2010) with 1000 replicate simulations for every interval of $0.5Ne$ generations. Values of a single quantitative trait were simulated using the expressions for expected variation outlined in the main text, coded in R. Simulation code is available in the materials for the book. Expected divergence of mean traits between species is $\mathrm{sqrt}(4T\sigma^2_m/\pi)$, whereas expected trait distance between two individuals within species is $\mathrm{sqrt}(8N_e\sigma^2_m/\pi)$. Hence, expected Birky index for a neutral additive phenotypic trait is $B_p = \mathrm{sqrt}(T/2N_e)$. Median (solid line) and upper and lower 95 per cent confidence limits (dashed lines) of phenotypic Birky index divided by genetic Birky index (derived in Box 3.1) are shown. For an organelle locus, $B_o = (2T+N_e)/N_e$, and for a neutral locus, $B_n = (T+2N_e)/2N_e$. Expected B_p for a phenotypic trait is therefore $\mathrm{sqrt}((B_o-1)/4)$ or $\mathrm{sqrt}(B_n-1)$.

differentiation, because divergence between species is constrained to be less than the expected divergence with neutrality. These predictions can be used to identify characters under divergent selection: with known neutral traits as a baseline, we can test for traits that are significantly more differentiated than neutral expectation. The rationale is equivalent to that used for the McDonald–Kreitman test on sequence data.

Although no morphological traits can be robustly assumed to be neutral, morphological differentiation can be compared with the molecular differentiation for silent variation in molecular markers. For example, Q_{st} in phenotypic traits is compared to F_{st} of neutral markers such as microsatellites in tests for adaptive differentiation among populations (Leinonen et al., 2008; Le Corre and Kremer, 2012). At broader taxonomic scales, phenotypic differentiation could be compared to differentiation at silent sites in a mitochondrial or plastid organelle marker or several nuclear markers.

For example, the expected Birky index for a neutral additive phenotypic trait outlined above can be compared to the equivalent index for a neutral organelle marker (Fig. 4.11): under neutrality, phenotypic traits should be less differentiated than molecular markers, especially haploid, maternal markers. A robust threshold for judging 95 per cent significance for divergent selection on a trait is a Birky index greater than the index for an organelle locus or twice the index for a nuclear locus (Fig. 4.11). In contrast, values below the 95 per cent confidence limits would indicate shared purifying selection on the traits. Standard terms and conditions apply. These predictions make simplifying assumptions of panmictic species, additive genetic traits, constant population size, silent variation strictly neutral, and so on. But they show a framework that could be used to infer patterns of selection across species pairs by combining phenotypic and standard molecular data.

The predictions can be expanded to a phylogenetic context across a whole clade by comparing rates of interspecific phenotypic divergence across a species tree to phenotypic variation with species, relative to expectations predicted from neutral genetic markers (Box 4.3). Units of divergent selection could be optimized by fitting a model with two rates of phenotypic change (one within species, one between species) and finding the delimitation yielding the maximum likelihood solution. An application in bdelloid rotifers is presented in chapter 6.

Beyond the standard caveats, however, there are several complications to analysing phenotypic traits in this way. The predictions apply to additive genetic variation, whereas traits also incorporate phenotypically plastic variation that can alter variation within each species and potentially divergence between species. For example, the range of environments experienced by each individual can increase variation, but phenotypic plasticity can also counteract additive genetic variation (e.g. countergradient variation (Phillimore et al., 2010)). Even without tackling this problem, comparisons of differentiation indices for multiple traits, referenced against molecular data, could be used to identify candidate traits that are consistent with uniform versus divergent selection—with due caution for alternative explanations.

The tests could be improved if we knew the genetic basis of the focal trait (e.g. from empirical measures of heritability or crossing and quantitative trail loci (QTL)-based

Box 4.3 Testing for diversifying selection on phenotypic traits across clades.

Assume that a phenotypic trait has been measured for samples of multiple individuals per species from a clade of three species (see the true species tree in Fig. 4.12A). Gene trees from an organelle (e.g. mitochondrial or plastid, Fig. 4.12B) and nuclear marker (Fig. 4.12C) are also available.

For simplicity, assume that every species has the same effective population size, N_e. Either the trait is assumed to have high heritability or breeding values have been inferred for each individual to remove the effects of phenotypic plasticity. Consequently, under a null model of neutral evolution, the expected variance within each species under standard neutral assumptions is $2N_e\sigma^2_m$, where σ^2_m is the rate of increase in variation due to mutation per generation. Assuming a model of neutral Brownian motion, the expected variance along species branch j is $\sigma^2_m t_j$, where t_j is the duration of the branch in number of generations.

The gene trees have branch lengths in units of per cent sequence distance, which are assumed to reflect neutral variation (e.g. estimated just from silent substitutions). The average pairwise genetic distance within each species for the organelle marker is $\theta_o = N_e\mu_o(1-1/i)$, where μ_o is the mutation rate per generation and i is the sample size per species. The

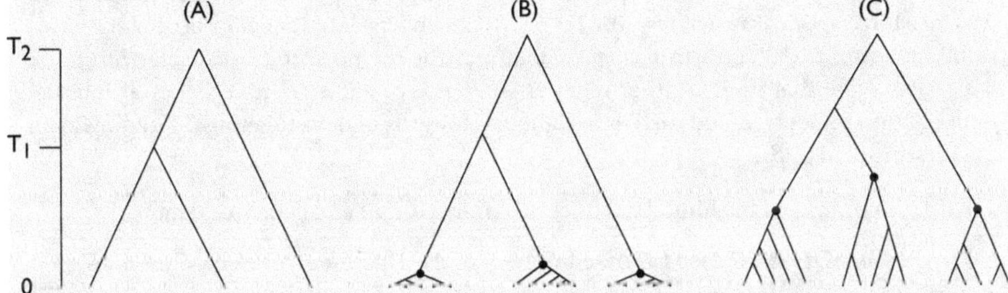

Fig. 4.12 Species tree of three species (A) with divergence times T_1 and T_2, and a reconstructed mitochondrial (B) and nuclear (C) gene tree, with MRCA nodes for each species denoted by black circles.

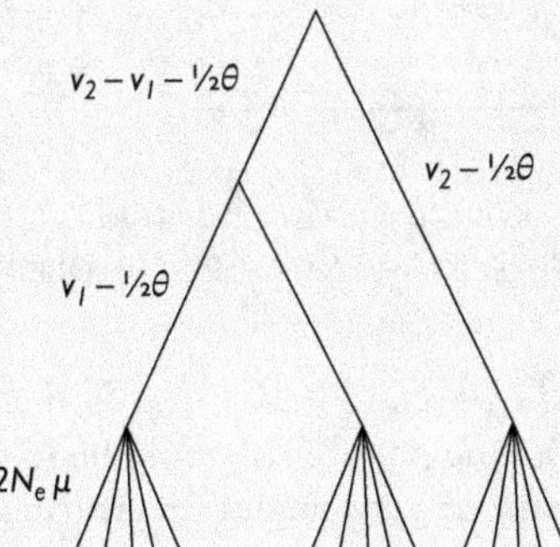

Fig. 4.13 Transformation of a gene tree to yield an expectation of uniform Brownian motion within and between species under a neutral model of trait evolution. v_k refers to genetic distance from each internal species divergence to tips in observed gene trees (Fig. 4.11).

Box 4.3 *Continued*

expected genetic distance back to species node k from the tips is $v_k = T_k \mu_o + \tfrac{1}{2}\theta_o$: the second term corrects for the effects of ancestral polymorphism on genetic divergence times (Box 3.1).

These results imply a transformation to the gene trees that would yield an expectation of uniform Brownian motion across both within- and between-species branches (Fig. 4.13). First, extract the molecular species tree and subtract $\tfrac{1}{2}\theta_o$ from each branch to estimate $T_k \mu_o$. The expected morphological variance per unit species branch length is σ^2_m/μ_o. Second, assign a polytomy with i tips within each species, with tip branch having length $2N_e\mu_o = 2\theta_o i/(i-1)$. Again, the expected morphological variance on those branches is σ^2_m/μ_o.

A similar transformation can be made for the nuclear marker to yield expected morphological variance on all branches of σ^2_m/μ_n. The only difference is that tip branches are lengths of $2N_e\mu_n = 0.5\theta_n i/(i-1)$, because $\theta_n = 4N_e\mu_n(1-1/i)$.

The null hypothesis of neutral evolution can be tested by comparing the likelihood of a model with a single Brownian rate parameter, $\beta = \sigma^2_m/\mu$, to the likelihood of a model fitting two rate parameters, β_b for between-species branches and β_w for within-species (Table 4.1). A trait under divergent selection between species should yield $\beta_b > \beta_w$ whereas $\beta_b < \beta_w$ might indicate a trait under selection for the same optimum in all species. Analyses of simulated data for a three-species case with $i = 10$, $T1 = 15$, $T2 = 20$ under neutral and non-neutral scenarios (1000 replicates each) confirm the applicability of the method (code available with book materials). Note that model fitting requires inference of ancestral trait values at internal nodes of the species tree, and there is a slight bias towards underestimating β_b, leading to an underestimate in β_b/β_w.

Table 4.1 Summary of likelihood analyses of simulated trait and gene tree data

Simulation treatment	Organelle locus as baseline			Nuclear locus as baseline		
	Mean β_b/β_w (95% C.I.)	Mean ΔAICc*	Power†	Mean β_b/β_w (95% C.I.)	Mean ΔAICc	Power†
Neutral: $\beta_w = 1$, $\beta_b = 1$	0.80 (0.02, 3.35)	0.25	0.05	1.02 (0.02, 4.50)	0.51	0.05
$\beta_w = 1$, $\beta_b = 2$	1.55 (0.03, 6.08)	0.25	0.05	1.83 (0.04, 8.45)	0.60	0.07
$\beta_w = 1$, $\beta_b = 5$	3.86 (0.07, 15.96)	2.93	0.18	4.28 (0.08, 19.85)	3.53	0.19
$\beta_w = 1$, $\beta_b = 10$	7.69 (0.24, 31.92)	7.96	0.42	8.29 (0.22, 35.16)	9.08	0.40

* Mean ΔAICc is the mean difference in Akaike information criterion corrected for small samples (AICc) between the single-rate model and the two-rate model.
† Power is the number of replicates with ΔAICc > 95 per cent value under the null model.

studies). We are unlikely, however, to have this information for many traits in many clades (Wood et al., 2016). One possibility would be to estimate heritability and the genetic structure of traits from genealogies reconstructed within species. A future study could sample multiple individuals within multiple species and sequence the whole genome of each individual as well as measuring trait variation (Ellegren, 2014). We could then fit alternative models for the coding of the trait that vary in their predictions for the correspondence between phenotypic variation and genomic marker variation (e.g. from the simplistic case outlined above). A model combining the

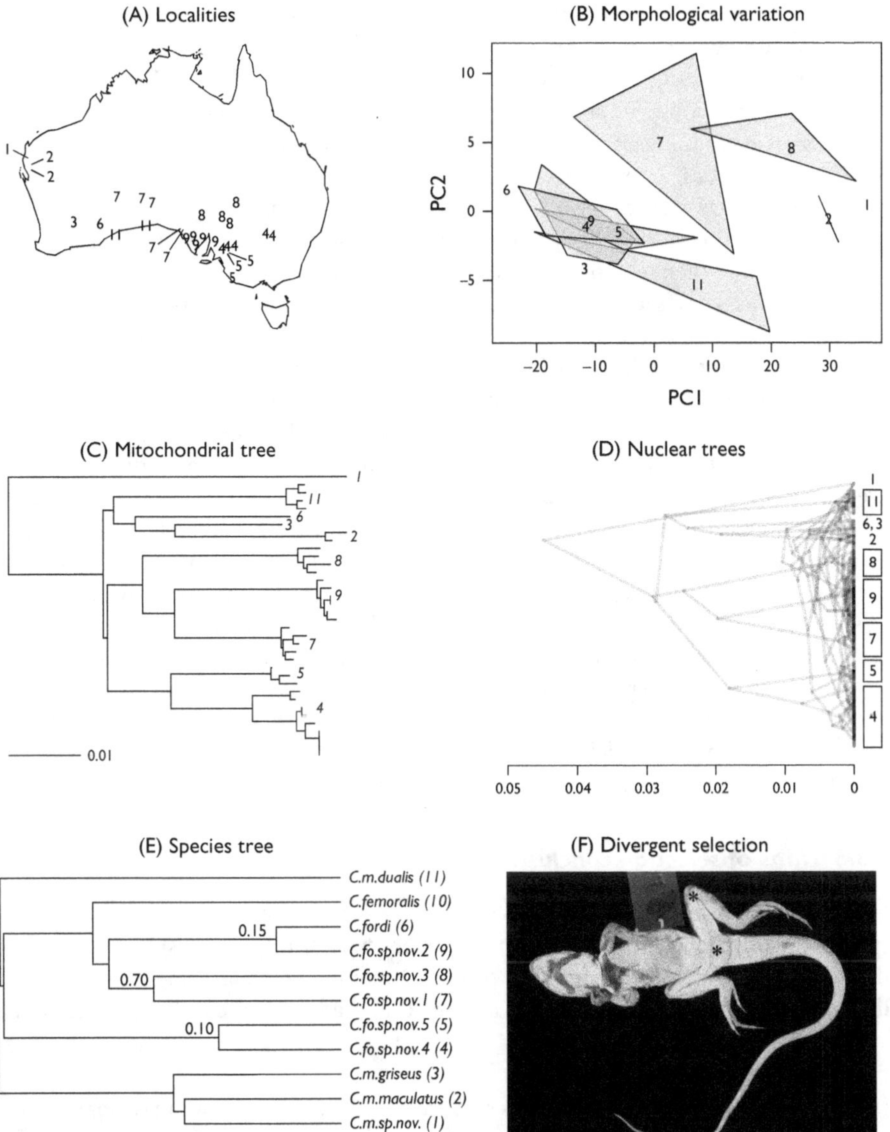

Fig. 4.14 Molecular and morphological variation in the *Ctenophorus maculatus* species complex of lizards (after Edwards and Knowles, 2014). Data trimmed to include only 36 males with all genes and morphology sampled. Numbers indicate species groupings assigned by the authors (species 10 trimmed here). (A) Collection localities of individuals. (B) Variation in 18 morphological characters along first two principal component axes: centroids and polygons connecting all measurements for each labelled species are shown. (C) Mitochondrial DNA tree from concatenated sequences of 16S and ND2 genes. (D) Overlaid trees from four nuclear markers: BACH1, GAPD, NTF3, and PRLR. A single allele per individual was retained (rather than both diploid alleles) for visualization. Branch lengths are in units of substitutions per site.

genetic basis of the traits and the pattern of selection on them could then be evaluated. Potentially, with a large enough sample of individuals, the causal genetic elements contributing to trait variation could also be modelled (Ellegren, 2014). Analyses such as these applied to whole clades are still far away, but there is great potential in assembling an integrative approach to phenotypic and genomic evolution across whole clades.

At present, few studies have truly integrated molecular and morphological data into species delimitation. Many studies that claim to be integrative look at multiple data sources and then use consensus or a subjective assessment to bring those sources together. Solis-Lemus et al. (2015) devised a method for delimiting species using combined multilocus molecular and phenotypic trait data, which they applied to Australian sand dragon lizards (Fig. 4.14). This method looks for genetic isolation—as evident from separate normal distributions of trait variation between species—rather than divergent selection per se, although power should be increased for traits under divergent selection and the outputs allow traits to be ranked by how divergent they are across the delimited units. The power of combining data sources, especially in the light of potential issues for phenotypic traits, remains to be evaluated. In the sand dragons, the same delimitation is obtained using just the mitochondrial marker, without adding in the five nuclear loci or 18 morphological traits. An early impression of such methods is that they tend to optimize delimitation of all hypothetical species assigned by the researcher, similarly to multilocus methods. Any lumps and bumps in the trait data (e.g. different growth stages, geographic structure) could lead to the maximum number of species being delimited simply to soak up variance that is inconsistent with the model assumptions (e.g. panmictic, additive genetics, normal variation within species), just as in DNA-based methods.

4.3.3 Units of selection conclusions

Estimating patterns of selection is an underexplored avenue for statistical methods of species delimitation. The observation that phenotypic traits often display a pattern of discrete clusters within clades (see section 3.2) might be seen as evidence for units of divergent selection, if in addition we assume that many of those traits are likely to be

Fig. 4.14 Continued

(E) Species tree recovered by multispecies coalescent method (BEAST*), with taxonomic names and number used for species grouping in parentheses. All putative species were delimited with full support using just DNA or DNA plus trait data in analyses with iBPP, which fits a multispecies coalescent to molecular data and separate normal trait distributions to morphological data (Solís-Lemus et al., 2014). Support values show nodes supported with < 1.0 posterior probability when using just trait data. (F) Testing for divergent selection with the method outlined in Box 4.3 using mitochondrial variation as a neutral baseline revealed one trait to be under significant divergent selection ($\beta_b > \beta_w = 6.74$; p = 0.05, correcting for false discovery rate): the number of femoral pores, which are present in males and secrete chemical signals used in mate choice and territory marking (Edwards et al., 2015). (Photo: *C. maculatus dualis*, species 11, online Digitized Type specimen from Western Australian Museum, with permission.) Start and end of femoral pores highlighted with *.

under strong selection (and hence optima must differ among clusters). Furthermore, correspondence of phenotypic clusters with measures of reproductive isolation further supports overlap with units of reproductive isolation. These data are limited at present, however, to cases where direct measurement of reproductive isolation is feasible. Returning to scenarios outlined in chapter 2, it is possible in some circumstances that units delimited by reproductive isolation do not experience divergent selection and that units of divergent selection do not exhibit reproductive isolation (i.e. in speciation with gene flow scenarios). Statistical evaluation of patterns of reproductive isolation and divergent selection are needed that combine genetic and phenotypic trait data.

4.4 Prospects for whole-genome data

Whole-genome data have revolutionized inferences on the history and causes of divergence between closely related species. It is possible to fit demographic models of how the effective population sizes of two species and their common ancestor, and levels of migration between the diverging species, changed over time (Fig. 4.15). It is also possible to identify genomic regions that exhibit excessive divergence between the species, interpreted as evidence of regions under divergent selection, or those that are unexpectedly uniform, indicative of ongoing gene flow or shared patterns of purifying selection. The number and distribution of such regions across the genome is a key metric for comparing models of divergence with or without gene flow (Nosil et al., 2009; Cruickshank and Hahn, 2014). Based on genome data alone it is not possible to attribute genetic variants to particular traits or consequences on fitness, although hypothetical connections are made from gene annotations. Further genetic crosses, reciprocal transplant and common garden experiments, and ideally genetic modification are used to confirm phenotypic and fitness effects of genetic variants.

Soon it will be feasible to sequence whole genomes from samples of thousands of individuals across large clades using long-read technology. How will these data help us to understand whether life diversifies into discrete species units and how reproductive isolation and divergent selection shape their formation? Because the field is moving so rapidly, I concentrate on what whole-genome data could be used for, rather than how to do it.

An immediate benefit will be ready access to thousands of loci with known patterns of linkage. Extending current methods for estimating reproductive isolation and gene flow will allow for a comprehensive evaluation of how these processes are patterned across clades. How often are there strict units of reproductive isolation rather than a gradual decline in gene flow between distantly related taxa (Liti et al., 2006)? What factors correlate with strict isolation versus ongoing gene flow, such as time since divergence, geographical proximity, and ecological similarity (c.f. Smillie et al., 2011)?

Beyond existing methods that focus on sequence variation among orthologous loci, whole-genome data will allow comprehensive use of structural genome changes, such as duplications and genome rearrangements, for inferring genetic isolation. For

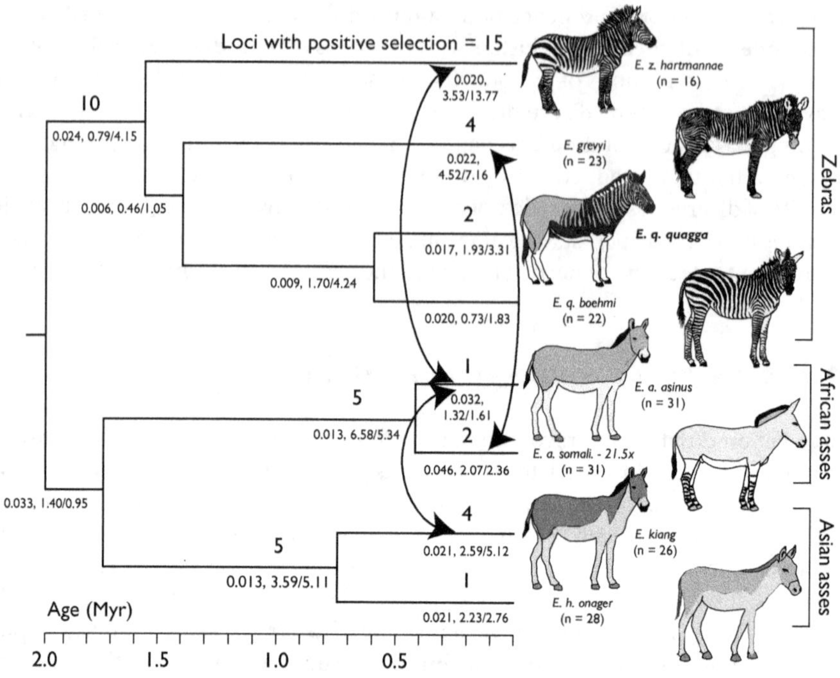

Fig. 4.15 Inferences on gene flow, structural genome changes, and selection during history of the *Equus* genus: asses and zebras (Jonsson et al., 2014, modified with permission). Dated phylogeny reconstructed from whole-genome sequences is shown, with numbers of loci inferred to be under positive (i.e. divergent) selection above each branch, and rates of gene loss and chromosome gains/losses below each branch. Numbers of chromosomes in each extant taxon are shown next to their names. Curved arrows indicate gene flow inferred from coalescent models. Gene flow is reconstructed even between taxa with different chromosome numbers.

example, it has long been posited that rearrangements might facilitate divergence and coexistence of species by protecting divergent genome regions from the homogenizing effects of recombination (Rieseberg, 2001). Comparative studies on organisms ranging from yeast (Hou et al., 2014) to rodents (Castiglia, 2014) have identified genome rearrangements as a correlate of reproductive isolation and/or range overlap (indicative of a pressure to preserve divergence against the action of gene flow; but see Fig. 4.15). By corollary, patterns of genome rearrangements such as inversions or the fusing or splitting of chromosomes could be used in future to infer both interacting gene pools and the extent to which mechanisms are required to protect against the homogenizing effects of gene flow.

Another possible use implied by studies of bacteria and yeast is to develop models of the probability of recombination between two individuals based on the similarities and differences between their genomes (Liti et al., 2006). Although the probability of interbreeding and recombination is unlikely to be simply a function of DNA sequence divergence in multicellular eukaryotes, as it has been modelled in bacteria and yeast, it will still be impacted by divergence at particular types of genes (e.g. those determining mating compatibility) and structural differences such as chromosomal rearrangements. Combining the signatures of isolation at orthologous loci described in section 4.2 with specific loci involved in mating and wider structural differences could allow for more powerful estimation of the probability of gene exchange, both between individuals within a clade and across genome regions (Fig. 4.16). Alternative 'species' models could be fitted to this network of probabilities, ranging from strict discrete units, to partial gene flow at particular genome regions between species, to declining gene flow with increasing divergence of particular genomic features.

The other major benefit from whole-genome data will be to infer the pattern of selection across individuals in the sample. For example, an initial goal would be to estimate the probability that each gene region and non-genic regulatory region is under uniform selection or selection for different optima between each pair of individuals. This probability depends on the variability at that region across individuals both within and between species units, and may also be affected by whether the region has experienced a chromosomal rearrangement. Assuming a network of shared versus divergent selection pressures can be reconstructed across individuals and genes (Fig. 4.17), alternative models could be fitted: the simplest species model is that conspecifics share the same selection pressures across their genome (even though there is variation due to mutation and the strength of selection may vary) but differ in selection pressures for at least several traits and genomic regions from other species. Applied simplistically, this criterion

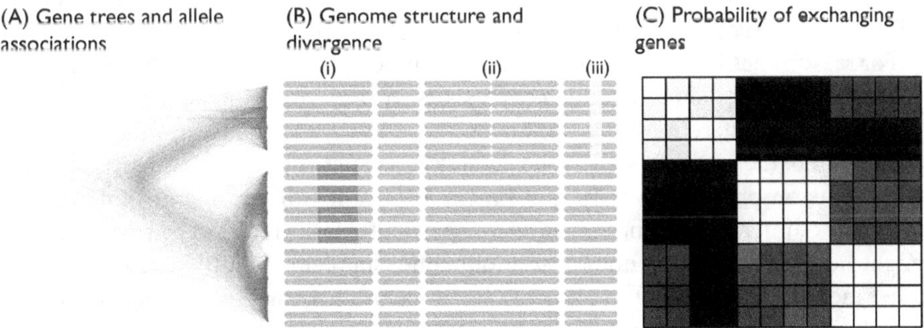

Fig. 4.16 Hypothetical illustration of estimating the network of probabilities of exchanging genes across a set of 12 individuals from whole-genome sequence data. Multiple information sources are integrated including: (A) the pattern of gene tree topologies and networks of allele association within diploid individuals across orthologous loci; (B) structural changes in the genome such as inversions (dark grey shading, i), chromosome fission (ii), and deletions (pale shading, iii), and sequence divergence across loci, and mating incompatibility loci in particular if known (not illustrated). Horizontal bars indicate chromosomes in diploid individuals arranged vertically. Together, these feed into prediction of (C) probabilities of exchanging genes between all pairs of individuals (portrayed as a heatmap: ranging from black (0 probability of exchanging genes) to white (1.0)), either averaged across whole genome as shown or reconstructed separately for different genome regions. In this case, three discrete and cohesive units of gene exchange are present containing four individuals each.

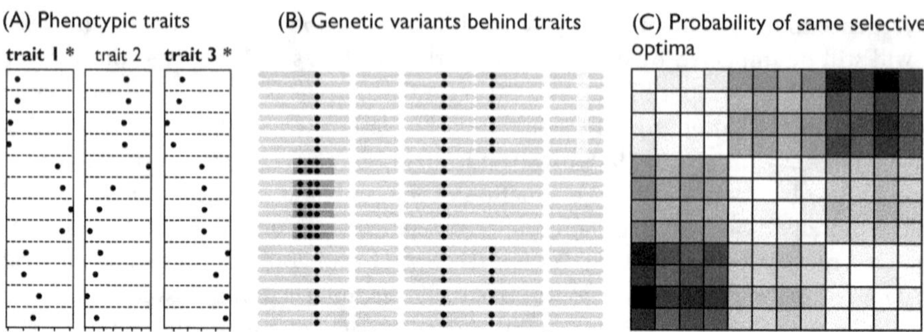

Fig. 4.17 Hypothetical illustration of estimation of the network of probabilities of 12 sampled individuals sharing the same selection pressures across multiple traits from phenotype and whole-genome sequence data. (A) Three traits are sampled and patterns of variation analysed relative to variation of neutral genetic markers to identify traits under divergent selection (e.g. trait 1 and 3). (B) GWAS-type analyses are used to infer gene regions associating with variation in the selected traits (black circles), supplemented with further functional knowledge where available. Multiple data sources are then integrated to estimate (C) the probability, averaged across traits and/or genome (as shown) or separately for different sets of traits or genes, that each pair of individuals share the same selective optima (shown as a heatmap: ranging from black (0, different optima) to white (1, same optimum)). In this case, different selective optima for each of the three discrete units of genetic isolation in Fig. 4.16 are apparent, although the bottom two units retain a moderate probability of sharing the same optimum (mid-grey shading).

would delimit males and females as separate species, so the method would need to be adjusted to take account of sexual dimorphism where present.

For reasons outlined above, there will be numerous challenges to realizing this goal. First, current methods rely on detecting shifts in rates of amino acid change, rather than detecting different optima, and hence have low power between more divergent taxa. Second, there remains a disconnect between patterns of DNA sequence variation and effects on phenotypic traits and fitness. Ideally, it will become possible to infer networks of interacting genes that underlie particular traits targeted by selection. That might be feasible from genome variation alone, supplemented by transcriptome or other molecular data (Koch et al., 2017), but it seems likely that integrating phenotypic trait data with genome data will be more productive over the shorter term. With a large enough sample of individuals, it may be possible to infer the contribution of genetic variants to traits using a genome-wide association study (GWAS) or pedigree analysis approach (Ellegren, 2014).

Assuming those challenges are met, the final step in comparing alternative models of diversification is to overlay the networks of gene exchange and selection—the two key parameters for models of species divergence—to map out the diversification landscape (Fig. 4.18). This can then be used to test for interactions between selection and gene flow. For example, how often are regions under divergent selection protected by low rates of recombination, either in cold-spots for recombination (Samuk et al., 2017) or associated with genome rearrangements? How important are genetic mechanisms for reducing gene flow versus direct selection for different optima in each species? Does the

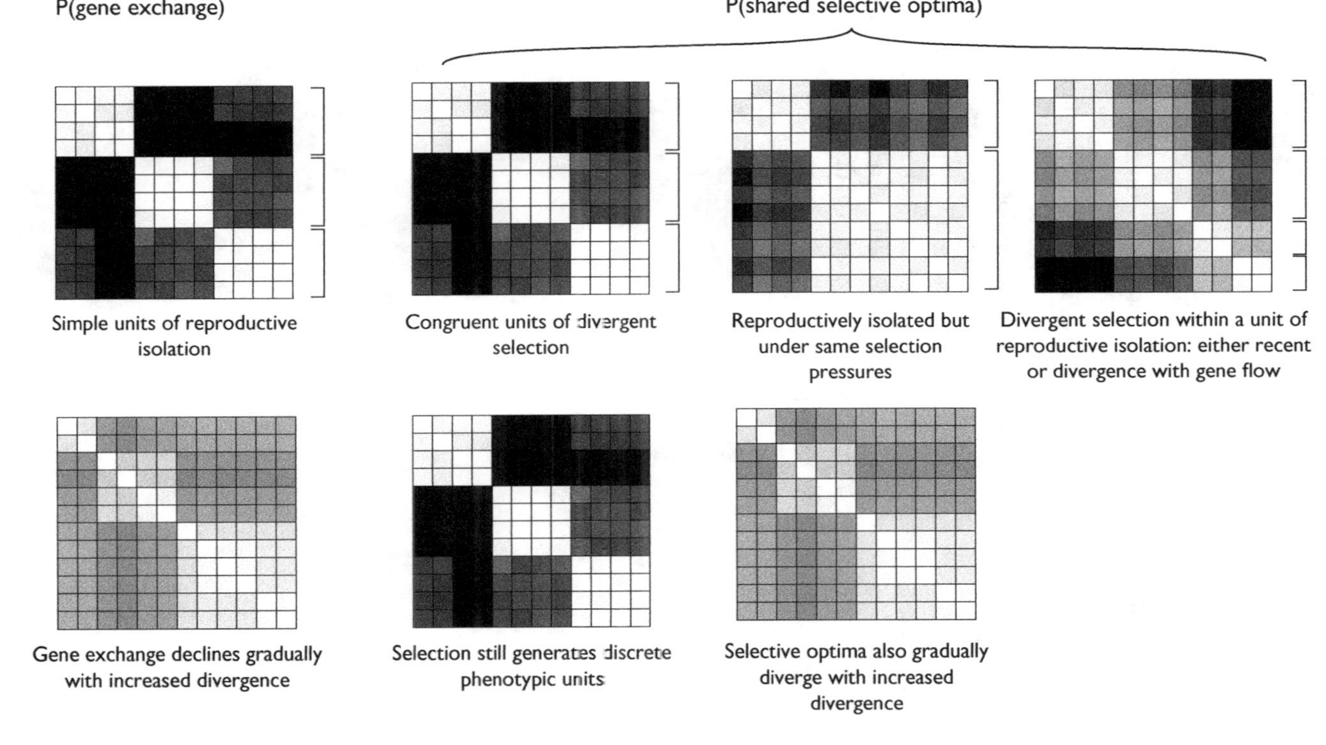

Fig. 4.18 Hypothetical comparison of the network of probabilities of gene exchange (column 1) versus probabilities of shared selective optima. Within each heatmap, rows and columns are ordered from individual 1 to individual 12 (white, higher probability). Two scenarios for probabilities of gene exchange are shown: discrete units of genetic isolation (row 1, column 1) versus declining probability of gene exchange with increasing divergence (row 2, column 1). For each of those scenarios, alternative scenarios of the network of selective optima among individuals are shown. Not necessarily all of these scenarios are equally plausible, but they illustrate alternative possible mappings between networks of gene exchange and similarity in selective optima.

simple species model of a single unit structuring diversity patterns apply generally across eukaryotes and prokaryotes, and, if not, what circumstances lead to departures from this model? I continue some of these lines of questioning in chapters 5 and 6.

4.5 Conclusions

A range of new data and analytical tools make it possible to survey the action of evolutionary processes across large samples of individuals. Rather than using descriptive and informal methods to identify diversity units, we can now estimate evolutionary parameters and test alternative hypotheses for the structure of diversity and its causes. There is clear evidence from single-locus data and from morphological surveys that coherent clusters exist in most groups as expected if discrete and independently evolving species are present. Nonetheless, departures from a simple 'species' model are also apparent, and these cases may be extended further as whole-genome data provide more resolved insights into levels of gene flow and patterns of selection across whole clades. In particular, there is great scope for using measures of divergent selection more explicitly in evolutionary analyses of species limits.

5

What causes speciation?

5.1 Introduction

The previous chapters considered forces that maintain cohesion within species and permit divergence and independent evolution between them. I now move to the formation of species—speciation. How does one cohesive species split into two or more independently evolving species? This process requires conditions to arise that favour splitting and then divergence of populations in response to those conditions. Understanding speciation therefore requires knowledge of when and why the conditions that favour divergence arise, as well as of the genetic responses once such conditions arise. There have been many excellent accounts of speciation that range from genetics of speciation to broad speciation patterns (Mayr, 1963; Coyne and Orr, 2004; Gavrilets, 2004; Butlin et al., 2012; Nosil, 2012; Seehausen et al., 2014). Arguably, less is known still about how genetic responses interact with the origin of geographical and ecological conditions favourable for speciation. I will explore evidence for these processes from analysis of speciation patterns across clades.

In the expanded view of diversity outlined in chapter 2, it is possible that discrete units are not present in some taxa, and the aim would then be to understand conditions favouring multiple causes of independence versus cohesion to arise. Present work on speciation, however, mostly assumes that species are present and correctly delimited. Chapters 2 to 4 provide enough evidence for species that we can continue with that assumption for now, although details may change when formal methods for investigating species boundaries are applied more widely.

5.2 Background

Speciation is a dynamic process made up of several steps (Fig. 5.1; Mayr, 1963; Allmon, 1992; Barraclough and Herniou, 2003; Dynesius and Jansson, 2014). Consider an ancestral species linked by cohesive mechanisms such as gene flow and shared selection pressures (Fig. 5.1A). First, conditions must change in order to break cohesion and promote divergence (Fig. 5.1B). Geographical restriction to gene flow might arise through the origin of an external barrier or loss of favourable habitat connecting separate parts of the species' range. Divergent selection might result from colonization of an area with new conditions, the emergence of new resources (such as a new

The Evolutionary Biology of Species. Timothy G. Barraclough, Oxford University Press (2019).
© Timothy G. Barraclough 2019. DOI: 10.1093/oso/9780198749745.001.0001

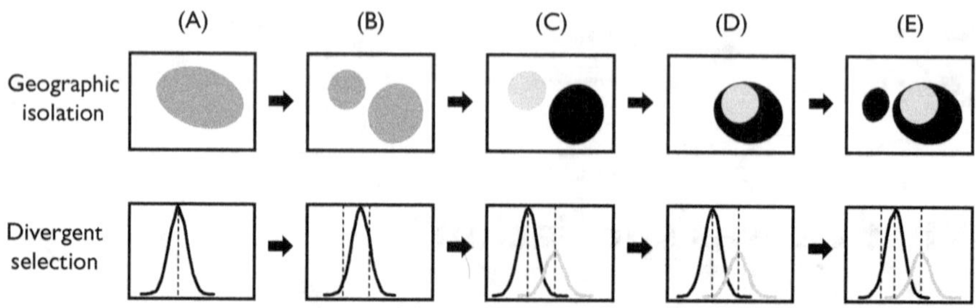

Fig. 5.1 Schematic diagram of stages of speciation shown for geographical isolation (top row) and divergent selection (bottom row). (A) Ancestral population with cohesive interbreeding in a single contiguous range (top panel) and selection towards the same optimum trait value (bottom panel, frequency of different trait values, dotted line showing optimum trait value). (B) Conditions arise that favour divergence (e.g. geographical isolation and divergent trait optima). (C) Populations diverge genetically (illustrated by shading in top panel) and phenotypically (bottom panel). (D) If emerging species come into contact, a sufficient balance of reproductive isolation and divergent selection is required to maintain coexistence. (E) Ongoing process requires descendent species to encounter new conditions favouring divergence, with one or both descendants cycling back to stage (B) onwards.

host for a plant-feeding insect, symbiont, or pathogen), or differential changes in the abiotic or biotic environment across the species range. Geographical and environmental changes are likely to go in concert, since different geographical areas will often experience different conditions. If so, geographic barriers need not be absolute, as long as the balance between gene flow and divergent selection shifts towards promoting divergence rather than cohesion.

Second, there needs to be a genetic response to the new conditions (Fig. 5.1C). In the case of two populations isolated by a strict barrier to gene flow, for example, genetic divergence will occur through drift (Schiffman and Ralph, 2018), through differential responses to parallel selection (Price, 2008; Schluter and Conte, 2009), or in response to divergent selection pressures (Schluter, 2009; Nosil, 2012). In the case of new conditions creating divergent selection in the face of gene flow, reproductive isolation would need to evolve concomitantly in order to permit genome-wide divergence. For example, pleiotropic effects on interbreeding (so-called magic traits (Gavrilets, 2004)) facilitate responses to divergent selection (as in models of speciation via changes in flowering time in plants (Devaux and Lande, 2008)).

Third, diverging populations must persist and coexist (Fig. 5.1D). Many related species continue to live in different geographical areas, and as long as geographical isolation persists, no special mechanisms for coexistence are required. If contact is re-established, then reproductive isolation is necessary to permit coexistence (see chapter 2 and section 5.3.4). Ecological differences are also required, otherwise one species should competitively exclude the other (see Fig. 2.4, unless they coexist neutrally, see chapter 10). It is a grey area whether to include ecological coexistence and persistence as an additional stage in speciation (Allmon, 1992), or whether to treat these as separate processes or part of extinction. For speciation events to be detectable, however, species need to survive long enough to be distinguishable and to either reach the present day or have left traces in the fossil record.

A final step that is required for ongoing speciation within a clade is the re-establishment of conditions for speciation in the species resulting from earlier speciation events (Fig. 5.1E). For example, speciation by geographical isolation leads to species with smaller population sizes and narrower geographical ranges than the ancestor (Fig. 5.2). Without subsequent expansion or movement of species ranges to encounter new geographical barriers, speciation events might cease or the average range size of species within the clade might decline, perhaps increasing extinction rates (Barraclough and Vogler, 2000; Pigot et al., 2010). Similarly, species resulting from ecological speciation and adapting to different niches or habitats would need to encounter further variation in environmental conditions for future ecological speciation events.

One goal of understanding causes of speciation is therefore to understand the relative importance of these different stages. For example, do the temporal dynamics of speciation events depend more on the origin of extrinsic conditions favouring divergence or on intrinsic genetic responses when faced with those conditions? The former is expected if genetic responses are rapid but the chance of encountering geographic barriers or new ecological conditions is low. At present, we lack broad enough data on genetic mechanisms, but especially on the dynamics of changing environments and species movements, to quantify such a model in any clade. There is also relatively little quantitative theory that considers all stages of the process (for example, see Gavrilets and Losos, 2009; Birand et al., 2012; Aguilee et al., 2018).

Another question is what are the relative roles of geographical isolation and divergent selection in promoting speciation? At one extreme, speciation might result when external circumstances keep populations apart for long enough that they establish independent evolutionary trajectories well before they re-establish contact again. At the other, the pattern of selection across members of a population might be structured (because of general features of the way organisms are distributed and environments vary) such that it tends to promote genetic divergence and reproductive isolation (the latter either by direct selection or as a by-product of ecological selection). To what extent do species result from selection acting at broad scales or from incidental breaks in the interconnectedness of populations?

I will explore evidence for the roles of geographical isolation and divergent selection in speciation, in each case contrasting the origin of conditions versus responses to those conditions. Answering such questions is hard, because evidence relies on observational inference and reconstructing the past. We cannot readily observe or manipulate the whole process, which takes place over many thousands to millions of generations. Fossil evidence is too sparse to observe speciation dynamics except in rare cases, such as planktonic foraminifera (Ezard et al., 2010). Phylogenetic studies reconstruct speciation events leading to a set of extant species with reasonable accuracy. But information content for conditions and responses during speciation decays over time because of other processes such as geographical range shifts, extinction, and phenotypic evolution (Barraclough et al., 1998b; Barraclough and Nee, 2001). Data on key parameters, such as dispersal, reproductive isolation, and ecological divergence, are scarce and often rely on surrogates for large-scale analyses. Different aspects are more tractable in some groups than others. I will emphasize these issues throughout and return to them at the end of the chapter to outline potential new avenues.

5.3 Geographical isolation

Geography is not as trendy in speciation studies as it once was, because of the expansion of tools for inferring gene flow and selection from genomes, but it remains a key factor linking speciation mechanisms and patterns. It influences not only gene flow but also the distribution of selection pressures among organisms and diversity patterns resulting from speciation. For this reason, I look at evidence for the geographical context of speciation in depth.

5.3.1 Evidence from geographical ranges of sister species

Closely related species or sub-specific taxa often have non-overlapping ranges. Dubbed Jordan's rule (Jordan, 1905; Fitzpatrick and Turelli, 2006), this pattern gained prominence from Ernst Mayr's taxonomic work on south-east Asian and Pacific island birds (Mayr, 1963). Cladistic approaches formalize the comparison of sister clades and demonstrated that overlapping (i.e. sympatric) ranges are characteristic of deeper clades (Lynch, 1989). With the advent of molecular phylogenies, comparisons focused on relative age and showed that young sister species pairs tend to have non-overlapping ranges, whereas range overlap is more likely between old sister clades (Fig. 5.2; Barraclough and Vogler, 2000). The hypothesis to explain this pattern is that speciation is often caused by geographical isolation, and sympatry only accumulates over time as species acquire sufficient genetic differences for reproductive isolation and ecological coexistence. Other metrics reveal other geographical patterns of speciation. For example, range adjacency provides evidence for parapatric speciation along large-scale environmental gradients (Fitzpatrick and Turelli, 2006) and range size asymmetry indicates that speciation often involves species with smaller ranges splitting from a more widely distributed ancestral species (Fig. 5.2; Barraclough and Vogler, 2000). Yet the primary focus of theory and evidence has been range overlap and Jordan's rule.

Clearly there are exceptions to the 'rule'. Some clades do have recently diverged species with overlapping ranges (e.g. *Rhagoletis*, Fig. 5.2; Rosser et al., 2015). Also, many species never come back into contact with close relatives and persist without any need for reproductive isolation or ecological coexistence. Evidence for the rule has focused on birds, plus other eukaryotes with complete species-level phylogenetic trees and good range data. There is limited evidence for micro-organisms. The assumption is that wide dispersal reduces the importance of geographical isolation in microbes, but the yeast *Saccharomyces paradoxus*, for example, displays a pattern of allopatric divergence and secondary contact among globally diverging populations (Kuehne et al., 2007).

How reliable are current ranges for inferring speciation patterns? The key problem is that geographic ranges move. We know this from fossil evidence of recent taxa, from the

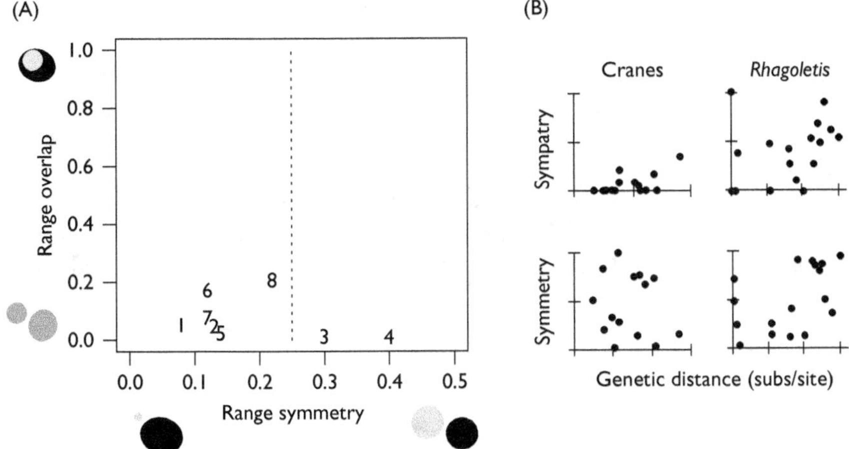

Fig. 5.2 Summary of patterns of geographical range overlap and symmetry in several animal radiations. *Sylvia* warblers (1), *Malurus* fairy wrens (2), Gruidae cranes (3), Alcidae auks (4), *Xiphophorus* swordtail fish (5), *Rhagoletis* fruit-flies (6), *Flexamia* leaf-hoppers (7), and *Cicindela* tiger beetles (8). (A) Y-axis indicates the Y-intercept of a plot between range sympatry of sister clades (0, no overlap, that is strict allopatry; 1, one range completely overlapped by other) and node age reconstructed from a molecular phylogenetic tree (= genetic distance). Most recent sister clades had non-overlapping ranges with growth of sympatry between older sister clades; hence low intercepts for most clades. X-axis is the intercept of a plot between range symmetry of sister clades (smaller range/sum of range sizes) and node age. Dotted line shows null expectation under a broken stick model of random subdivision of the ancestral range: there is an overall tendency towards significantly more range asymmetry than expected under the null model (Barraclough and Vogler, 2000). (B) Example plots for cranes and *Rhagoletis* fruit-flies: note very recent speciation event with sympatry = 1.0 in latter. (Redrawn with permission from Barraclough and Vogler 2000).

speed of changes in range distributions over observed timescales, and from sister taxa with partially overlapping ranges (Barraclough and Vogler, 2000)—a pattern that is not predicted by any model of speciation (even parapatric speciation involves divergence of adjacent ranges, rather than partially overlapping ranges). The above approaches assume that younger species pairs have shifted less since speciation, but ranges move over the timescale of glaciations and sea-level changes, which is far shorter than typical divergence times between recently diverged species of, say, 1 million years (Myr).

How could we solve this problem? One way is to model range movements and speciation simultaneously and coestimate parameters specifying each process. Models of Brownian motion of range edges or stochastic jumps to random locations have been developed. Initially, these were used to demonstrate 'null expectations'. For example, a pattern of allopatry can arise for small ranges occupying a large continental area if ranges move a lot, irrespective of the geographic mode of speciation (Barraclough and Vogler, 2000). Similar models have been used to guide inferences (Phillimore et al., 2008), but there have been few parametric approaches. This might be due to the complexity of the process: range movements are a factor of dispersal abilities, changing environmental conditions, and competitive exclusion between competing species. My instinct is that the nature of the processes and lack of temporal data mean inference is just too hard. Moderately precise estimates of the rate of range movement are feasible, but all but the lowest rates of range movement would obscure the geographical mode of speciation entirely. Or perhaps we have not applied the same rigour to macro-evolutionary inference as, for example, to theoretical population genetics.

Skeels and Cardillo (2019) used an approximate Bayesian computation framework to assess the frequency of different geographic modes of speciation among 30 animal and plant clades (Fig. 5.3). By comparing pattern metrics to those obtained under a

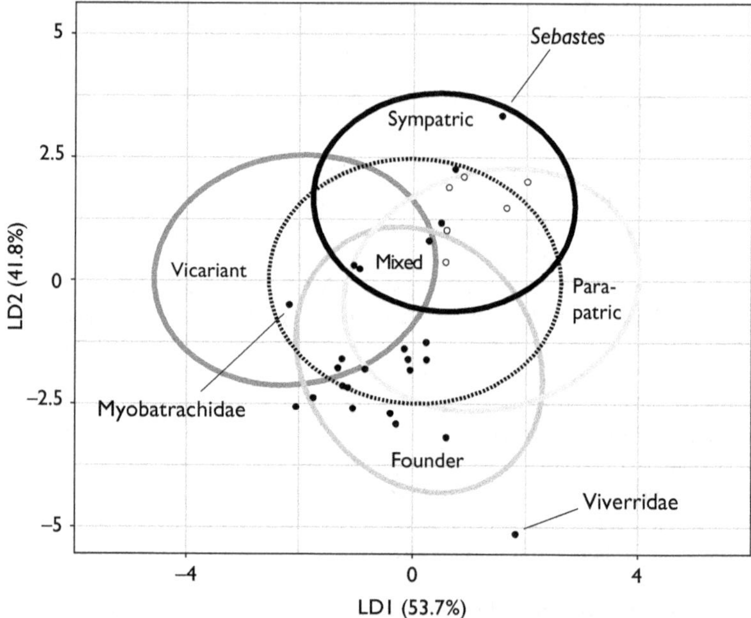

Fig. 5.3 Comparison of speciation patterns between empirical datasets (filled dots, animal clades; open dots, plant clades) and simulated datasets under alternative geographical models of speciation (ellipses contain 90 per cent of simulated datasets) by Skeels and Cardillo (2019). The first two axes of a linear discriminant (LD) analysis are shown, summarizing variation in 14 metrics of the geographical pattern of speciation, such as sister species overlap versus divergence time, range size asymmetry, and skew and bimodality of those metrics across multiple branching events within the focal clade. Simulations incorporated speciation, environmental change, range shifts, and extinction across a range of parameter values. Clades with the strongest signal for sympatric speciation (rockfish in the genus *Sebastes*; previous accounts recognize speciation along a depth gradient within overlapping geographical regions (Ingram, 2011)), founder event speciation (civets and genets in the family Viverridae), and vicariant speciation (frogs in the family Myobatrachidae) are highlighted. Plant clades lie in the sympatric–parapatric range of simulated metrics, but still overlapping with a mixed mode of speciation, which might also reflect 'indecisive', for example due to range shifts obscuring initial mode of speciation. (Reprinted and simplified from Skeels and Cardillo (2019) with permission.)

range of simulation models with different parameter values, they inferred a tendency for speciation of small, isolated founder populations in animals—a pattern also implicated by patterns of range asymmetry in earlier studies (Barraclough and Vogler, 2000)—and for sympatric speciation in plants. The results implied that mode of speciation is detectable from phylogenetic and species range data even after millions of years of divergence. The crux of these approaches is in the specification of the simulation models—do they incorporate all relevant processes? There are in principle multiple ways to obtain a given pattern. For example, overlapping ranges of close relatives could result from sympatric speciation or from a tendency towards environmental niche similarity and hence convergence in range distributions following range shifts after an initial isolation event. The ability to distinguish alternative models depends on robust specification of the probabilistic nature of each process involved.

Despite these challenges and the potential for improved methodology, the impression from many phylogenetic studies is of an allopatric world. Sad but true, perhaps most speciation events are boring—geographical isolation coupled with environmental differences between areas permits and drives genetic divergence. Many species never come into contact until they are so divergent that reproductive isolation and ecological coexistence are already in place. A different impression is given by case studies of speciation, but case studies look at interesting taxa—why study a pair of consistently allopatric species that diverged 9 million years ago? Case studies are vital for understanding mechanisms, but should be used with caution to determine the most prevalent mechanisms causing species to split.

5.3.2 Evidence from oceanic islands

An alternative to comparing geographical distributions of sister species is to take advantage of natural experiments, namely situations where range shifts are limited. Coyne and Price (2000) looked for evidence of within-island speciation of birds on remote oceanic islands. These islands are too small to hypothesize geographical isolation because even a flightless bird could saunter across them in a day. Consequently, within-island speciation would indicate sympatric divergence, for example to exploit distinct niches. By corollary, if sympatric speciation were feasible for birds, then *in situ* divergence into multiple species should occur on some islands. In fact, across 46 oceanic islands that include at least some endemic birds, no cases were found of sister species on the same island. A few cases of congeneric species found on the same island resulted from repeated colonization rather than *in situ* speciation, as each species had a different close relative elsewhere. The conclusion therefore was that sympatric speciation is of negligible frequency in birds. Since then, a case of divergence of finches within Tristan da Cunha has been presented (Ryan et al., 2007) and of allochronic speciation—divergence into a hot and cold season breeding species—in storm petrels from the Azores (Friesen et al., 2007), although the dominant pattern for co-occurring hot and cold season breeders on the same island across the genus was for independent origins rather than divergence *in situ* (Wallace et al., 2017).

The same rationale has been used to demonstrate sympatric divergence in other cases. For example, the Lord Howe Island palms have diverged onto two partially overlapping soil types since the origin of this volcanic island (Fig. 5.4A; Savolainen et al., 2006). Subsequent work demonstrated further plant radiations on the island adapting to different altitudinal bands (Fig. 5.4B,C; Papadopulos et al., 2011, 2013). While there is some geographical structure to divergence, it is argued that gene flow via pollen flow occurs more broadly across the island, at least in the wind-pollinated palms.

Another canonical example comprises the cichlid fish that diversified in crater lakes both in east Africa and in Nicaragua (which represent 'islands' of freshwater). These cases rely on evidence that species diverged in areas that are too small for geographical barriers to gene flow and that repeated colonization from outside the 'islands' is unlikely (Schliewen et al., 1994; Barluenga et al., 2006). In the crater lakes, movements are constrained because no streams flow into or out of the lake, but since at least one gravid female must have arrived in the lake, the probability of two genetically distinct colonizations is small, but not zero. It is tricky to distinguish single colonization and *in situ* divergence from repeat colonization plus some subsequent gene flow: there is evidence for secondary colonizations in the African crater lake Barombi Mbo (Martin et al., 2015), but how much this contributed to ecological, sexual, and morphological divergence in the lake, instead of in situ divergence, remains unclear (Richards et al., 2018).

The island approach was extended to identify the minimum scale for speciation in *Anolis* lizards (Losos and Schluter, 2000). Across Caribbean islands of different sizes, within-island speciation only occurred on islands above 3000 km^2. Species living together on smaller islands resulted from repeated colonization instead. By inference, smaller islands are ecologically heterogeneous enough to allow multiple species to coexist but lack suitable conditions for divergence. Kisel and Barraclough (2010) examined the same pattern across multiple taxa worldwide (Fig. 5.5). They found that the probability of speciation scaled with island area in most taxa, but the scale varied among taxa. Land snails speciated even on the smallest islands, whereas carnivores failed to speciate on islands smaller than Madagascar. The minimum area for speciation scaled with the typical scale of genetic differentiation within populations of each taxon—used as a measure of dispersal and gene flow. Speciation depends therefore on the interaction between the environment (island area) and organismal traits affecting responses (dispersal ability). In the original study, ferns were an exception to the general trend, with little correspondence between scale and the frequency of speciation. The authors speculated that sympatric speciation by polyploidy might account for this. Since then, however, purported cases of fern speciation on small islands have been shown to result from repeated colonization, meaning that ferns too follow the general trend (Igea et al., 2015).

The use of islands and island analogues offers the chance to estimate key determinants of speciation in more controlled conditions than in 'messy' continental areas. But these are still natural experiments rather than true experiments, and so the possibility of confounding factors remains. For example, scaling of speciation with area does not necessarily mean geographical isolation is the cause. Larger islands might

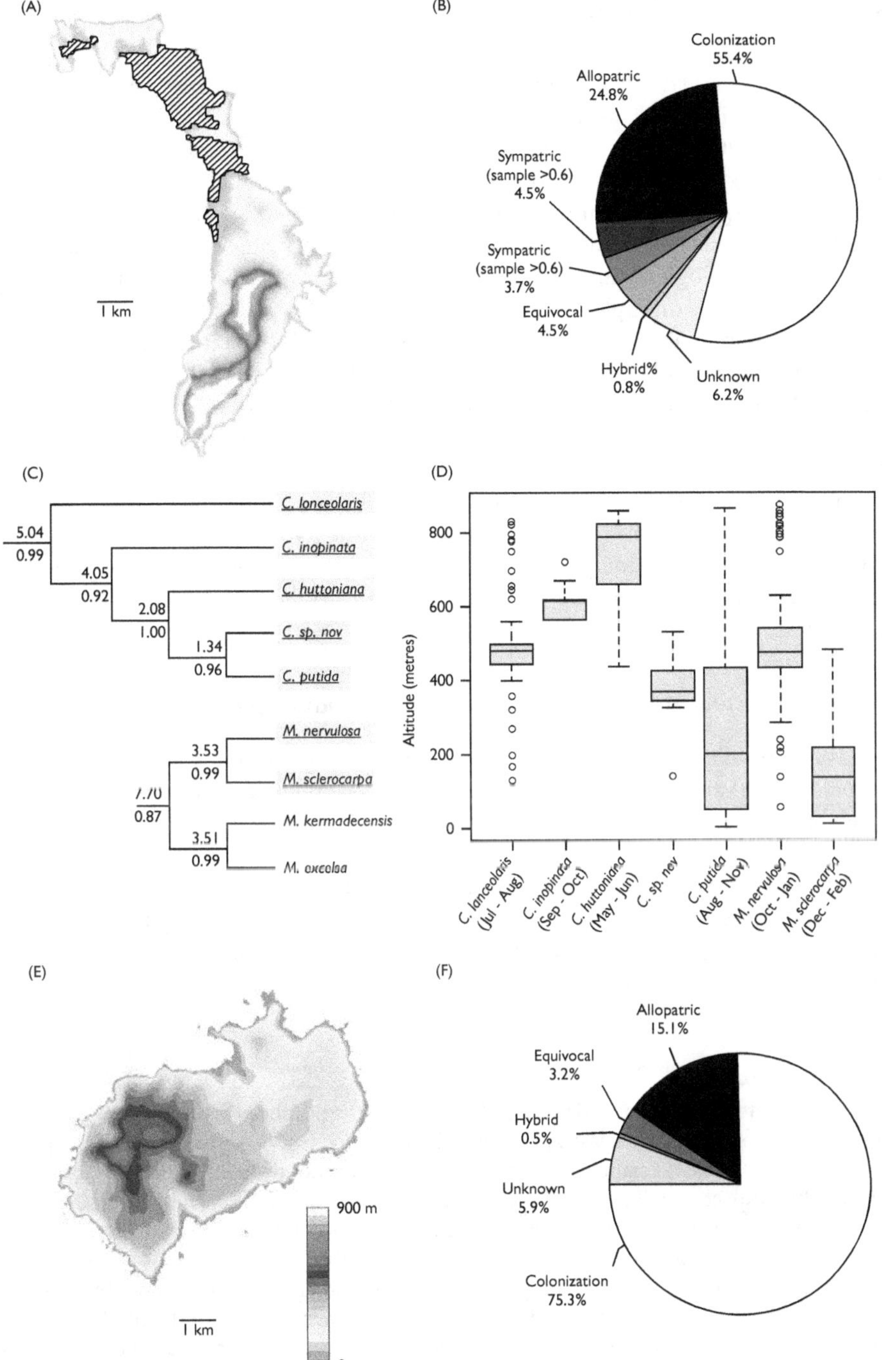

Fig. 5.4 Evidence for modes of speciation of flowering plants on isolated oceanic islands. (A) Lord Howe Island lies 600 km off the east coast of Australia and is home to two endemic sister species of Lord Howe palms that diversified since the island formed (7 Mya): one grows on

offer greater opportunity for divergent selection if there are more environmental niches and habitat types, as well as greater scope for geographical isolation. Kisel and Barraclough (2010) included altitude as a surrogate for environmental heterogeneity and found it explained less variation in speciation rate than area per se. Wagner et al. (2012, 2014) explored determinants of endemic cichlid diversity within African lakes, finding that depth or area plus solar energy and lake age explained most of the variation. Broad-scale ecological data are hard to obtain, however, because species vary ecologically across multiple axes.

Remote islands also provide an interesting but underexploited test-case for the assumptions of microbial biogeography. Current efforts to survey microbial diversity with molecular markers have not yet tackled the question of endemism on remote islands—more work has considered geographical isolation among hot springs (Papke et al., 2003). One exception is that an endemic thrush on Sao Tome diverged from a mainland relative, yet the composition of cultivable bacteria on the two birds was the same (Lobato et al., 2017). Other natural experiments aside from islands are also informative, such as the comparison of different taxa responding to the same geographical barrier in the case of marine life isolated in the Caribbean and Pacific by the emergence of the Isthmus of Panama (Lessios, 1998).

5.3.3 Geographical opportunity versus organismal responses

The island results shed light on the interplay between opportunity (large enough geographical area) and organismal responses (dispersal ability and hence the effects of a given barrier). More detailed evaluation is required, however, if we are to understand geographical speciation sufficiently to predict the timing and pattern of speciation from first principles. In particular, how does the rate and number of speciation events in a clade depend on the rate of origin of new barriers to gene flow, the strength of isolation resulting from the barrier, and the rate at which organisms diverge in response and evolve reproductive isolation?

Fig. 5.4 Continued

calcareous (hatched regions) and volcanic (rest of island) soils, the other just on volcanic soils. (B) Across the entire flora, at least 4 per cent and potentially up to 8 per cent of endemic species result from sympatric speciation, judged from evidence for monophyletic radiations within the island with respect to related species found elsewhere. For example, the genera *Coprosma* (C, five endemic species, upper 95 per cent confidence limit of node age above branches, support values as posterior probabilities below branches) and *Metrosideros* (C, two endemic species) have diversified along altitudinal gradients (D). Variation in flowering periods (phenology) among species might represent a pleiotropic response to adapting to local altitudinal conditions that helps to maintain reproductive isolation. In contrast, on Cocos Island off the Pacific coast of Central America (E), no uniequivocal within-island divergence was recorded (F): all endemic plant species are most closely related to species living elsewhere. Cocos lacks the range of altitudes and availability of different soil types (not shown) of Lord Howe. (Reprinted from Papadopulos et al. (2011) and Igea et al. (2015) with permission.)

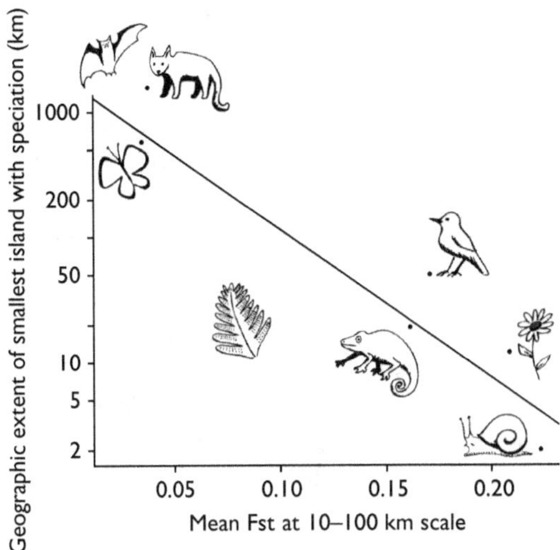

Fig. 5.5 Minimum island size for speciation across eight animal and plant taxa (bats, carnivores, butterflies, ferns, birds, lizards, flowering plants, and snails) scales with summary measures of degree of genetic differentiation between populations (Fst values compiled from molecular ecology literature). Clades in which populations differentiate more at 10–100 km scale—indicative of low rates of dispersal and gene flow—are able to speciate on smaller islands, whereas broadly dispersing clades only speciate on much larger islands (e.g. Madagascar is the smallest island displaying carnivore speciation). (Reprinted from Kisel and Barraclough (2010) with permission.)

Geographical opportunity depends on the dynamics of geographical isolating barriers over space and time. Correlations between net diversification rates (i.e. number of extant species remaining after a given period of time since the origin of a clade) and surrogates such as topographic complexity provide evidence for the importance of geographical opportunity; this is further supported by evidence for species separated by topographic barriers (Hazzi et al., 2018). Going beyond this is tricky, however. Even high-profile scenarios such as the role of forest fragmentation in Amazonian speciation are hard to pin down because direct, independent evidence for historical changes in connectivity is rare (Ribas et al., 2012). The timing of some isolating barriers, such as the Isthmus of Panama affecting marine organisms, is relatively well known and provides controlled conditions to explore variation in organismal responses (Lessios, 1998).

The key processes determining the chance of encountering a barrier and the resulting strength of isolation are dispersal and ultimately gene flow. Dispersal encapsulates the rate at which organisms cross barriers. Gene flow also depends on whether those organisms contribute genetically to future generations: they may not do if maladapted to local conditions. Broad surveys show that surrogates of dispersal ability correlate with diversification rates (Phillimore et al., 2006; Claramunt et al., 2012), but dispersal cannot easily be reduced to a single number. In the island examples, taxa need rare long-range dispersal to reach the island and encounter conditions for isolation, but low rates of local movement to experience geographical isolation on the island once there. For example, land snails presumably arrive carried by birds or by surviving salinity, but move short distances on land. So-called fat-tailed dispersal kernels over evolutionary time, which means that rare long distant events occur, should favour speciation by maximizing the rate of encountering barriers plus the response to those barriers. The 'sweet spot' for maximizing speciation rate should depend on the spatial structure and dynamics of geographical barriers. Evidence for the shape of dispersal functions is rare for ecological timescales and even rarer for evolutionary timescales (i.e. the probability of a long-range colonization event per million years; Pinsky et al., 2017). An alternative is to estimate how gene flow within species varies with scale and across geographic features. Some continental examples have tested for a correlation between gene flow measures and diversification. For example, levels of population differentiation within species correlate with speciation rates across clades in New World birds (Harvey et al., 2017). In contrast, population structure within species of Australian lizards did correlate with traits expected to be proxies for dispersal, but these did not predict variation in speciation rates across the sampled clade (Singhal et al., 2018); this is similar to a result found in Central American orchids (Kisel et al., 2012).

The final component identified above is the rate of genetic divergence and evolution of reproductive isolation in populations that experience geographical isolation. Genetic divergence depends on the mutation rate and generation time. Organisms with shorter generation times or higher per generation mutation rates might diverge faster than organisms with slower mutation rates when exposed to the same isolating conditions. Evidence is equivocal at present and tends to focus on marker genes that are not themselves directly responsible for traits involved in ecological divergence or reproductive isolation (Barraclough et al., 1996; Barraclough and Savolainen, 2001). Genetic divergence also depends on the strength of selection acting in isolated populations, which combines both the environmental pressures and organismal responses, for which there are currently little data across whole clades. Finally, for predicting whether species will persist if the geographic barriers are removed or dispersal reoccurs, the dynamics of speciation through geographical isolation also depends on the evolution of reproductive isolation. This topic merits its own section.

5.3.4 *Geographical patterns of reproductive isolation*

Co-occurring sexual species require barriers to restrict gene flow (see chapter 2), even if some residual level of gene flow remains. These barriers can act at multiple stages of the life cycle (Coyne and Orr, 2004). Pre-zygotic barriers reduce the chance of mating between members of different species—such as via different flowering times, different mating preferences, or spatially segregated mating grounds—or the chance of successful fertilization. Post-zygotic barriers reduce viability, survival, and fertility of hybrid zygotes. Barriers are also classified as intrinsic, namely those arising from genetic incompatibility, and extrinsic, namely those arising from ecological incompatibility between the species. Pre-mating barriers can be intrinsic (genetic differences in phenology of mating) or extrinsic (ecological niche separation reduces interspecific encounters), as can post-zygotic barriers (intrinsic: hybrid inviability or sterility; extrinsic: reduced survival of hybrid offspring in each parental niche).

Genetic studies have quantified the strength of each type of barrier in recently diverged species pairs (Coyne and Orr, 2004). Comparing patterns across taxa can then determine the relative importance of alternative isolating mechanisms (Baack et al., 2015; Lackey and Boughman, 2017; Turissini et al., 2018). Mechanisms that evolve more rapidly between diverging taxa should be more important than those that evolve later after reproductive isolation is already established. There is a general trend for pre-mating isolation to evolve faster than post-zygotic isolation. Variation across taxa can also reveal the evolutionary forces driving the emergence of isolating mechanisms. One possibility is that selection directly promotes reproductive isolation. If so, the strength of reproductive isolation should be greater between species that come into contact compared to those in allopatry, since selection on reproductive isolation can only act if there is a potential for interspecific crosses. Alternatively, reproductive isolation might evolve as an incidental by-product of genetic divergence. If so, there should be no systematic difference in levels of reproductive isolation between sympatric and allopatric species pairs, assuming rates of genetic divergence are equivalent in sympatry or in allopatry.

The classic study addressing these questions is Coyne and Orr's meta-analysis of data from crossing experiments across pairs of taxa (species or sub-specific) of *Drosophila* (Fig. 5.6A; Coyne and Orr, 1989, 1997). Pre-zygotic isolation evolved at a faster rate than post-zygotic isolation, consistent with its pre-eminent role in restricting gene flow between species. The strength of isolation from pre-zygotic barriers tended to exceed the level from post-zygotic barriers between the most recently diverged taxa. The effect was only found, however, between taxon pairs living in sympatry in nature. This is therefore consistent with direct selection on reproductive isolation. Either divergence occurred in sympatry and selection promotes reproductive isolation (for example, concomitant with adaptation to use different resources) or taxa initially formed in allopatry experience selection for reinforcement of incomplete reproductive barriers. Intriguingly, the effect is observed among the most recently diverged taxa, not older taxa as they come into secondary contact—could this be a signature of sympatric speciation? The same pattern has been observed in mushroom-forming fungi—a useful study taxon because of the widespread availability of crossing data from fungal taxonomy (Fig. 5.6B; Giraud and Gourbiere, 2012).

Genetic crosses are not feasible in all organisms and so relatively few studies have emulated these results. In other taxa, traits involved in mate choice or reproductive compatibility can be used as surrogates to infer pre-zygotic isolation. Darwin (1871) first proposed that sexual isolation was an important cause of speciation based on the observation that closely related species often differed most conspicuously in sexual traits. Notable cases include birds of paradise, in which male plumage used to attract females differs spectacularly among related species whereas females barely differ, and the use of genital morphology to distinguish insect species that are otherwise identical. The rationale is the same as described in section 5.3.1 for geographical

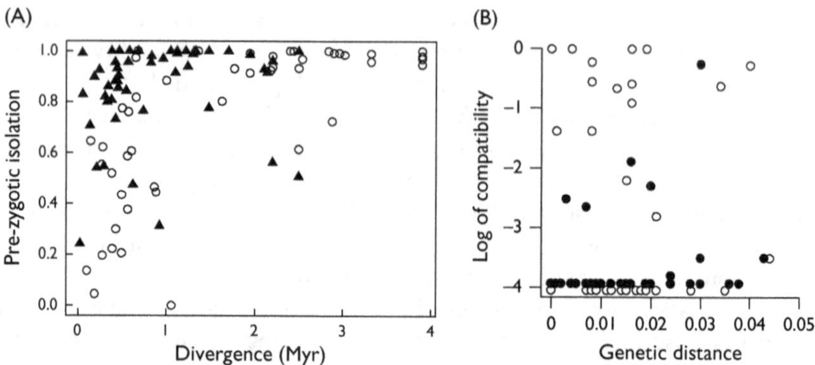

Fig. 5.6 Speciation patterns in the evolution of reproductive isolation over time. (A) Classic result of Coyne and Orr (1989) (updated by Coyne and Orr (1997)). Pairs of *Drosophila* taxa evolve pre-zygotic isolation faster in sympatry (black triangles) than in allopatry (open circles), consistent with species reinforcement—that is, selection to reduce gene flow between sympatriots—and with predictions of models of speciation with gene flow. (B) Equivalent pattern in basidiomycete fungi (modified from Giraud and Gourbiere (2012) with permission); note Y-axis shows compatibility rather than isolation, hence low value indicates reproductively isolated. Sympatric taxon pairs (black circles) show stronger incompatibility than allopatric pairs of equivalent divergence (open circles). No evidence for reinforcement was found in comparison of ascomycete fungi.

isolation—causes of speciation are inferred from characteristics of recently diverged species. Darwin and others theorized that strong mating preferences could drive speciation (Lande, 1981). Sexual selection tends to promote rapid evolution and elaboration of traits—sexual traits might evolve more rapidly than other types of traits. Also, divergence in mating preferences and traits between populations leads directly to reproductive isolation.

These hypotheses can be tested using phylogenetic methods to compare the rate of divergence of sexual traits versus ecological characters. For example, sexual signalling traits evolve faster than ecological traits among weakly electric fish in the genus *Paramormyrops* (Arnegard et al., 2010) and cichlids of Lake Tanganyika (Gonzalez-Voyer and Kolm, 2011). The role of species reinforcement can be tested by comparing sexual trait divergence to the degree of range overlap as a surrogate measure of co-occurrence (Hopkins, 2013). For example, sympatric species pairs of monkeyflowers display greater differences in floral morphology (but not vegetative morphology) than allopatric species pairs (Grossenbacher and Whittall, 2011). In contrast, finding that reproductive character values are more similar between overlapping species is consistent with convergence to adapt to local conditions rather than character divergence to reduce gene flow. This pattern was found in the tribe Sinningiae in the Atlantic forest of Brazil (Perret et al., 2007).

So how important are these mechanisms in shaping speciation patterns across taxa? One possibility is that the rate of acquiring reproductive isolation is a limiting step for speciation, and the dynamics of these responses determines the probability of speciation. Taxa might vary in their rate of evolving each isolating mechanism. Alternatively, reproductive isolation might evolve sufficiently rapidly that speciation depends more on the origin of circumstances promoting genetic divergence rather than the rate of response. Furthermore, if speciation with gene flow is common, isolating mechanisms might be less important than the strength of selection acting on genomic regions behind ecological adaptation (see section 5.4).

Early studies focused on surrogates for the ability to evolve reproductive isolation, such as measures of the strength of sexual selection by female choice. In passerine birds, clades with males and females that differ in colour tended to have significantly more species than their sister clades with monomorphic colouration (Barraclough et al., 1995). The argument is that these clades experience stronger sexual selection, which provides a basis for reproductive isolation between isolated populations. Similar results have been reported in insects, fish, and other birds, but not in other cases (Ritchie, 2007). Even the stronger cases explained little variation in diversification rates across taxa, which instead are explained more by ecological factors (see chapter 10).

Recently, more direct tests have been performed on the importance of isolating mechanisms on the accumulation of species within lineages. Rabosky and Matute (2013) compiled data on levels of reproductive isolation as a predictor of net diversification rates in *Drosophila* and birds. They found no evidence of a correlation between metrics of post-zygotic isolation and diversity. The conclusion was that speciation depends less on genetic mechanisms of the accumulation of reproductive isolation and more on the timing and rate that isolating barriers appear. Data availability is currently a constraint on such studies. Few taxa have been subject to experimental measurement of reproductive isolation, and intrinsic reproductive isolation might not be the key organismal response—for example, in birds, pre-zygotic isolating mechanisms are believed to be more important for maintaining species boundaries. Coordinating a comprehensive survey of reproductive isolation across taxa would be a valuable endeavour to complement and extend insights derived from particular case studies.

Other traits could increase the chance of evolving reproductive isolation. For example, chromosomal translocations facilitate genetic divergence by providing barriers to recombination for parts of the genome (Rieseberg, 2001). Similarly, polyploidy can generate both reproductive isolation and phenotypic differences instantaneously. Systematic variation in the probability of such events arising or persisting within populations (e.g. at a faster rate in annual than perennial plants; Baack et al., 2015) could lead to higher speciation rates. Testing these hypotheses with phylogenetic data is hard—a clade with more species might have a greater range of chromosome numbers or accumulated number of translocation events, but is that a cause of speciation or a consequence of more 'lineage time' for accumulating them?

Together these examples highlight many of the difficulties of investigating causes of speciation: often, we are restricted to the use of surrogates for variables of real interest, we lack detailed historical data or direct records of speciation over time, and tests often focus on a single mechanism or hypothesis rather than evaluating multiple interacting mechanisms.

5.4 Divergent selection

I now turn to the second major explanation—speciation is caused by different subsets of an ancestral species adapting to different ecological niches or environments. Heterogeneity in physical and biotic conditions means that selection often acts in different directions for different populations or sets of individuals within a population. Organisms faced with fluctuating conditions over time and space should evolve a generalist or plastic genotype that maximizes fitness across variable environments (Lande, 2014). But sustained divergent selection—for example, following colonization of a new host, expansion into a new habitat, or environmental differences in distinct geographical areas—can drive divergence and speciation (Schluter, 2009; Nosil, 2012). Ecological speciation can occur with or without geographical barriers to restrict gene flow. With gene flow, mechanisms are needed to prevent genetic homogenization due to random mating (Gavrilets, 2004). Typically, these are reproductive barriers, but, in theory, selection might be strong enough to maintain differentiation of ecological traits while gene flow occurs across other genomic regions (see chapter 2).

5.4.1 Ecological divergence and reproductive isolation

The emphasis on ecological speciation and divergent selection over the last few decades came from theory and laboratory experiments showing that genetic divergence and reproductive isolation between isolated populations only evolve rapidly when they adapt to different local environments (Schluter, 2001). For example, Rice and Hostert (1993) reviewed selection experiments on *Drosophila* showing that only populations kept in cages with different food supplies or physical environment evolve appreciable levels of reproductive isolation. Genetic drift or differential responses to selection under the same conditions (parallel selection) also lead to genetic divergence between isolated populations, but these are much slower processes. Although these results indicate that divergent selection is the more powerful force, it does not necessarily follow that it plays the dominant role in nature. The conditions needed for slower divergence by drift or parallel selection might arise more frequently than those for ecological speciation, enough to counter-balance the faster rate of response. We need to consider the whole dynamic.

Surveys of reproductive isolation across taxa strengthen the view that ecological divergence is an important driver of speciation. For example, the primary mechanisms preventing gene flow between pairs of stickleback species relate to assortative mating in different habitats and maladaptation of hybrids to either parental habitat (Lackey and Boughman, 2017). Similarly, Ostevik et al. (2016) measured components of reproductive isolation between sunflower ecotypes on sand dunes versus non-dune habitats. Ecological selection against immigrants and hybrids, and distinct pollinator communities in each habitat were the main barriers to gene flow. Once again, however, we must take care in extrapolating from organisms initially chosen as models of ecological speciation. Broad tests of ecological causes of speciation across taxa are needed.

Funk et al. (2006) extended the Coyne and Orr approach to show that reproductive isolation evolved more rapidly between species pairs that were more ecologically divergent, across 500 species pairs of plants, invertebrates, and vertebrates. Either ecological divergence promotes reproductive isolation or ecological divergence is only permitted once reproductive isolation in place or (most likely) the two processes evolve in concert. A limiting factor for such studies is the availability of data on reproductive isolation and especially ecological divergence, which were fairly coarse. Detailed data on resource use and habitat requirements are lacking for most organisms, and different kinds of traits are associated with speciation in different taxa.

5.4.2 Ecological traits and speciation

A similar approach can be taken as outlined above for geographical and reproductive isolation. Ecological traits associated with aspects of resource or habitat use that are repeatedly targeted by divergent selection causing speciation should tend to vary between closely related species (Barraclough et al., 1999; Barraclough and Nee, 2001). More specifically, they should vary more between recently diverged species than expected under a null model of the even accumulation of trait differences over time (that is, along branches on a phylogenetic tree; Webster et al., 2003; Jiggins et al., 2006; Magnuson-Ford and Otto, 2012) or relative to other traits. The opposite pattern is expected for key innovation traits that open up diversification into a new adaptive zone (Schluter, 2000). Those traits will tend to be phylogenetically conserved and shared among close relatives occupying the same adaptive zone.

Schnitzler et al. (2011) used these predictions to investigate alternative hypotheses for speciation in four clades from the Cape flora of South Africa. The Cape is home to over 10,000 species of flowering plants, packed into a small region with Mediterranean-type climate. Hypothesized drivers of speciation include pollinator shifts (expected to cause reproductive isolation), specialization on alternative soil types, and traits associated with fire regeneration strategy (Cowling and Holmes, 1992; Linder, 2003; Barraclough, 2006). Specialization on different soils types displayed the greatest variability between closely related species in three clades and shifts in fire-regeneration strategies in one clade. Pollinator shifts were phylogenetically conserved in these clades, rejecting their role in recent speciation but consistent with a potential role opening up new adaptive zones for subsequent radiation.

This approach implicitly assumes that conditions are constant over time. Patterns among recently diverged species reflect speciation, whereas patterns between deeper branches reflect long-term persistence and coexistence. An alternative explanation for differences between recent and deeper branching events is that different causes of speciation have operated over time. For example, niche-filling models assume that early speciation events partition one niche axis and later speciation events then partition alternative niche axes. Which assumption is made can change the interpretation of a given pattern. For instance, among Himalayan songbirds, recently diverged species tend to occupy different elevational bands, whereas deeper divergences reflect changes in body size and shape (Price et al., 2014). Assuming a niche-filling model, which the authors favour, this pattern is interpreted as showing that diversification first involved filling different body size and shape niches, then different elevational bands. Assuming constant conditions and turnover, however, it would be interpreted as showing that speciation tends to involve divergence on an elevational gradient, whereas long-term persistence requires accumulation of body size and shape differences. These alternatives are quite hard to distinguish without direct evidence of extinction, which is lacking in most clades.

Whatever method and assumptions are used, this kind of evidence is observational and shows a correlation between changes in a trait and speciation events, not causation. Some rapidly evolving traits might diverge late in speciation after the main action has already happened, for example, or as an outcome of conditions (such as reduced population sizes caused by sub-division) rather than a cause.

5.4.3 Spatial pattern of ecological divergence

The relationship between ecological divergence and geographical overlap can provide further insights into what types of selection pressures promote speciation. Community ecology recognizes two types of ecological niche traits: α and β. α-niche traits permit coexistence of species within an area by partitioning resource use (Ackerly et al., 2006). They vary between co-occurring species. β-niche traits allow organisms to colonize a particular environment and therefore tend to be shared by co-occurring species. These types of traits are detected by testing whether observed variation within a region is greater or less than expected relative to a suitable null model—the set of species in an area varies more (α) or less (β) in their traits than expected based on random assembly from a continental pool. Shifts in either type of trait could drive speciation: α-niche traits could result from sympatric speciation, β-niche traits from speciation driven by adaptation to the environment in different regions.

While correlations might discriminate some scenarios, they cannot definitively identify a mechanism of speciation for the same reasons as highlighted for geographical isolation above: changes since speciation can also affect these patterns. For example, increased ecological divergence of co-occurring related species could indicate ecological character displacement in species coming into secondary contact (to reduce competition) or an ecological assembly rule that only ecologically divergent species are able to coexist (Davies et al., 2007; Pigot and Tobias, 2013).

A study of Coryciine orchids in South Africa examined speciation patterns relating to geographical isolation, reproductive isolation, and ecological divergence (Fig. 5.7; Waterman et al., 2011). These orchids are found across the Cape region, which is typified by high levels of endemic plant diversity, and comprise 60 described species. Close relatives tend to be found in different geographical areas, supporting the role of geographical isolation in speciation. In contrast, local guilds of up to 15 orchid species comprise more distantly related species. Interactions between two types of partner—pollinating bees and mycorrhizal fungi—were compared in order to infer patterns of reproductive and ecological divergence. Pollinator shifts are hypothesized to lead to reproductive isolation, whereas mycorrhizal fungi might mediate adaptation to the mosaic of soil types found in the Cape and thereby facilitate ecological speciation, but should have no direct influence on gene flow.

Bees from the genus *Rediviva* pollinate the orchids while collecting oil from specialist appendages on the flowers using their long legs. Pollen is deposited in a sac called a pollinarium on a specific part of the bee's body, which varies with species. Orchids within a region are pollinated by the same species of bee, whereas different bee species are found in different areas associated with different soil types found there (Fig. 5.7B). Orchid species in a local guild place pollen on different parts of the bee's body, however. Field experiments confirmed that this strategy minimizes pollen flow and hybridization between co-occurring species. Further experiments showed that 'foreign' flowers suffer reduced visitation and pollen delivery when transplanted to areas with different bee species present. Speciation is therefore associated with adaptation to new pollinators in different areas, coupled with a shift in pollinarium attachment site to reduce pollen interference and hybridization between co-occurring species. As a result, pollinator type and attachment site are highly variable between closely related species (Fig. 5.7)—to a significantly greater extent than a null model shuffling trait values among species.

Mycorrhizal fungi belonged to five main clades and, in contrast to the relationships with pollinators, sister species of orchids tended to share similar fungal partners (Fig. 5.7A). There was no evidence at the level of resolution available that colonization of a new area (and associated shift in soil type) required a change in fungal partner. Instead, fungal partner was conserved within deeper clades of orchids. These associations were conserved even when orchid seeds were experimentally germinated in different areas. Furthermore, co-occurring orchid species associated with significantly increased range of fungal partners than predicted under a null model of random association. A putative explanation is that fungi provide access to alternative resources in the soil and hence permit coexistence of multiple orchid species. With respect to theories of speciation, the example showed that adaptation to different conditions in different areas was associated with speciation, but notably for traits with a direct impact on reproductive isolation. Shifts in fungal partners were associated with coexistence of species belonging to different clades rather than species divergence.

Islands have been used to help investigate the interaction between geographical isolation and ecological divergence. In Darwin's finches from the Galapagos, sister species tend to be isolated on different islands, and tend to adapt to different food resources on those islands (Grant and Grant, 2009; Farrington et al., 2014). Incipient divergence associated with food resources is observed within a single species on a single island (Hendry et al., 2009), but the conditions for divergence might not typically persist long enough for speciation. Spiders in the genus *Tetragnatha* from Hawaii comprised two main radiations, one of which speciated on separate islands consistent with geographical speciation, whereas the other diverged into different habitats within islands (Gillespie, 2004; see also Fig. 5.4 for an example in plants).

There is further scope for natural experiments focusing on ecological divergence as a driver of speciation. For example, analogous to the Isthmus of Panama for geographical isolation, there are places with steep environmental gradients that can be

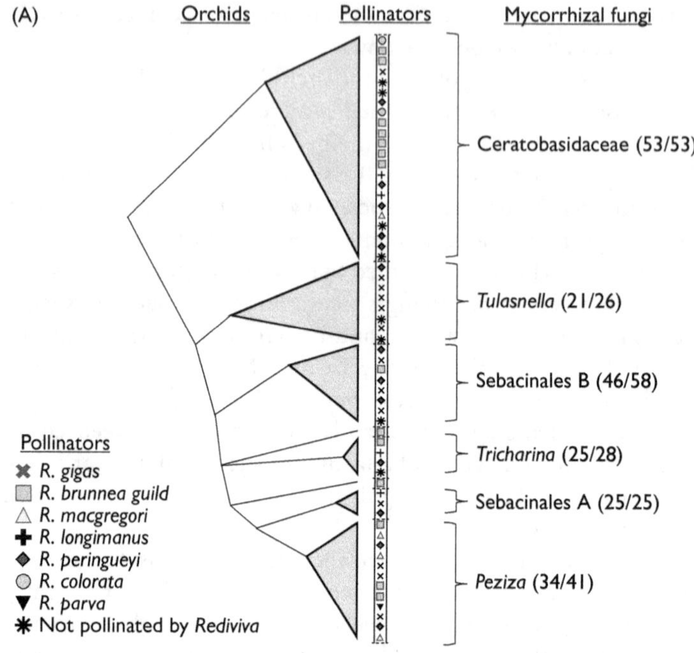

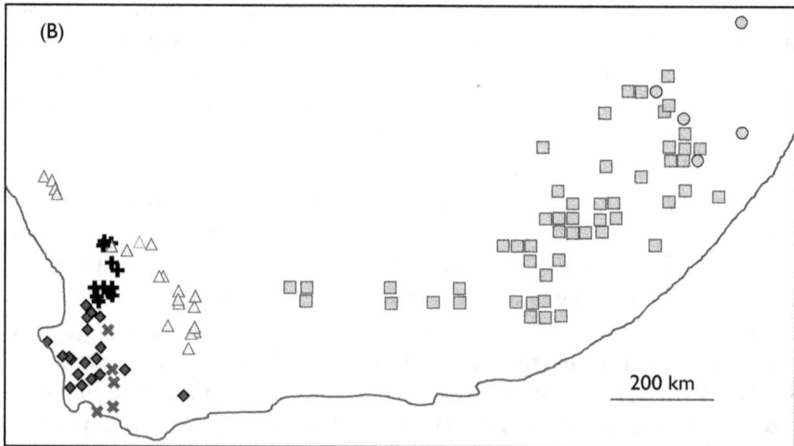

Fig. 5.7 Speciation patterns of Coryciine orchids in South Africa with respect to pollinating bee species (genus *Rediviva*) and mycorrhizal fungal associations. (A) A phylogenetic tree summarizing relationships among sub-clades of orchids shows that related species tend to be highly variable in pollinator type (symbols) and pollinator attachment point (not shown). In contrast, associations with different clades of mycorrhizal fungi are highly conserved within sub-clades (numbers of samples belonging to predominant partner clade shown). (B) Pollinator species and guilds are separated geographically. (Modified from Waterman et al. (2011) with permission.)

used to compare responses of multiple taxa. 'Evolution Canyon' in Israel has been proposed as such a case, with mesic European conditions on one slope and dry African savannah conditions on the other slope (Yablonovitch et al., 2017). The difficulty is that these regions tend to coincide with contact zones between faunas and floras, and therefore there is a signature of historical events, dispersal, and secondary contact, as well as *in situ* divergence.

5.4.4 Ecological opportunity versus organismal responses

The conditions favouring ecological speciation relate to the frequency of encountering new environments, habitats, or resources. Populations need to establish in places with contrasting environments or to be able to invade and use a new niche—heterogeneous niches will often already be occupied by other species, thereby preventing occupation. Cases of major ecological shifts driving speciation might therefore be relatively rare compared to subtle divergence and adaptation in different areas or restricted to expansion into empty areas. This could explain the observation that recent speciation rates are higher in temperate regions—in areas recently recolonized following glaciation—than in the tropics (Weir and Schluter, 2007; Schluter, 2016).

The rate of response of organisms to selection depends on the generation time, genetic variation for focal traits, and the genetic architecture of the trait: the variance and covariance among traits. The response to divergent selection requires not only that genetic architecture favours evolution in a particular direction but also that it permits evolution in two contrasting directions (Dochtermann and Matocq, 2016). Furthermore, in the presence of gene flow it requires that selection can promote reproductive isolation to arise at the same time. This explains why mechanisms that lead directly to reproductive isolation as a consequence of ecological divergence are favoured. It is unclear, however, whether genetic architecture is limiting over time periods involved in speciation—selection can act on the architecture itself. Architectures can be conserved between closely related species, but perhaps selection is conserved as well (Schluter, 1996). The big challenge is to separate conditions from outcome, especially evidence for ecological opportunity that is independent from how organisms responded to it.

5.5 Towards a dynamic and integrated model of speciation

The examples discussed in this chapter demonstrate the role of particular processes in particular taxa. However, no single study has quantified all of the processes outlined in section 5.2. At the broad level, we have at best qualitative understanding of the relative importance of each process. Can we progress towards a quantitative and dynamical understanding of speciation? The first step is to imagine what complete understanding of speciation dynamics would constitute.

Consider a clade diversifying into separate species within a geographical region (Fig. 5.8). The chance of encountering conditions that favour divergence depends on

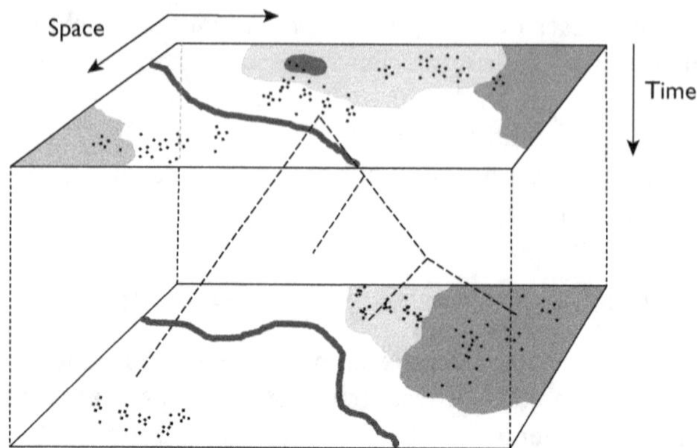

Fig. 5.8 Illustration of a clade diversifying within a geographical region over time. Shaded areas at each time point indicate spatial variation in environmental conditions (e.g. habitat types). A river crosses the region (thick dark grey line). Dots represent localities for members of the clade. A single species at time zero experiences geographical isolation and divergent selection between habitat types and diversifies into three extant species at the later time. Probability of speciation events and persistence to later time point depends on a combination of ecological and geographical conditions encountered, and responses of organisms, mediated by traits such as dispersal ability, ecological tolerance, rate of adaptation, and rate of acquiring reproductive isolating mechanisms, where required.

the environmental and geographical contexts, and how organisms respond to them. First measure how the physical environment and resources vary over space and time. This could constitute spatial patches with a certain probability of changing state, or environmental gradients, or both. The frequency spectrum of spatial and temporal variation will have a large impact on whether organisms evolve to generalize, or diversify into species or short-lived ecotypes or clines.

Then quantify the effect of environmental variation on organismal fitness—that is, selection pressures as a function of genes/phenotypes and environmental conditions (including physical environment, resource availability, parasitism, etc.). This allows calculation of the strength of diversifying or unifying selection on traits and genes at each time point. Some combinations of environment and genetic structures are more conducive to divergence than others—for example, when selection aligns with existing genetic covariances among traits and the resulting changes also act to promote reproductive isolation.

Next, measure geographical isolation across the landscape in terms of dispersal probabilities between different locations, and how that changes over time. Dispersal depends on the combination of physical environment and organismal traits. There might be strict barriers to dispersal or patches of habitat that are more/less resistant to movement. Their effects will depend on dispersal kernels: rare, long-range dispersal might be important for colonizing new areas, without maintaining ongoing gene flow with source populations. Dispersal rates might evolve in response to the spatial distribution of available habitat or resources. If so, structures that might superficially appear conducive to geographical isolation might lead to evolution of high dispersal abilities and reduction in speciation rates over time.

For sexual organisms, finally measure the probability of interbreeding and the probability of producing viable offspring between pairs of individuals. Together with the fitness of hybrid offspring in a given environment, this will determine the degree of reproductive isolation between diverging species that encounter one another. For prokaryotes, measure instead the probability of gene transfer between individuals, which might vary across genes and could involve multiple mechanisms for gene transfer such as plasmids or transformation.

Now we observe dynamics in these quantities over time, recording when speciation events occur. We then model the probability of speciation as a function of the geographical context, environment, organismal traits, and interaction among these factors. This model takes account of not only conditions when speciation occurs but also conditions when it does not. We measure the speed of different genetic responses (e.g. ecological divergence and the origin of reproductive isolation) in order to determine whether the rate of origin of new conditions or the rate of genetic responses limits the overall rate of speciation. We compare how often conditions for alternative mechanisms arise and coincide or oppose each other. Finally, we test our understanding by manipulating parameter values and observing whether the probability of speciation and component processes increases or decreases as predicted.

5.5.1 Future avenues

The first requirement for achieving this 'perfect knowledge' is new theory that incorporates the entire process. Most speciation models focus on genetic responses when an ancestral population faces a static set of environmental conditions. Gavrilets and co-workers constructed simulation models of speciation within landscapes, which were applied to several empirical studies of speciation (Birand et al., 2012). These yielded some general predictions for matching to patterns of diversification (Gavrilets and Losos, 2009). Aguilee et al. (2018) developed an individual-based simulation model of these processes in order to generate predictions for the main determinants of speciation probability (Box 5.1).

The next requirement is new empirical approaches to quantify the dynamics of the process, encompassing conditions, genetic responses, and multiple interacting factors. The best current evidence for temporal dynamics of speciation events comes from exceptional fossil records such as planktonic foraminifera. Ezard et al. (2011) used statistical models to identify environmental and biotic conditions correlated with probabilities of speciation and extinction (Fig. 5.10). The probability of speciation was greatest when diversity was low, consistent with the hypothesis that ecological opportunity controlled speciation rates. In contrast, extinction rates correlated with environmental fluctuation, namely periods of rapidly changing temperatures. At present, however, mechanisms of speciation in foraminifera (forams) are unknown in terms of reproductive barriers, levels of gene flow, and ecological traits associated with divergent selection. It would be profitable to fill these gaps through culture

Box 5.1 Determinants of speciation rate in a dynamic landscape.

Aguilee et al. (2018) developed a model of organisms diversifying in a heterogeneous landscape experiencing fluctuations in environments and the degree of geographical isolation (equivalent to the landscape visualized in Fig. 5.8). The model incorporates resource competition, genetic divergence, and reproductive isolation via genetic incompatibility. Three stages were observed during the radiation of a clade arriving in an empty landscape.

Stage 1: Speciation rates are high as the clade adapts to geographically isolated patches with different abiotic environments. The speciation rate depends mainly on the time spent in geographical isolation, resource abundance (determining population size and the strength of selection to local optima), and the rate of origin of post-zygotic isolation. Extinction rates are low and diversity increases rapidly.

Stage 2: Speciation now involves divergence for coexistence of species in the same area. The speciation rate depends mainly on the rate of landscape fluctuations, which exposes different parts of an ancestor to local adaptation, and reduced interspecific competition, which allows species to co-occur in overlapping sites. Extinction rates increase as ranges are subdivided, population sizes decrease, and new species are outcompeted by or hybridize with phenotypically similar species. Diversity continues to increase but at a decelerating rate.

Stage 3: Speciation rates fall further, because each species is restricted to a very small geographical area and narrow range of environmental conditions. There remains turnover of species (i.e. speciation and extinction) at a rate determined by the rate of origin of post-zygotic isolation and pace of landscape dynamics. Steady-state diversity is reached.

The model shows how intrinsic and extrinsic factors interact to determine the rate of speciation—especially in stage 2, speciation rate is decoupled from the rate of origin of reproductive isolation, as observed in the study of Rabosky and Matute (2013).

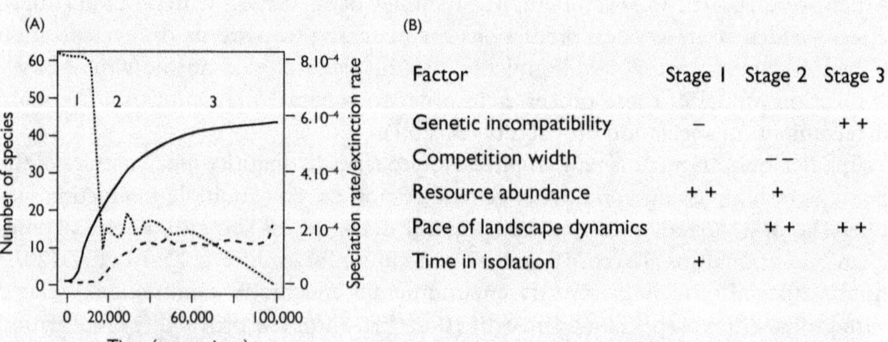

Fig. 5.9 Summary of simulation results of Aguilée et al. (2018). (A) Number of species increases to a stationary level (solid line). Speciation rates start high and fall over time (dotted line). Extinction rates start low and rise to moderate level (dashed line). Three stages are numbered. (B) The main factors that increase (+) or decrease (−) speciation rate vary among the three stages. (Figure adapted from Aguilée et al. 2018 under creative commons license).

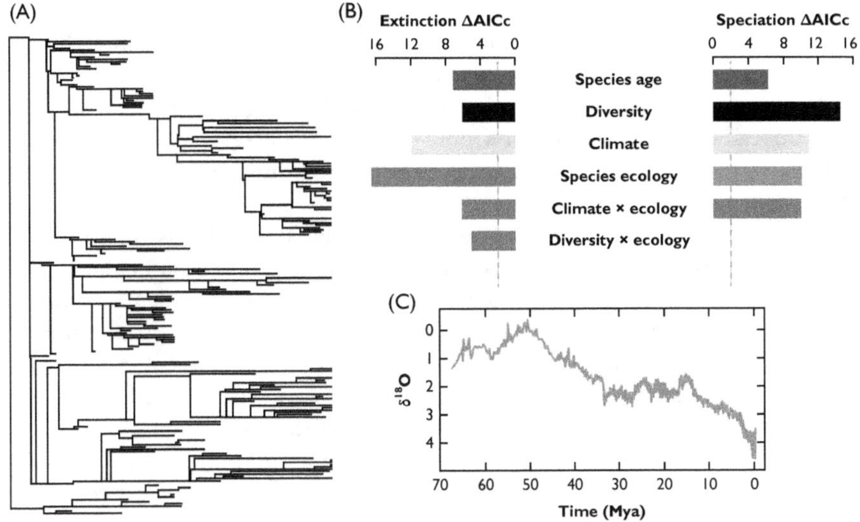

Fig. 5.10 Tracking probabilities and causes of speciation and extinction in the fossil record. A lineage phylogeny of planktonic foraminifera (A, Aze et al. (2011)) based on fossil evidence was used to calculate per lineage speciation and extinction rates and model their correlation with explanatory variables (summarized in (B)), such as a proxy of global temperatures (inversely related to $\delta^{18}O$, the ratio of ^{18}O to ^{16}O in carbonate sediments (C)). Speciation rates increased when diversity was lower, interpreted as ecological opportunity unconstrained by competition from other species. Extinction rates depended more on abiotic conditions and species ecology (habitat and morphology). (Reprinted from Barraclough (2015) with permission, using material from Ezard et al. (2011) and Aze et al. (2011).)

experiments and/or genomic studies. Another limitation of forams as a model system is that they lack much geographical signal. Species have global ranges, which is a rare characteristic that partly explains their good fossil record. Forams do segregate along depth gradients, however, which could provide extrinsic barriers to gene flow and impose divergent selection on populations.

The key requirement for phylogenetic and genomic studies of extant clades is a more systematic compilation of data on key parameters outlined in this chapter. We need comprehensive data across whole clades of reproductive isolation and population genetic/genomic estimates of gene flow and selection (see chapter 4). These data should be compiled in a way that allows estimation of parameter values and mathematical relationships between different parameters—for example, how does gene flow, reproductive isolation, and the pattern of selection on genomes change with phylogenetic distance, and with geographical and environmental distance? The most promising avenue is whole-genome sequencing, since it can be applied irrespective of study taxon, and statistical methods are well advanced for estimating the history of gene flow and selection across genes. These methods are not magic and are still affected by limitations with historical and observational inference. Care is needed not to overestimate confidence. For example, Bayesian and approximate Bayesian computation methods estimate parameters in complex models, but the outputs depend on the choice of model and priors. These approaches are less conservative with respect to replication than traditional statistics—a single datum outside the range expected for a simpler model can lead to strong support for a more complex model (even if the extreme value of that datum was caused by different processes than those specified in the model under consideration).

Assembly of systematic ecological and reproductive trait datasets is also required (see chapter 4). Multivariate analyses comparing variation to environmental variation and other data (e.g. measures of reproductive isolation) could then identify traits with strong associations with components of the speciation process. As discussed in chapter 4, new methods are needed to investigate the genetic basis of trait evolution across large scales, and coordination across multiple laboratories is needed to compile comprehensive data at a large enough scale to answer these questions.

Another gap at present is identifying conditions that do not promote speciation. Extending sampling to multiple individuals and locations within species could allow estimation of the frequency that different conditions arise, including those that do not typically result in speciation. A snapshot in time through the complex process in Fig. 5.8 might allow quantification of the typical distribution of gene flow and divergent selection within species. How many species are cohesive versus experiencing early stages or partial signatures of divergence? If we then make the assumption that the frequency of conditions across species represents the frequency of occurrence over time, does a simulation model with those frequencies generate speciation dynamics consistent with phylogenetic evidence? Such an approach would effectively relax the focus on species as units as discussed in chapter 2 and quantify the history of diversifying forces and resulting responses across the clade.

For organisms with fast generation times it might even be feasible to re-estimate conditions after 100 or 1000 generations to gain direct evidence of temporal changes. For animals and plants, we have no hope of observing repeated speciation events, but we could estimate trajectories around a short time-slice in Fig. 5.8. How consistent are patterns of isolation and selection detected from a single snapshot in time? Can we extrapolate from dynamics operating over short time-slices to overall patterns, or do we need to incorporate rare events? Clades in a relatively contained geographical area that can feasibly be sampled would be needed. Annual plants in the Cape Floristic region come to mind: repeated endemic radiations within a small geographical region with multiple environmental gradients and physical barriers. Alternatively, phenotypes and genome sequences could be sampled over ~ 100 years from museum specimens of comprehensively sampled clades (Holmes et al., 2016). For some organisms, such as viruses, it is possible to track macroevolutionary dynamics live (Neher et al., 2016). There has not been much work looking at diversification per se, as most focuses on resistance evolution, but viruses do display distinct genetic clusters indicative of independently evolving sets (Herniou et al., 2015).

Large samples are needed to identify general trends, but replication brings further challenges. As more clades, geographical areas, and ecological lifestyles are amalgamated, sub-clades vary in their responses to particular conditions (Davies et al., 2004a). As replication increases, the complexity of model needed to explain variation in speciation dynamics therefore also increases. Is there enough replication on earth to tease apart all of the interacting factors that determine speciation dynamics across taxa? This raises a chief limitation with approaches described in this chapter, namely that they are observational. As well as the problem of inferring causation, with observations we can compare how different taxa respond to the same conditions, but not how the same taxon responds to different conditions. Organisms adapt locally to their region, and therefore related taxa do not have exactly the same traits in different regions even when they belong to the same wider clade.

A final avenue for improved understanding is manipulative experiments of speciation in fast-evolving organisms, specifically microbes. These are considered in detail in chapter 6.

5.6 Conclusions

Speciation results from the interaction between environmental conditions and organismal responses, and is driven by both geographical isolation and divergent selection. The 'sweet spot' occurs when divergent selection and geographical isolation align. Yet, we currently lack integrated, quantitative understanding of the dynamics of speciation. Speciation plays out over timescales that have not been directly observed in any organism, making it hard to track. To progress, theoretical models need to be coupled with determined efforts to quantify each set of processes contributing to the whole phenomenon, encompassing the dynamics of conditions favouring divergence as well as responses (e.g. estimates of selection and gene flow from genome sequences). Coordinated efforts targeting whole clades or regions are needed to achieve this synthesis—going beyond haphazard investigation of different components of speciation in different taxa. An integrative approach with expanded theoretical models, environmental data, genomics, and experimental evolution offers exciting new potential to fill the gaps in knowledge.

6

Species and speciation without sex

6.1 Introduction

Recombination and reproductive isolation play such a central role in theories of speciation that it has been argued that 'species' and 'speciation' are properties exclusive to sexual organisms. This view makes sense if species are defined as reproductively isolated groups. As argued in chapter 2, however, recombination is not the only cohesive force and reproductive isolation is not the only mechanism leading to independent evolution. So, what is the expected pattern of diversification in strictly asexual organisms? Comparing diversity patterns between obligate out-crossers and strictly clonal organisms might shed light on the relative importance of different mechanisms behind diversification (Maynard Smith and Szathmary, 1995; Coyne and Orr, 1998; Barraclough et al., 2003). This chapter outlines the theory behind diversification in asexual clades, before applying methods described in chapters 3 and 4 to test those ideas in bdelloid rotifers. It then considers how intermediate and alternative forms of recombination might influence diversity patterns, looking in more detail at bacterial species.

6.2 Theory of diversification in asexuals

Barraclough et al. (2003) synthesized theory for the expected pattern of diversity in sexual and asexual clades faced with the diversifying forces of geographical isolation and divergent selection. By asexual, I refer to the strict case of obligate clonal reproduction: mixed strategies are discussed in section 6.5. In short, faced with the same diversifying forces, asexual clades should diversify in qualitatively similar ways to sexual clades (Fig. 6.1). With geographical isolation, independent evolution in asexuals occurs because new genotypes arising in one population cannot disperse into another isolated population (Fisher, 1930). After a sufficient period of time, genetic clusters arise equivalent to those expected in sexual populations facing the same conditions (see chapter 3). All other things being equal, the degree of differentiation in the sexual versus asexual case would depend on the ploidy level and the population size. For example, if additional females replace the lost males, a mitochondrial marker would take longer to differentiate in the asexual populations than in the sexual case because of the larger effective population size (Chapter 3). The main qualitative difference between

The Evolutionary Biology of Species. Timothy G. Barraclough, Oxford University Press (2019).
© Timothy G. Barraclough 2019. DOI: 10.1093/oso/9780198749745.001.0001

sexuals and asexuals in this scenario is the structure of variation within populations. In asexuals, the same genealogy is expected for all loci and so variation across multiple loci is hierarchical right to the level of individual. This gives the appearance of discrete rather than continuous variation within a population—apparent as the 'fractured' appearance of phenotype distributions in Fig. 6.1—but any discrete sub-clades within each population are transient and the total amount of variation is limited by drift and selection. Significant clustering of samples from the isolated populations and their genetic and phenotypic divergence over time is still expected, as described in detail in chapter 3.

Divergent selection can also drive the emergence of independently evolving asexual species. An asexual population can diverge to specialize on two discrete ecological niches, unencumbered by the gene flow that limits sympatric speciation in sexuals (Fig. 6.1, bottom row). After sufficient time, this can result in genetically and ecologically

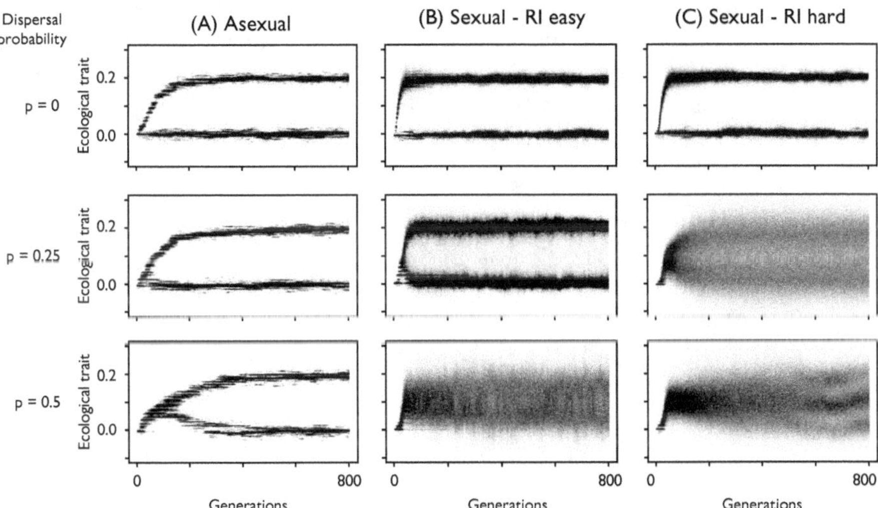

Fig. 6.1 Simulations of diversification of (A) obligate asexual, (B) obligate sexual with easy evolution of reproductive isolation (RI), and (C) obligate sexual with hard RI. Model details for the sexual populations are described in Box 2.2 and Fig. 2.3: individuals were assumed to be hermaphrodite. In each case, two patches were specified with distinct optima for a continuous phenotype of 0 and 0.2, respectively, coded for by ten additive, diploid loci. Distribution of phenotypes in the populations is shown: darker intensity indicates higher frequency of individuals with that phenotype. Mating in sexual cases occurred randomly within each patch. Probability that offspring dispersed to the other patch varied from 0.0 (complete geographical isolation) to 0.25 to 0.5 (completely random settlement between patches). Easy RI assumed that probability of reproduction is a function of similarity of ecological trait (i.e. complete pleiotropy between reproductive trait and ecological trait, Fig. 2.3D), whereas hard RI set the probability of reproduction as depending on a separate reproductive trait encoded by ten separate loci. Sexual reproduction speeds up the rate of divergence when dispersal rates are low, but diversification in sexuals is limited by increasing dispersal rates.

distinct species that are limited independently—the survival of offspring from one species does not affect the chance of survival of offspring in a separate species. New mutations that convey advantages in one niche are not beneficial in the second niche, and hence the genotype that contains them only spreads to fixation in its own niche (Fig. 6.2A). This is Cohan's ecotype theory of speciation in clonal bacteria (Cohan, 2001; Koeppel et al., 2008), where selective sweeps occur separately in populations adapted to distinct ecological niches, leading to distinct ecological and genetic clusters referred to as ecotypes.

Additional assumptions are required to generate a standing pattern of discrete genetic clusters within asexual clades. First, ecological niches must be stable over time periods sufficient for genetic differentiation (see chapter 3): selection pressures that fluctuate over time would not maintain consistent divergent selection on each species. Second, selection on all loci and traits must act independently on each cluster, or the response to a common selection pressure must be independent. Imagine that one trait has the same optimum in all species (e.g. optimum temperature for growth), whereas other traits have different optima among species (e.g. resource use). Then the shared optimum shifts in all species (e.g. there is global warming). The set of species could respond in two different ways (Majewski and Cohan, 1999). Either a new genotype adapted to

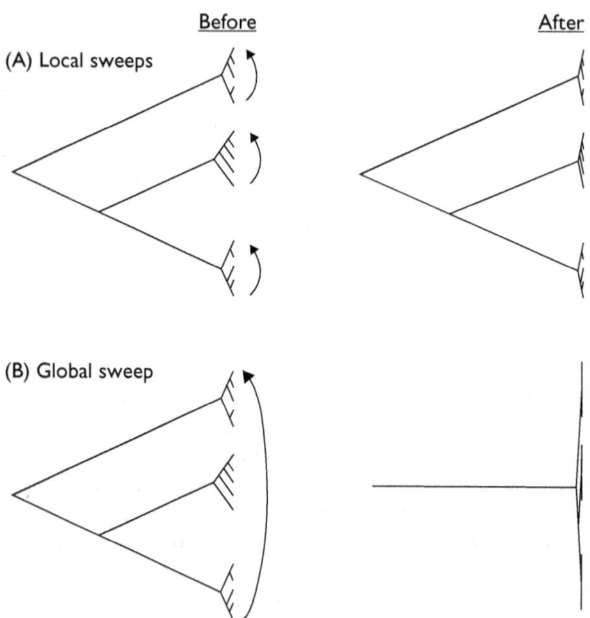

Fig. 6.2 Two forms of response to a change in shared environmental conditions in a set of asexual species adapted to different niches. (A) Beneficial selective sweeps occur in parallel in each species, strengthening the pattern of discrete clustering. (B) New beneficial genotype arises in one species and sweeps through all species, removing species variation. Subsequent local sweeps might re-establish genotypes adapted to the separate local niche differences.

higher temperature originates and spreads to fixation in every species (Fig. 6.2A). Or a new genotype arises in one species, spreads to replace genotypes from all other species (i.e. a global selective sweep, driving those other species extinct, Fig. 6.2B), and its descendants then re-evolve the other ecological differences among species (for example, they specialize to use different resources). Which outcome occurs will depend on the phenotype–fitness map (i.e. is the largest reduction in fitness caused by having the wrong resource genotype or the wrong temperature genotype?) and the chance order of mutations. The standing diversity and its turnover over time in an asexual clade therefore depend on the shape of selection pressures acting on its members. I return to these ideas in chapters 7 and 8.

Although qualitative responses to diversifying forces are expected to be the same, sexuals and asexuals should still differ in their propensity to diversity in two main ways. First, asexuals should diversify more readily than sexuals in the face of gene flow, because they do not need special mechanisms for reproductive isolation (Fig. 6.1, bottom row, and Fig. 6.3A; Felsenstein, 1981). This assumes that reproductive isolation is a limiting factor for speciation in sexuals—a view supported by the classic emphasis on reproductive barriers in the speciation literature (Coyne and Orr, 2004; Gavrilets, 2004). If, instead, reproductive isolation is able to evolve easily and rapidly between diverging populations—for example, as a correlated by-product of ecological

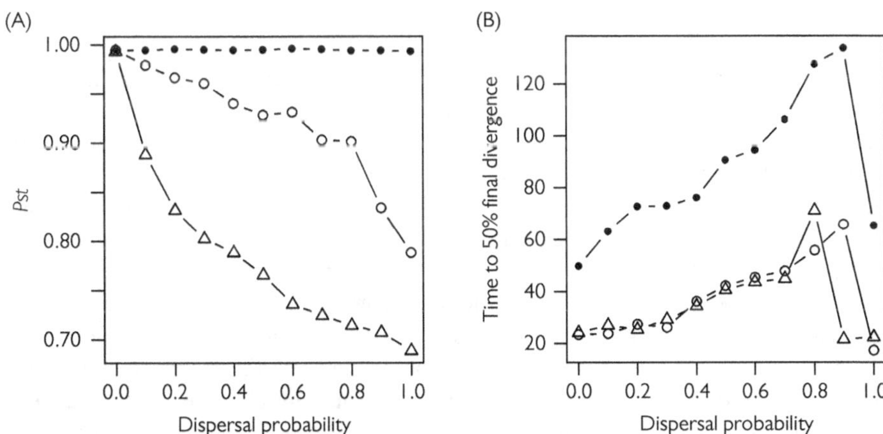

Fig. 6.3 Trends in phenotypic divergence across a range of dispersal rates. The model in Fig. 6.1 was simulated across a range of dispersal rates for asexual (black circles), sexual with easy reproductive isolation (open circles), and sexual with hard reproductive isolation versions (open triangles) (each point equals mean of five simulations, and standard errors are same size as the point). (A) Final level of differentiation into two phenotypic clusters, expressed as *Pst*; proportion of total phenotypic variation attributed to being between the two clusters, fitted by k-means clustering after simulation run of 800 generations. (B) Time in generations taken to reach 50 per cent of final phenotypic divergence between patches. Asexual populations differentiate more than sexual ones (*Pst* is higher), especially at high dispersal rates, but they take longer to diverge due to slower rates of multilocus adaptation.

divergence (Fig. 6.1B)—the difference between sexuals and asexuals might be less, at least for intermediate levels of gene flow.

Second, whenever diverging populations are protected from the homogenizing effect of gene flow and recombination (either geographically or by emerging reproductive barriers), sexual populations might diverge more rapidly than asexuals (Fig. 6.1, top row, and Fig. 6.3B). Recombination is well-known to speed up the spread of beneficial gene combinations under certain conditions (Otto and Michalakis, 1998), which might allow sexuals to adapt more rapidly to new conditions (Burt, 2000). This could lead to a more differentiated pattern of diversity in a sexual clade than an asexual clade facing the same conditions, if diversifying conditions originated recently or change regularly.

Comparing diversity patterns in asexuals and sexuals might therefore reveal the relative importance of reproductive isolation and ecological divergence in shaping diversification. If reproductive isolation is limiting in sexuals, we predict greater ease of speciation in asexuals than sexuals, but if rates of ecological divergence are limiting, we predict the opposite pattern. The relative propensity of sexuals and asexuals to diversify into discrete, independently evolving species via geographical isolation and divergent selection is therefore an empirical question.

6.3 Evidence for asexual species—bdelloid rotifers

Testing the ideas in section 6.2 requires examples of strictly asexual clades, ideally with comparisons to related sexual clades. In principle, there are numerous origins of obligate asexuality to choose from, even in multicellular eukaryotes. The problem is that most asexual lineages are recently derived or maintained by the ongoing origin of asexual clones from a related sexual population. The pattern of variation in these lineages tends to be amorphous (Sepp and Paal, 1998). There are myriad variants, often given taxonomic names (called agamospecies in plants), but without clear differentiation into discrete genetic clusters consistent with independently evolving groups as defined in section 6.2. This might indicate that speciation does not occur in asexuals—perhaps the conditions are generally not met for consistent selection to occupy discrete ecological niches or they adapt too slowly to those opportunities. Or it might indicate that asexuals tend not to persist long enough to have the chance to diverge into separate species. The evolutionary disadvantages of asexuality, such as failure to keep pace with coevolving parasites and accumulation of deleterious mutations (Bell, 1982; Burt, 2000), mean that asexuals are usually destined to early extinction before subsequent speciation can occur. What pattern of diversity emerges in those rare cases when asexuals escape their normal doom and persist for millions of years?

Bdelloid rotifers offer the best case among animals or plants for answering this question (Fontaneto and Barraclough, 2015). Bdelloids are microscopic animals that live in moss, ponds, soil, and any other habitat with occasional or permanent freshwater (Fig. 6.4). They are remarkable among animals because no males have ever been recorded in over 300 years of looking and with many 10^5 animals observed (Birky, 2010), in contrast to some other putative ancient asexuals in which occasional males are

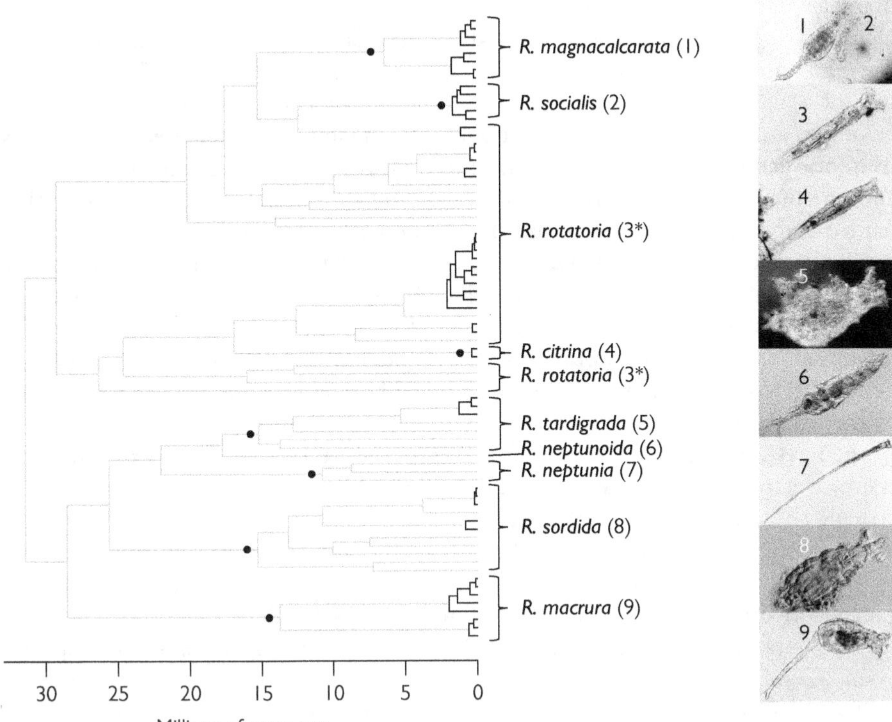

Fig. 6.4 Genetic relationships among a sample of 76 clones of bdelloid genus *Rotaria* based on the mitochondrial *cox1* gene and nuclear 28S rRNA gene. Branch lengths are in units of millions of years calibrated using a generic invertebrate molecular clock for *cox1* of 1.76 per cent sequence change per million years. Taxonomic identifications are shown and most named species comprise monophyletic clades (indicated with black circles on branch, all with 1.0 posterior probability support in Bayesian phylogenetic analysis), with the exception of the pelagic *R. rotatoria*, which comprises a wide range of distantly related genotypes. Black branches indicate clusters identified by GMYC analysis for genetic clustering; grey branches indicate branches delimited as between-cluster branching. (Redrawn with data from Fontaneto et al. (2007). Photograph for species 1 and 2 by T.G. Barraclough; all other photographs by Michael Plewka (plingfactory.de) with permission.)

found (Wilson et al., 2018). Despite this, bdelloids have persisted for over 50 million years (Tang et al., 2014b) and diversified into over 450 taxonomically recognize species and 19 genera. This might seem to show already that bdelloids have diversified into species but, although morphologically diverse, these taxa could constitute names applied to an amorphous spread of variation equivalent to agamospecies in plants. Statistical evidence of genetic and phenotypic variation using methods outlined in chapters 3 and 4 is needed to test for independently evolving groups.

Birky et al. (2005) provided the first evidence of genetic clustering in bdelloids, by demonstrating the existence of clusters with four times the divergence between them than variation within them—a level of differentiation argued to be consistent with reciprocal monophyly (see Box 3.1 for discussion of this method for delimiting species). Fontaneto et al. (2007) subsequently applied the GMYC method (Box 3.2) to test for a significant pattern of clustering relative to a null model that individuals belong to a single interacting super-population of clones. They sampled individuals from one genus, *Rotaria*, for one mitochondrial gene (cytochrome oxidase I (*coxI*)) and one nuclear gene (28S ribosomal RNA). Individuals belonging to each of nine named taxonomic species were mostly monophyletic (Fig. 6.4). More critically for testing the nature of asexual diversification, significant genetic clusters were apparent, separated by longer internal branches. Many traditional species included several genetic clusters, but it is not unexpected to observe cryptic species, especially for understudied microscopic animals like bdelloids. There was significant evidence therefore for independently evolving sub-groups of bdelloids, as predicted by theories of asexual speciation.

The next step was to determine the causes of clustering. Were genetic clusters a result of geographical isolation or did they result from divergent selection, as in ecotype theories of asexual speciation? Natural history supports ecological differences among species. Some of the *Rotaria* species live in desiccating habitats such as dry ponds or moss on trees and can survive desiccation through the process called anhydrobiosis (another remarkable feature of bdelloid rotifers). Others live permanently in water in ponds and do not survive desiccation. Moreover, two species living as epibionts on the body of the water louse *Asellus aquaticus* partition space neatly between different body regions: the chest area (*R. magnacalcarata*) versus the base of the legs and the anus (*R. socialis*). These observations, however, do not constitute evidence for divergent selection driving differentiation. Instead, tests for divergent selection as outlined in chapter 4 are needed.

The DNA data alone could not answer this question. *coxI* is under purifying selection in bdelloids (Barraclough et al., 2007) and 28S codes for ribosomal RNA and is not amenable to tests for selection. Instead, we turned to morphological variation, measured as morphometric variation in the bdelloids' jaws (called trophi), which are the only hard part of the animal. If the size and shape of the jaws evolved neutrally, the ratio of interspecific to intraspecific variation should be proportional to the equivalent ratio for neutral molecular changes. Alternatively, if the bdelloids experience divergent selection on jaw morphology, there should be a greater level of interspecific variation relative to intraspecific than for neutral characters (see chapter 4). Using silent substitutions in mitochondrial DNA as a baseline for neutral expectations, we found that the ratio of morphological variation between species versus within species was significantly higher than expected from neutrality (Fig. 6.5). There has been adaptive divergence of bdelloid jaw morphology—presumably associated with different food or feeding strategies, although the exact functional significance of shape variation remains obscure. Interestingly, this pattern was apparent between taxonomic species but not between the genetic clusters (i.e. cryptic species). It is possible that other traits have diverged between the genetic clusters, or some of the clustering could reflect geographical isolation. In any case, the qualitative pattern of diversity

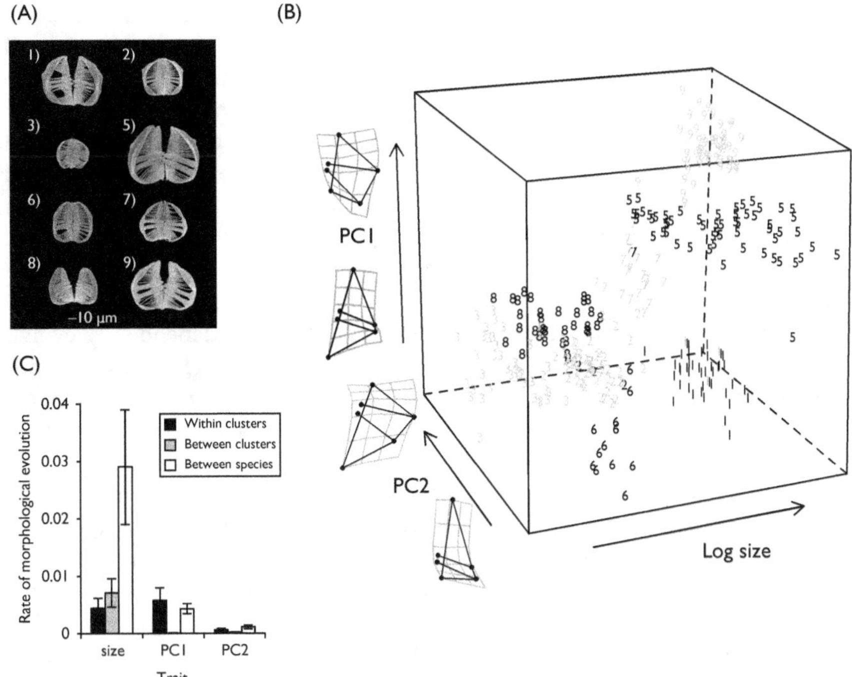

Fig. 6.5 (A) Scanning electron micrographs of jaws (called trophi) of *Rotaria* species. Numbers correspond to species in Fig. 6.4. (B) Morphometric variation of trophi size and first two principal component (PC) axes representing shape (schematics shown for gradient in shape along each axis) reveals morphological clustering of each species. (C) Mean rates of morphological evolution mapped onto the DNA tree and estimated separately on branches within genetic clusters, between genetic clusters within named species, and between taxonomic species. Rate of size evolution is significantly higher between named species, consistent with divergent selection operating between these clades. (Redrawn from Fontaneto et al. (2007) under creative commons license.)

matched patterns found in sexual clades, where cryptic species are also relatively common, especially in groups that are so poorly known.

The next question is whether the pattern of differentiation in bdelloids is stronger or weaker than in equivalent sexual organisms. The nearest relatives of bdelloids that live in partly overlapping habitats are the monogonont rotifers. Comprising ~ 1500 named species, monogononts live primarily in permanent water bodies and are cyclical parthenogens that have a sexual stage in their life-cycle (like in *Daphnia*). Incidentally, males are well known in monogononts and commonly observed in both the laboratory and field. We therefore have a picture of what bdelloid males might look like (reduced version of the female with visible sperm under the microscope) though they have not been found. A disadvantage of bdelloids' ancient status is that it permits only one phylogenetically independent comparison: between bdelloids and monogononts that diverged over 300 Mya (Tang et al., 2014b). Aside from reproductive mode, there are other differences between the clade; in particular, monogononts disperse by resting eggs, whereas bdelloids disperse as dried-up adults during anhydrobiosis—they can be recovered following rehydration from wind-socks. Nonetheless, the two clades co-occur in some habitats (ponds and streams) and so remain broadly comparable (unlike the nearest obligate sexual relatives of the bdelloid rotifers—the macroscopic acanthocephalan parasites or the three species of seisonid rotifers that live commensally on marine Crustacea).

Tang et al. (2014b) therefore compared the degree of genetic differentiation in *cox1* between representative clades of bdelloids and monogononts. They reported that bdelloids differentiated into more genetic clusters, that is, independently evolving units, within equivalent time periods than monogononts did. The net diversification rate per lineage per unit time was 50 per cent higher in bdelloids than monogononts. As a result, however, bdelloid species were less distinct than monogonont species, because they were more closely related to their nearest sister species (Fig. 6.6). The results support the hypothesis that diversification is less constrained in obligate asexuals because there is no requirement for reproductive isolation, but consequently the resulting pattern of diversity is less distinct than in sexuals. If classifying organisms into species represents a model to describe the pattern of genetic variation across a clade, 'species' explains a stronger shift in pattern in monogononts than in bdelloids.

Does this help us understand the role of sexual reproduction in differentiation? The remaining critical question is whether bdelloids are really strictly clonal without any recombination between different individuals (Fontaneto and Barraclough, 2015). A failure to observe males does not necessarily mean they do not exist. The first published genome of bdelloid rotifers appeared to find positive evidence for the lack of meiosis, because homologues that would represent diploid copies in a sexual animal were sometimes found on the same chromosome (Flot et al., 2013). This finding was not replicated, however, in other species (Nowell et al., 2018) and may represent an artefact of genome assembly or a peculiarity of the particular laboratory clone. Another approach was to search for genes involved in meiosis or sperm production in bdelloid genomes, which are present (Flot et al. 2013; Nowell et al. 2018), but animals lack a clear set of sexual markers that only function in sexual reproduction—these genes could be performing other functions (Hanson et al., 2013). At the time of writing, genomic evidence for lack of sex in bdelloids remains indecisive in either direction.

Even if bdelloids lack conventional meiosis, they might possess some alternative mechanism for recombination, which might still limit their utility as a case of asexual diversification. Gladyshev et al. (2008) discovered unexpectedly high levels of foreign gene uptake in bdelloid rotifers from bacteria, plants, fungi and other non-metazoa, which have since been shown to comprise up to 10 per cent of expressed genes in multiple species (Boschetti et al., 2012; Flot et al., 2013; Nowell et al., 2018). Gene uptake is hypothesized to occur when bdelloids repair double-stranded breaks in their DNA caused by desiccation—occasionally incorrect pieces of DNA in the environment get incorporated by mistake. Horizontal transfer contributes to genetic differences between related bdelloid species in different habitats (Eyres et al., 2015),

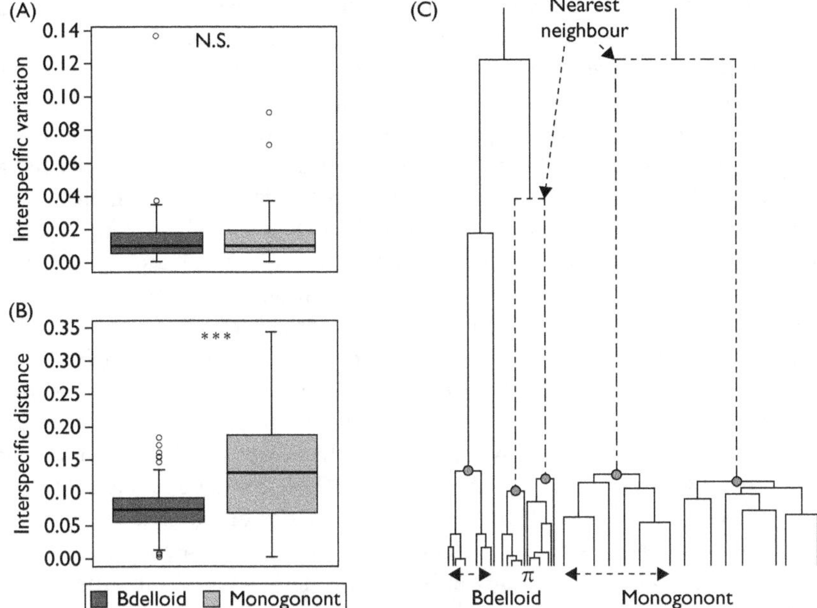

Fig. 6.6 Comparison of pattern of differentiation between six representative bdelloid rotifer clades (assumed to be obligate asexuals) and seven monogonont rotifer clades (assumed to be facultatively sexual and asexual). (A) Boxplot of average pairwise genetic distance within genetic clusters identified using the GMYC algorithm. Box shows interquartile range, circles show outlier points. Significance in linear models is shown at top of plot. (B) Boxplot of average distance to nearest neighbour belonging to a different genetic cluster. (C) Schematic representing the main result that variation within clusters is similar between bdelloids and monogononts, but bdelloid clusters (i.e. hypothetical species) tend to be more closely related to their nearest sister cluster than observed in monogonont clades. (Reprinted from Tang et al. (2014b) under creative commons license.)

and so might introduce new genetic functions equivalent to horizontal gene transfer among unrelated lineages in bacteria. However, for theories of the evolution of sex and speciation, the key question is whether a similar phenomenon could promote recombination among different individuals within bdelloid populations, and whether barriers to exchange exist between species.

Debortoli et al. (2016) reported evidence of allele sharing within and between bdelloid rotifer species from a single location, which they interpreted as evidence for horizontal gene transfer. The hypothesis was that bdelloids take up DNA from the environment and incorporate it via homologous recombination, a similar mechanism to natural competence and transformation in bacteria. If occurring at a high enough rate, this process could lead to homogenizing effects of gene flow within populations and create conditions requiring isolating mechanisms for divergence to occur in sympatry. The result was later shown, however, to reflect accidental contamination of tubes with DNA from multiple rotifers rather than a true signal of gene exchange (Wilson et al., 2018).

A second study reported evidence suggestive of an alternative mechanism for gene exchange in bdelloids. Signorovitch et al. (2015) found evidence of the assortment of whole multilocus haplotypes in a small sample of individuals of *Macrotrachela quadricornifera*. The authors propose rare, unorthodox meiosis without crossing over, similar to that found in the evening primrose, *Oenanthera*, as an explanation. At the time of press, this has not been found in other rotifers and its wider significance remains to be elucidated. Even if present, such a system might limit outcrossing to the shuffling of combinations of whole-genome level haplotypes: bdelloid genomes would behave as a single diploid locus, a phenomenon hitherto restricted to some very simple genetic models of speciation and arguably constraining rather than facilitating their capacity to adapt to spatially and temporally variable habitats.

It remains possible, however, that some form of recombination occurs between bdelloids of the same species cluster, based on what we know about uptake of foreign DNA. If this occurs frequently enough, there might still be a role for 'reproductive isolation' in bdelloid species divergence—in this case, barriers to uptake and incorporation of genes from another population. These could be ecological based on habitat preference or due to decreasing probability of homologous recombination with genetic divergence, as observed in bacteria and yeast (see section 6.5).

How much recombination among individuals would be needed in order for bdelloid clusters to be explicable by conventional ideas used in sexually reproducing organisms? Theory for bacteria says that a rate of recombination per site above the rate of mutation per site is the threshold level for transitioning from 'asexual' to 'sexual' dynamics for neutral markers (Fraser et al., 2007). So far there is no evidence for a rate in bdelloids above the level needed for sexual-style dynamics. Knowledge of bdelloid genetics and evidence for recombination continues to develop apace and this conclusion might change in future.

6.4 Other asexual eukaryotes

A chief limitation of bdelloid rotifers as a study system is that repeated comparisons cannot be made between recombining and non-recombining clades—they arose just once. Oribatid mites contain multiple apparently asexual clades, which might offer an alternative system for exploring these questions in animals (Domes et al., 2007). The presence of males across multiple widely distributed taxa in the clade raises the possibility, however, that the remaining species also have unobserved male production and cryptic sex or transitions to asexuality were recent. Fungi display a wide range of mating systems and offer an underutilized source for evolutionary studies (Billard et al. 2012). Care is needed because many apparently asexual fungi have been demonstrated to be sexual either by connecting an asexual 'anamorph' to a sexual 'teleomorph' by DNA sequencing, or by finding signatures of recombination (see chapter 4). Ozkilinc et al. (2018) found positive evidence for both differentiation of species and clonal evolution in the section Porri of the ascomycete genus *Alternaria* using multilocus sequencing. The different species are specialist pathogens on different host plant species, consistent with the ecological speciation mode outlined for strict asexuals in section 6.2. In contrast, a second section, Alternaria, had less distinct, generalist species with a wide host range and positive evidence for recombination. The authors highlighted that the difference in patterns of differentiation could result from recent divergence or differential sampling of the two sections as well as from the frequency of recombination.

Another potential source of multiple comparisons of sexual and asexual lineages comes from single-celled eukaryotes, but the same issues of demonstrating true asexuality arise. Colpodean ciliates were proposed as an extremely ancient asexual clade—around 900 Myr old—but sex has been demonstrated in one deeply embedded species. This could either result from a reversal to re-originate sex after nearly 1 billion years of abstinence, or the apparent lack of sex in other species could be an artefact of the difficulty of observing sex in these tiny organisms. The presence of micronuclei—a feature typically associated with sexual reproduction in ciliates—and genes required for meiosis further supports the likelihood of widespread secretive sex in this clade (Dunthorn and Katz, 2010; Dunthorn et al., 2017). Similarly, a traditional view of amoeba as anciently asexual has been challenged by recent phylogenetic reconstructions, which show several recent transitions from sexuality to asexuality. These could be amenable for a comparative evaluation of speciation in sexuals and asexuals (Lahr et al., 2011; Doerder, 2014).

6.5 Speciation across a continuum of recombination rates—bacteria

An excellent system for exploring the effects of varying recombination rates on diversification is bacteria—unsurprisingly, since bacteria constitute a major part of the diversity of life. Although all bacteria reproduce clonally, and are often deemed 'asexual' in theories of speciation, in fact there is wide variation in the degree and type of recombination. Some bacteria display undetectable rates of recombination using likelihood models applied to multilocus sequence data for house-keeping genes, whereas others have per nucleotide rates of recombination up to 60 times the per nucleotide mutation rate (Fig. 6.7; Vos and Didelot, 2009). How should this variation affect diversity patterns across those clades?

Most models of divergence in bacteria consider the effects of homologous recombination, which involves the copy and paste of DNA from the environment into a bacterial genome and requires regions of homologous sequence at each end of the pasted DNA. Although this differs from the shuffling of two genomes during reproduction observed in sexual reproduction, the effects of recombination on divergence are similar.

In strictly clonal models, divergence occurs via the processes outlined in section 6.2. A combination of geographical isolation and/or adaptation to distinct ecological niches could lead to the establishment of independently evolving and genetically distinct entities. Cohan and colleagues have argued that adaptation to distinct ecological

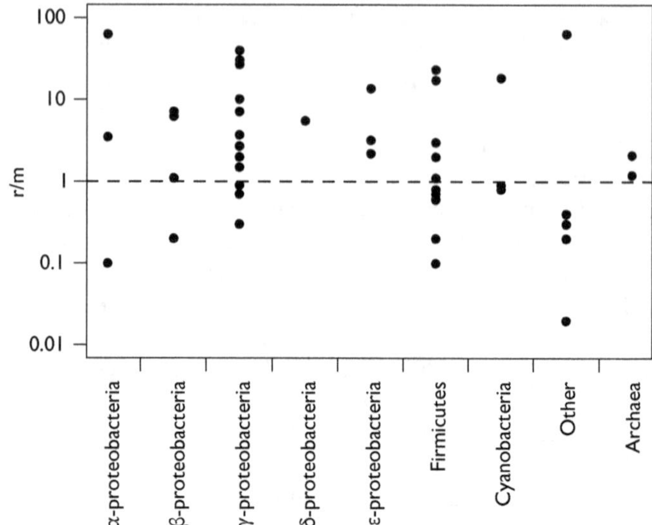

Fig. 6.7 Survey of per nucleotide recombination rates (r) relative to mutation rates (m) inferred across 46 bacterial and two archaeal species by Vos and Didelot (2009). Each point represents a bacterial species, with multilocus data available for multiple isolates analysed with ClonalFrame, which estimates the ratio *r/m*. Data are grouped by phylum; 'other' includes Bacteroidetes, Spirochaetes, and Chlamydiae. Dashed line shows the threshold of equal recombination and mutation rates that represents transition from asexual to sexual-type dynamics (Fraser et al. 2007). (New figure drawn with data from Vos and Didelot (2009).)

niches is the main explanation for the existence of bacterial species, which they call ecotypes to distinguish from species defined by reproductive isolation in sexual organisms (Cohan, 2001; Cohan and Perry, 2007; Koeppel et al., 2008). Strict linkage means that regular selective sweeps reduce variation across the whole genome of a clonal niche-specialist while creating genetic divergence between ecotypes. Occasionally new mutations that are advantageous across a wider clade might arise that reduce diversity back to a single genetic population. Observing a standing level of diversity within a wider clade would therefore indicate that niches are sufficiently stable for multiple discrete genetic clustering to persist. Geographical isolation is sometimes argued not to play a major role because it is assumed that tiny bacterial cells can disperse widely—but this may vary depending on whether they form resistant spores. Evidence from hot springs supports geographical isolation as playing a role in bacterial speciation (Papke et al., 2003), but the sheer diversity of bacteria means that we still lack comprehensive evidence of how important dispersal limitation is for speciation in bacteria.

As rates of homologous recombination increase, a threshold is crossed above which the recombination rate is sufficient to cause cohesion of neutral markers within populations (Fraser et al., 2007)—roughly when recombination rate exceeds mutation rate per nucleotide (dashed line in Fig. 6.7). Genome regions targeted by divergent selection will diverge more than neutral regions, but as in sexual models, there is a limit to divergence set by the balance of selection and the influx of locally maladapted genes via gene flow (Fig. 6.8). The limit to divergence increases as the number of loci coding for ecological differences increases (Friedman et al., 2013). Consequently, divergence of bacteria with high recombination rates requires mechanisms to limit gene flow that are analogous to reproductive isolating mechanisms in sexual eukaryotes (Polz et al., 2013; Bobay and Ochman, 2017). This does not affect a small minority of taxa: 53 per cent of taxa surveyed by Vos and Didelot (2009) exceeded the theoretical threshold for 'sexual-type' dynamics of $r/m > 1$, and 21 per cent were highly recombinogenic with $r/m > 10$.

Several mechanisms have been proposed for restricting gene flow in recombinogenic bacteria. An early proposal was that the probability of homologous recombination decreases exponentially with the percentage of sequence divergence because similarity is needed for the pasting mechanism to work. Initial models showed that this could help to reduce gene transfer between diverging populations and reinforce divergence (Fraser et al., 2007). The rate of decline in recombination with genetic distance seems to be too shallow in nature for non-recombining clusters to arise in strict sympatry (Shapiro and Polz, 2014). The mechanism could still help to maintain clusters formed

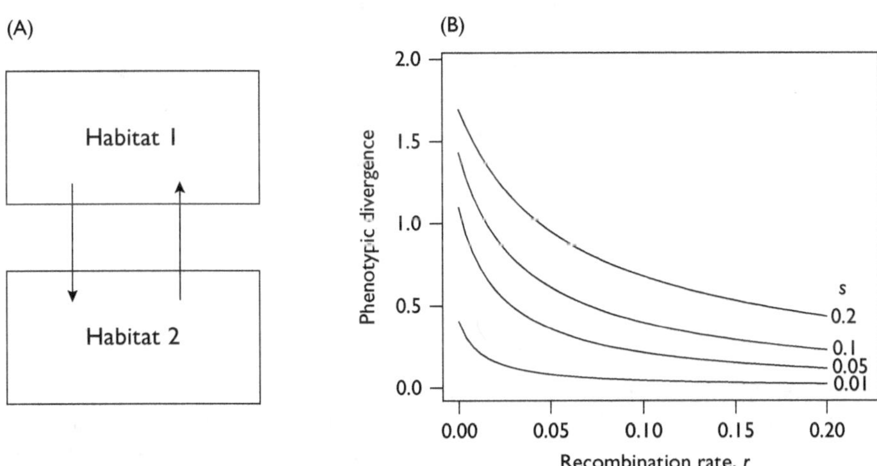

Fig. 6.8 Phenotypic divergence between two bacteria populations adapting to habitats with different phenotypic optima is constrained by gene flow via homologous recombination. Phenotype is encoded by two loci with equal additive effects, and has an optimum of +1 in habitat 1 (coded by genotype AB) and −1 in habitat 2 (coded by genotype ab). It is assumed that DNA flows freely between the two habitats and so the species share an external gene pool of secreted DNA, but that cells do not move between habitats. Just as in models of sexual populations, increasing per locus recombination rate (and hence rate of influx of maladapted genes arriving from the other habitat) reduces the equilibrium level of phenotypic divergence and hence mean fitness of populations. Each line shows results for a different selection coefficient, *s*, against maladapted genotypes (i.e. genotype ab in habitat 1 and genotype AB in habitat 2, respectively). (Reprinted from Schmutzer and Barraclough (2019) under creative commons license.)

during a period without contact. It has been rejected as a truly general mechanism, however, because Archaea show similar patterns of genetic clustering (Cadillo-Quiroz et al., 2012), yet homologous recombination does not decline markedly with genetic distance with the molecular mechanism they use (Polz et al., 2013).

A second class of genetic isolating mechanisms involves specific molecular machinery to prevent heterospecific gene transfer. One such mechanism uses pheromones. Bacteria communicate with one another via secreted molecules, which is called quorum-sensing. Carrolo et al. (2009) showed how a quorum-sensing pheromone called competence stimulating peptide (CSP) restricted recombination between two co-occurring groups of *Streptococcus pneumoniae*—the same clade found to contain multilocus clusters (Fig. 4.8 Fraser et al., 2007). Bacteria only take up DNA from their environment in a physiological state known as competence. Accumulation of CSP in the medium induces competence (i.e. increasing the rate of uptake of DNA) and also triggers lysis of non-competent cells (i.e. increasing the release of DNA fragments to be taken up by the competent cells). Most strains produce one of two varieties, CSP-1 and CSP-2, and cells with one variety do not respond to the signalling peptide of the other variety. Consequently, Carrolo et al. (2009) showed that two populations with CSP-1 and CSP-2 in turn corresponded to two distinct multilocus sequence typing (MLST) clusters that lacked exchange of house-keeping genes between them. This constitutes an isolating mechanism in these bacteria equivalent to mating preferences in animals or floral structures in flowering plants. Other authors have argued that restriction enzymes that cut specific sequences of DNA might play a role in reducing transfer of heterospecific DNA (Jeltsch, 2003). More work is needed to ascertain how widespread these specific mechanisms for restricting gene transfer are.

An alternative to genetic mechanisms for limiting gene transfer is the hypothesis that niche-specific gene pools facilitate ecological divergence in bacteria (Polz et al., 2013). If the structure of the environment means that bacteria from one habitat or utilizing a particular resource supply are unlikely to encounter DNA from bacteria using a different habitat or resource, gene flow between those populations will be low even if recombination rates are high. This mechanism is analogous to models of habitat-based assortative mating in sexual organisms (Diehl and Bush, 1989) and is one of the easiest mechanisms for ecological divergence of geographically connected populations. Because it depends on general features of the environment and dispersal range of DNA fragments and organisms, it is potentially general across all microbes.

Schmutzer and Barraclough (2019) investigated this hypothesis together with an alternative genetic hypothesis inspired by models of divergence with gene flow in sexual organisms. Because the level of divergence depends on the number of loci underlying adaptation to distinct niches, another possible response to divergent selection is for genetic effects to become concentrated into fewer loci. Moreover, differential gene loss—a well-documented phenomenon in bacteria—might provide a further mechanism to protect diverging genomes from the homogenizing effects of gene flow. They therefore investigated the effects of evolving genetic architecture, in terms of locus effect sizes on traits and the loss and gain of loci, on bacterial divergence, with and without the presence of niche-specific gene pools and at varying levels of recombination rate.

The model confirmed earlier work that transfer of DNA via a global gene pool can limit the divergence of bacterial populations with realistic levels of recombination. Strictly clonal simulations (Fig. 6.9, top left, black lines) were able to diverge to a greater extent than any simulation with homologous recombination and a global gene pool (Fig. 6.9, top left, grey lines), when the ecological trait was coded by a fixed number of loci. With a local, niche-specific gene pool, recombining populations diverged to the same extent as the clonal populations (Fig. 6.9, top right). The effects of global gene transfer diminished over time, however, when genetic architecture was able to evolve (Fig. 6.9, middle left). In particular, with gene loss, populations evolved towards having a single, private locus to encode the ecological trait, which therefore removed the scope for influx of maladapted genes to the population (Fig. 6.9, bottom left). Gene loss and gain are well known to shape variation among bacteria, and indeed rates of turnover sometimes exceed thousands of events per 1 per cent amino acid divergence (Nowell et al. 2014). Loss is typically interpreted in terms of reducing costs of maintaining and replicating DNA as bacteria specialize to particular conditions, but an alternative hypothesis is therefore that differential loss serves to protect local adaptation from the homogenizing effects of gene transfer.

The conclusion from these theories is that recombination should play a role in bacterial divergence among those taxa with high rates of homologous recombination. Other mechanisms for gene transfer in bacteria are also relevant for models of speciation. Similar arguments to those for homologous recombination would apply if transduction mediated by phage has a generally cohesive impact on genomes, as has been found among *Escherichia coli* isolates (Dixit et al., 2015). Different effects will arise if gene transfer brings in new genes. For example, if a plasmid carries genes that convey local adaptation in a particular environment, this will have consequences on whether two forms with divergent chromosomal genomes can coexist or not in that area (see chapter 7). Furthermore, integrons and virus-mediate transduction provide mechanisms for transferring operons or blocks of genes into new chromosomal backgrounds. A major role for these mechanisms is bringing in new functional capabilities and allowing lineages to occupy new habitats or hosts (Niehus et al., 2015). The role of these mechanisms for restricting gene flow to particular traits is discussed further in chapter 7.

Much of the empirical evidence for bacterial species at broad scales comes from marker-gene or multilocus sequence-type data, described in detail in section 4.2.2. To recap, genotypic clusters are apparent both in clades with high homologous recombination rates (e.g. *Streptococcus pneumoniae* complex) and in effectively clonal lineages (e.g. some *Bacillus*). Typically, the genome can be partitioned into core and non-core genes, with the former showing variation restricted to within-species clusters, whereas the latter transfer more broadly.

Whole-genome surveys of pairs of diverging species provide further evidence for the role of recombination and isolating mechanisms in bacterial species. For instance, Shapiro et al. (2012) studied divergence in *Vibrio cyclitrophicus*, a bacterium adapting to specialize on inhabiting two different particle sizes in the sea while undergoing high rates of homologous recombination (1000× the mutation rate across most of the

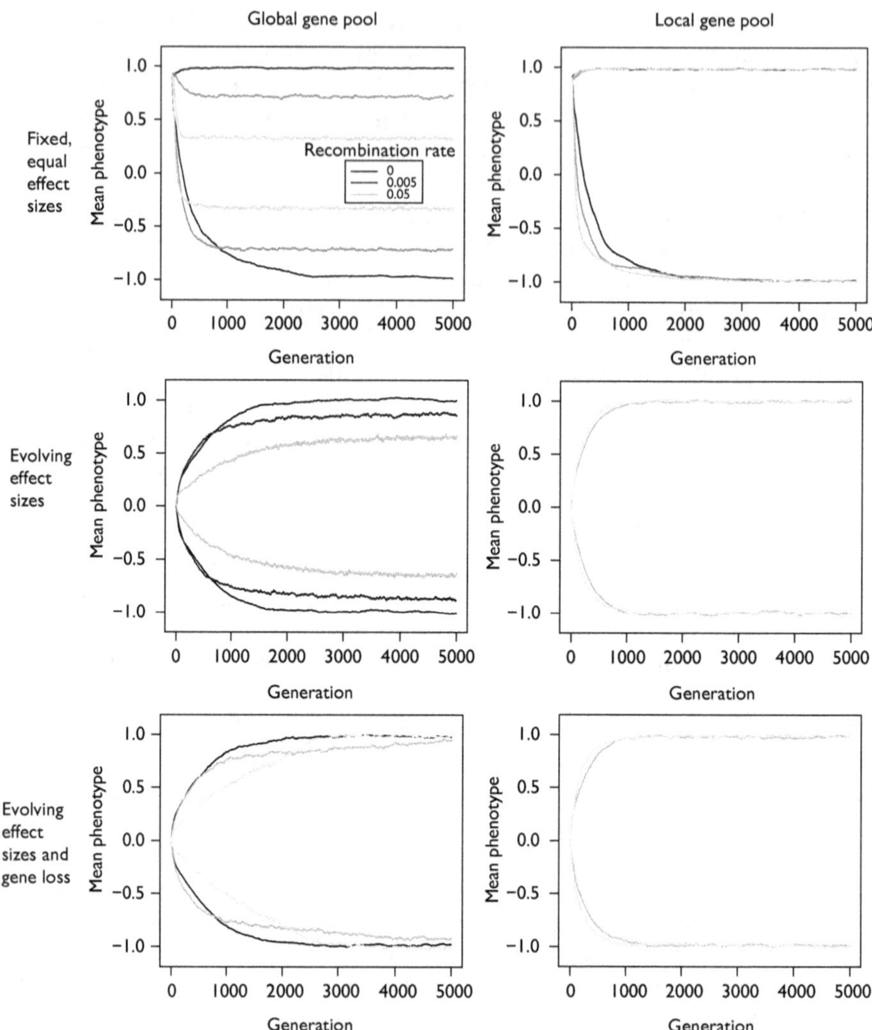

Fig. 6.9 Simulations of bacterial populations diverging to adapt to niches with different phenotypic optima (+1 and −1) under different scenarios and mechanisms for restricting gene flow. Left-hand column indicates results assuming a global gene pool, whereas right-hand column indicates results assuming a local gene pool within each habitat. The local, niche-specific gene pool always permits divergence to the phenotypic optima, whereas in the global gene pool models, recombination permits gene flow and thereby reduces divergence. This effect is minimized, however, over the longer term if effect sizes of the ten loci underlying the trait are allowed to evolve (middle row)—which results in the trait being coded primarily by a single locus—and especially if there is gene loss—which results in the trait being coded by a separate private locus in each niche-adapted population. (Reprinted from Schmutzer and Barraclough (2019) under creative commons license.)

genome). Differentiation between the two forms was restricted to 11 genome regions, and over 80 per cent of single nucleotide polymorphism was found in just three regions. The rest of the genome was intermingled between the two forms. This observation matches predictions of Friedman et al. (2013) and Schmutzer and Barraclough (2019) that divergence with gene flow is facilitated by concentration of divergence into relatively few loci. It tends not to support niche-specific gene flow, however, as then recombination would be restricted at all loci, not just those involved in adaptation.

Cadillo-Quiroz et al. (2012) reported a different pattern in the archaean *Sulfolobus islandicus* diverging into two groups of strains, called 'red' and 'blue', in hot springs in Russia. Across the genome as a whole, recombination rates were estimated to be 1.3–13 times the mutation rate, which is just above the threshold for observing recombinational cohesion at neutral regions in the models of Fraser et al (Fig. 6.7). Recombination both at homologous regions and via transfer of non-core genes between strains occurred more frequently within the red and blue sets than between them (Fig. 6.10). The genomic landscape of differentiation differed from the previous example, however, because most of the genome was significantly differentiated between the red and blue sets, except for three genomic islands that were significantly conserved. This pattern fits more closely with a niche-specific gene flow model—leading to differentiation across the genome, except for one or two regions under strong selection to retain the same genotype—or perhaps the fact that rates of recombination are too low to strongly constrain divergence with gene flow in this case. The latter explanation was favoured in a study of hot spring bacteria, which reported evidence of recombination that was restricted between ecologically differentiated sets, but the study argued that this was a consequence rather than a cause of differentiation (Melendrez et al., 2016). For this reason, it is still disputed whether recombination has the same primary role in bacteria as it does in sexual eukaryotes.

Additional evidence would be needed to distinguish these alternatives. For instance, confirmation of the niche-specific gene flow requires direct evidence that DNA does not transfer between different patches (e.g. sampling free DNA in the environment). Otherwise, apparently low rates of transfer between diverging forms might simply result from selection against locally maladapted genes. Furthermore, understanding the joint action of selection, gene flow, and recombination on divergence requires elucidation of the phenotypic and fitness effects of loci behind divergence, including interactions among loci, that is currently hard to obtain.

A systematic survey of diversity patterns across clades with varying levels and rates of recombination is now needed (Bobay and Ochman, 2017). If recombination does shape diversification, we predict a gradient in diversity patterns along the gradient of recombination rates found among different clades. With whole-genome data for dense systematic samples of species and strains within species, more complex models of divergence could be tested: for example, a model with a gradual decline in the probability of recombination (e.g. as a function of sequence divergence) compared to a model of discrete units (e.g. associated with pheromonal barriers to gene exchange or adaptation to distinct niches). Significant outlier genes or chunks of genomes that differ from underlying core patterns could then be identified.

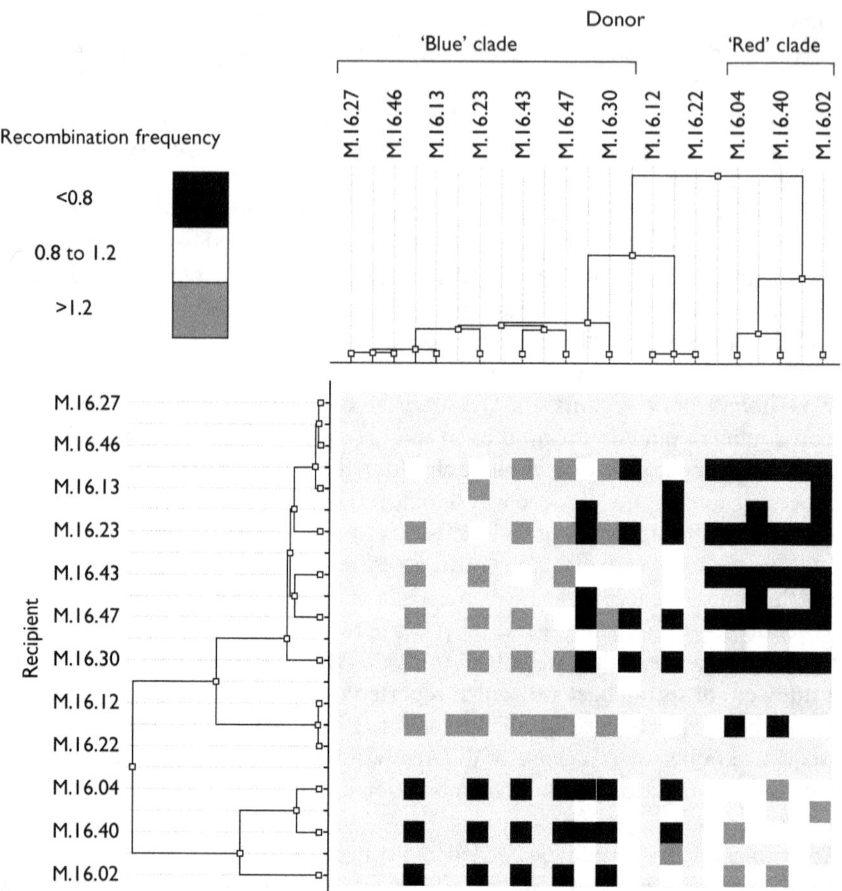

Fig. 6.10 Recombination rates are higher within 'blue' and within 'red' clades of *Sulfolobus islandicus*, an archaean found in a hot spring from Kamchatka in Russia, than between those clades. Phylogenetic relationships based on core genome phylogeny are shown, and the matrix shows observed frequency of recombination events from donor to recipient branches relative to the expected number under a uniform model of recombination probability across the tree. Excess recombination is found between 'blue' strains and between 'red' strains. The two intermediate strains received a high input of transfers from the 'blue' group, but did not donate at a significantly high rate to either 'blue' or 'red'. (Redrawn from Cadillo-Quiroz et al. (2012) under creative commons license.)

6.6 Speciation experiments in microbes

It would be extremely useful to have more manipulative experiments to test the theories outlined in chapters 5 and 6. Bacteria and other microbes offer great potential and are currently underexploited. Classic experiments on *Pseudomonas fluorescens* showed that spatial mixing inhibited divergence into distinct ecomorphs (Rainey and Travisano, 1998). The ecomorphs differ in single or a few mutations (Bantinaki et al., 2007). Similarly, bacteria grown on mixed resources diverge in sympatry to specialize on different metabolites (MacLean et al., 2005); for example, *E. coli* diverges into a glucose specialist and a generalist growing on both glucose and acetate, underpinned by three key genetic changes (Herron and Doebeli, 2013). In a more complex setting, *E. coli* inoculated into germ-free mice diverged into three mutant types that coexisted and differed along an axis of stress tolerance to bile salts, nutritional specialization, and motility (De Paepe et al., 2011). The role of recombination in such phenomena is starting to be uncovered (see chapter 9).

A study of bacteriophage lambda, a virus, specializing on different *E. coli* hosts provided a direct test of the speciation theories outlined in section 6.5 (Meyer et al., 2016). Culturing the phage on hosts with different receptor types, used by the virus during host attachment, led to improved binding to the local host and reduced binding to the alternative host in every replicate case. In contrast, when phages were cultured on both host types in sympatry (i.e. the same flask), they often diversified into specialists adapted to each host, but not quite as effectively as in allopatry. Recombination of the phage occurs within a host cell and therefore ecological adaptation directly led to assortative recombination (another example of niche-specific gene flow facilitating divergence); in addition, the fact that hybrids were inviable indicated a genetic incompatibility resulting from adaptation to different hosts. Theories for the effects of recombination on divergence could therefore be applied readily to phage lambda, which is definitely not a conventional sexual organism.

There is scope for more work: genetically engineering bacteria to vary in recombination rates and dispersal abilities (e.g. expression or not of motile flagella), in concert with varying the strength of divergent selection and physical barriers to dispersal. These experiments could also be used to test methods for observational inference: can we reconstruct past changes based on genetic analysis of the final time-point? Classic work investigated phylogenetic reconstruction methods this way (Hillis et al., 1992); equivalent experiments testing methods of genomic reconstruction of alternative past histories of gene flow would be equally useful. In addition, there is scope for experimental validation of observational inferences on bacterial species. For example, a prediction that ecological differences restrict gene exchange between a pair of bacterial species based on genomic inference can be tested with culture experiments in the laboratory (Cadillo-Quiroz et al., 2012). Culture conditions can be successfully predicted for bacteria based on their genomic repertoire, which could be extended to predict culture conditions promoting or inhibiting the coexistence and isolation of multiple species.

Perhaps the lack of exact correspondence to speciation of higher eukaryotes, and uncertainty about the nature of species in microbes, explains the relative scarcity of speciation experiments in microbes compared to the wealth of work on adaptation in single populations. Yeasts have been used as a model for reproductive isolation in sexual organisms (Kuehne et al., 2007). Mirroring earlier work on *Drosophila*, reproductive isolation has been shown to arise through adaptation to distinct environments (Dettman et al., 2007). Gray and Goddard (2012) elaborated on these experiments to investigate the effects of gene flow on divergence in both sexual and asexual strains of yeast (the latter produced by genetic modification). They found, contrary to predictions outlined in section 6.5, that a combination of sex and higher rates of gene flow enhanced adaptation to local conditions in different flasks: gene flow facilitated rather than constrained divergence in sexual treatments. This was because super-generalists evolved in these conditions that performed better in both environments than the specialists that evolved in other treatments. Whether this outcome could arise generally or is a quirk of the particular selective conditions and genetic responses that were implemented remains to be tested.

Mostly, experiments set up a given fixed diversifying force and then track the response. Potentially, more elaborate experiments could be constructed where the probability of diversifying versus unifying selection and levels of isolation versus gene flow vary over time (see section 5.5). The genetic architecture for reproductive isolation and selection can also be manipulated by gene knock-outs or prior episodes of selection. These could provide direct experimental evidence for the interaction between the frequency of changing conditions and rate of genetic responses (see chapter 5) that could be compared to observational evidence for speciation dynamics both in wild yeast and in other organisms.

6.7 Conclusions

Theories of species developed with sexual organisms in mind. As a result, they emphasize a key aspect for explaining the emergence and coexistence of sexual species, namely reproductive isolation. Applying the same simple population genetic theories for divergence used in sexual organisms to asexual populations shows that similar outcomes occur in asexuals and organisms with alternative modes of recombination. Both geographical isolation and divergent selection can generate discrete genetic clusters of individuals that evolve cohesively and independently from other sets, and diverge genetically over time. This pattern is not observed frequently in strictly asexual eukaryotes because the majority of asexual lineages are unable to persist for long periods. It is observed across bacteria, Archaea, and indeed viruses (Herniou et al., 2015), but much work is still needed to elucidate the role of recombination and isolating mechanisms in that process.

There are broader implications of comparing diversity patterns in organisms with different forms and rates of recombination that lead into topics of later chapters. First, diversity patterns are, in part, a consequence of how selection pressures are distributed across organisms and different traits and genome regions (e.g. Fig. 6.2). Ideally, to predict the pattern of diversity within a clade would require partitioning of selection acting on multiple populations into different components with an estimate of their relative strengths. This depends on the nature of the external environment as well as how that maps to individual fitness. For example, the models described in sections 6.2 and 6.5 assume that discreteness exists in the environmental conditions that organisms encounter. If underlying opportunities and resources were continuous, any emerging pattern of discreteness would depend on constraints on the width of resources or environmental conditions that a specialist genotype is able to monopolize, as well as the distribution of resources and conditions.

Second, in organisms where recombination rates and isolating mechanisms display variation and are under selection, the optimal solution to enhance individual fitness depends again on the pattern of selection pressures and gene flow they experience. Classic theory shows that recombination enhances the efficacy of selection in temporally changing environments. However, if there are high levels of gene flow between populations that experience different ecological conditions, this conclusion might change. For example, this might partly explain why tiny organisms are more likely to be asexual: higher dispersal rates lead to high rates of potential gene flow between differentially adapted populations, which in turn favours the protection of multilocus genotypes more than rapid adaptation to temporal changes.

Third, some of the processes elaborated here for strictly asexual organisms might also operate in sexual organisms. Consider the scenario in Fig. 6.2. Sexual species also compete for limited resources and in principle innovation in one species could lead to replacement of a second species. The pattern of diversity in a sexual clade also depends therefore upon the shape of selection pressures acting on its members. In a clade occupying a mosaic of habitats with temporally shifting environments, does adaptation to a change in environment, such as global warming, occur independently in each species? Or does it occur in one species (perhaps one already pre-adapted to the new conditions) that then spreads and replaces the other species and eventually diversifies to specialize on their niches? In sexual organisms, independent evolutionary responses might be more likely than in asexuals: each species can adapt faster and more efficiently to multiple optima across multiple traits than asexuals can, and rediversifying to fill empty niches is harder. But the alternative dynamics are also possible. This idea is pursued further in chapter 10.

7

Species boundaries and contemporary evolution

7.1 Introduction

Much of the evolutionary study of species is retrospective and reconstructs the past processes leading to extant diversity. Yet the nature of species and extent of diversity have profound implications for adaptation to ongoing environmental and biotic change. Concepts of independent evolution and cohesion incorporate evolutionary fate as well as history (see chapter 2). Many studies of contemporary evolution aim to predict evolution on a species-by-species basis. Selection pressures are characterized in a single population and responses predicted from genetic architecture. All organisms coexist with many other species, however, and the nature of boundaries between them will affect how they adapt to change. Responses will depend on the extent of genetic interactions (such as hybridization or horizontal gene transfer (HGT)). Even with strict barriers to gene flow, responses will depend on ecological interactions, which in turn relate to the forces behind the formation and persistence of species in the first place.

This chapter considers the importance of species boundaries in shaping evolutionary dynamics. Evaluating alternative models of species presented in chapter 2, I explore the consequences of different types of species boundaries and genetic interactions on how organisms within a clade or region evolve as environments change.

7.2 Background

Consider the simple case of two species in an enclosed region. It could be two endemic plants on an island or two bacteria found in a host. The environment is specified by values of a set of physical conditions ($E = e1, e2, e3, \ldots$) and by input rates of a set of resources ($R = r1, r2, r3, \ldots$). Stable coexistence of the species implies reproductive isolation. There might be some gene flow but not enough to erode genetic differences promoted by divergent selection. Stable coexistence also implies ecological differences. Assume species coexist by partitioning their use of different resources (R). Other aspects of the physical environment (E) are experienced by both species: a particular set of shared traits is required in order to be able to colonize the region (e.g.

The Evolutionary Biology of Species. Timothy G. Barraclough, Oxford University Press (2019).
© Timothy G. Barraclough 2019. DOI: 10.1093/oso/9780198749745.001.0001

optimum temperature, pH). Niche partitioning traits are called α-niche traits, whereas those required to occupy the location or region are β-niche traits (see section 5.4.3).

Now what happens if the environment changes? There are several possible scenarios. One is that no evolution occurs and species shift in abundances but do not adapt (Fig. 7.1A). One or both species might dwindle to extinction. Another scenario is that species adapt to the changes independently and maintain existing niche differences (Fig. 7.1B). This matches the implicit assumptions of much existing work on contemporary evolution that concerns single species. This is plausible for a change in β-niche axis, such as when multiple species adapt to a gradual increase in temperature. The scenarios of interest here are when interactions between the species alter evolutionary responses. First, there might be genetic interactions: for example, new mutations arising in one species might spread into the other species, or the environmental change might disrupt or create barriers to gene flow (Fig. 7.1C). Second, ecological interactions might alter evolutionary trajectories. For example, one species might be

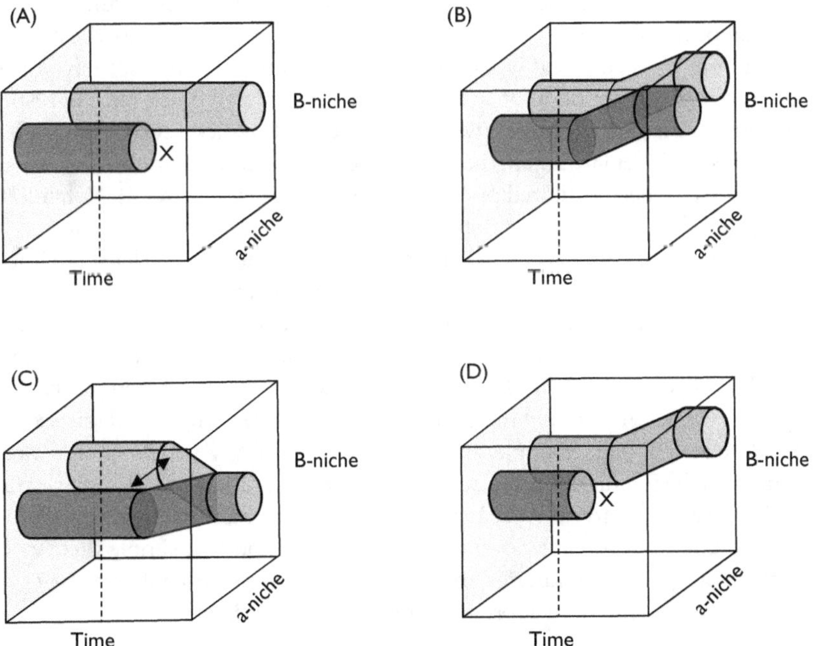

Fig. 7.1 Hypothetical examples of interactions between two species faced with environmental change. Species partition an α-niche axis but share the same β-niche environment, which changes at time indicated by dashed line. (A) Neither species evolves—dark grey species goes extinct. (B) Both species adapt independently. (C) Species interact genetically and exchange genes, in this case collapsing back into one species as a result of the change in environment, but they could both persist. (D) Competition constrains evolution of dark grey species, resulting in its extinction, even though it could have adapted to the change in environment in absence of light grey species.

pre-adapted to new conditions or evolve faster to use the new conditions, which disrupts coexistence (Fig. 7.1D). The second species could go extinct (even if it could have adapted to the environmental change in isolation) or evolve to re-establish niche partitioning and coexistence.

In the latter two cases, the presence of the other species changes the trajectory of adaptation and determines the fate of the lineage: that is, whether species are able to persist or not. Chapter 8 focuses on ecological interactions shaping evolutionary responses. But first, this chapter looks at the effects of genetic interactions—in other words, leaky species boundaries.

Regarding genetic interactions, there is a possible tension on closing the species barrier. Mechanisms that restrict gene flow to zero could in some cases reduce fitness and increase the risk of future extinction—by preventing beneficial mutations from crossing the species boundary. But with no restriction to gene flow, adaptation to the distinct environments or niches would be limited in the first place (see chapters 2 and 5). We can envisage different types of species boundary: strict boundaries with no gene flow; incomplete boundaries with some gene flow but also with some divergence and evolutionary independence; and the lack of species boundaries, resulting in gradual variation over space (clines) or among habitats (ecotypes). What are the dynamics of adaptation in each case and what consequences do those dynamics have for patterns of diversity? Most models of speciation have considered divergence in response to static conditions that promote divergent selection (Gavrilets, 2004)—are different outcomes expected if environmental conditions change through time?

These ideas have been explored previously in bacteria (see chapter 6). Cohan (2001) developed a model of local versus global selective sweeps, where some mutations have benefits within species units whereas others offer benefits across a wider set of lineages (see section 6.5). This was originally proposed for clonal taxa rather than recombining taxa: local sweeps generate distinct ecologically based clusters, whereas global sweeps occasionally reduce diversity in the clade back to a single cluster. But in recombining taxa, similar outcomes apply to different genes—some evolve independently in separate species, while others might rarely spread between them. This could arise solely as a result of selection—for example, in hybrid zones, globally beneficial and neutral genes can cross the interaction zone, whereas differentially adapted genes are restricted to the parental areas (Harrison and Larson, 2014). Or mechanisms of gene flow might facilitate movement of some genes but not others. I will present simple theory of the forces acting on species boundaries and their consequences for contemporary evolution, and review evidence and approaches for testing such theories.

7.3 A model of genetic interactions between species

The general concepts can be demonstrated with a simple model inspired by classic models of sympatric speciation by Felsenstein (1981) and Diehl and Bush (1989). Consider a population comprising two types of patches, such as two soil types of equal prevalence (Fig. 7.2A).

Genotype	Patch 1 fitness	Patch 2 fitness	Early mating pool	Late mating pool
ab	$(1 + s)^2$	1	r	1–r
Ab, aB	1 + s	1 + s	0.5	0.5
AB	1	$(1 + s)^2$	1–r	r

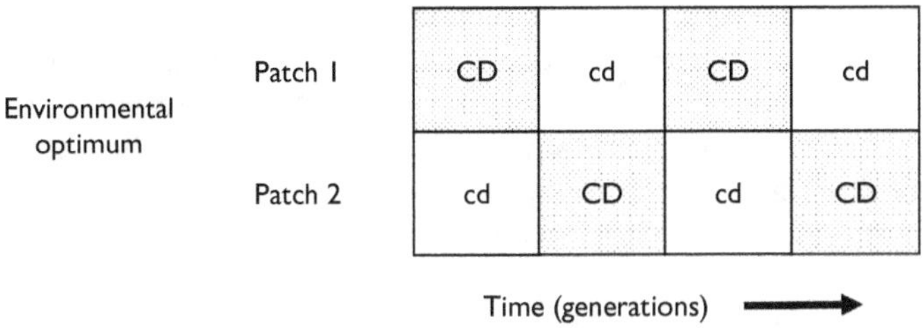

Environmental optimum					
Patch 1	CD	cd	CD	cd
Patch 2	cd	CD	cd	CD

Time (generations) ⟶

Fig. 7.2 Summary of a population genetic model of the effects of fluctuating selection pressures on species boundaries. (Top) Hypothetical plants occupy two patches (e.g. different soil types) with different optimum genotypes at two loci (a and b) and multiplicative fitness effects. Loci also determine proportion of plants entering the early versus late mating pool each year, which are non-overlapping in time. (Bottom) Plants also adapt to the physical environment, which fluctuates over time between conditions favouring different optimum genotypes at two other loci (c and d) that also display multiplicative fitness effects. An out-of-phase fluctuation between the two patches is shown. See text for more details.

Individuals are hermaphrodites and by default there is random mating: each individual is equally likely to fertilize any other individual irrespective of soil type. It could represent a plant population with random pollen transfer. Each female produces X propagules that settle in their mother's patch type. Selection then acts on viability: N individuals survive on each soil type from the total pool of NX propagules. For simplicity, the organisms are haploid and the probability of survival depends on two unlinked loci with two alleles. Alleles a and b improve survival on patch type 1, alleles A and B improve survival on patch type 2, and the effects are multiplicative, so that ab and AB have highest fitness in patch type 1 and 2, respectively (Fig. 7.2).

With random mating, the outcome is a stable polymorphism at both loci, with allele frequencies of 50 per cent and an excess of ab and AB that depends on the strength of selection. Ab and aB genotypes are selected against but continually produced through recombination. Adding assortative mating reduces the amount of gene flow between ab and AB genotypes. Felsenstein (1981) considered assortative mating controlled by similarity at a third locus. Instead, to stack the odds in favour of speciation, I consider a so-called magic trait model (Gavrilets, 2004) in which the similarity at the same loci determines assortative mating. Specifically, I assume there are two mating pools, 1 and 2. A parameter r controls the strength of assortative mating by determining the probability that a given genotype enters each mating pool. A fraction of r individuals with the ab genotype enter mating pool 1 and a fraction $1-r$ mate at random (i.e. there is a 50:50 chance of entering either mating pool). Similarly, a fraction r of the AB individuals mate with other AB individuals, and $1-r$ mate at random. All other genotypes mate at random. If $r = 1$, ab and AB individuals enter separate mating pools and the only gene flow between them occurs if Ab or aB genotypes arise by mutation and then mate at random. The two groups could represent early and late phenology groups, with intermediates mating at random with either early and late breeders. For example, Lord Howe Island palms (see chapter 5) have diverged on two soil types associated with divergent but partially overlapping timing of pollen production (Fig. 5.4, Savolainen et al., 2006).

Now the population can diverge into two independently regulated populations with limited gene flow between them, constituting ab and AB genotypes, which represent species. Gavrilets (2004) discusses variations of this model and the factors that determine whether reproductive isolation evolves or not, such as costs to assortative mating. The question here is under what conditions does selection favour complete closure of the species boundary (i.e. $r = 1$) versus remaining open to partial gene exchange? To investigate this question, I assume r is controlled by a modifier locus with a value constrained between 0 (random mating) and 1 (full assortative mating of ab and AB). Mutation increases or decreases the value of r and selection on r acts solely via its indirect effects on producing aB and Ab genotypes with low fitness.

In a model with consistent selection as described above, selection always pushes r towards 1. What scenarios favour a residual level of interbreeding between ab and AB genotypes? Inspired by the Cohan models of bacterial species (see chapter 6), I added in two unlinked loci each with two alleles—c, C and d, D—that code for an independent trait reflecting an additional environmental axis (Fig. 7.2B). The optimum genotype for that axis is either the same between patch types, or different, and can be constant over time or fluctuate. In contrast, I assume that the optima for locus 1 and 2 are different between patches (i.e. ab and AB) and remain constant over time, and only these loci affect assortative mating. Hence, locus 1 and 2 code for species traits, whereas locus 3 and 4 code for environmental traits that can be shared or divergent and constant or changing over time.

Starting with fully isolated species adapted to each patch type, if the optima for environmental traits are constant over time or fluctuating but always shared, selection favours the maintenance of a strict species boundary with $r = 1$. If the environmental optimum is divergent between species but fluctuates, however, then selection favours a lower average value, permitting 1–2 per cent of ab and AB genotypes to mate at random. When the environment switches, for example from favouring abcd and ABCD to favouring abCD and ABcd, interbreeding between species is initially favoured until the new optimal genotypes begin to spread, then selection returns to favour strict boundaries (Fig. 7.3A). The effect is stronger when selection is stronger on the environmental trait than the species trait, because interbreeding also breaks up favourable ab and AB combinations (Fig. 7.3B). This scenario is analogous to co-evolutionary models of hosts

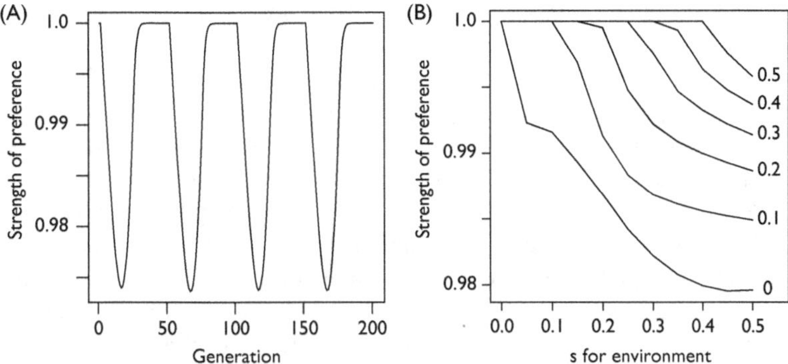

Fig. 7.3 (A) Example of how strength of preference, *r*, evolves with a change in physical environment every 50 generations between the two patches as shown in Fig. 7.2. Selection coefficients for adaptation of the divergent trait (namely soil type) is 0.1 and to the fluctuating trait (namely physical environment) is 0.2, where *s* is the fitness advantage of a heterozygote relative to a maladapted homozygote (Fig. 7.2). (B) Average strength of preference, *r*, that evolves under a combination of divergent and fluctuating selection acting on different traits. Lines show mean results for a range of selective coefficients for the divergent trait (i.e. adaptation to soil type), whereas X-axis shows selective coefficient on fluctuating trait (i.e. adaptation to physical conditions). If selection on fluctuating trait exceeds strength of selection on divergent trait, then marginally lower levels of assortative mating evolve, leaving open residual gene exchange between the two species.

and parasites in relation to the evolution of sex: interbreeding between the two species is favoured when there is negative linkage disequilibrium: that is, newly optimum genotypes are scarce and produced by recombining previously optimal genotypes from the separate species. It is also somewhat analogous to models of selection for higher mutation rate in more rapidly changing environments.

A model of alternating optima between species is not particularly realistic, but there are other scenarios that create these conditions. Asymmetric models in which one species gets a head start in adapting to a new optimum—for example, because of a higher beneficial mutation rate, stronger selection for the optimum, or a lag in onset of new conditions—would also favour interbreeding. In a stochastic model, one species might by chance acquire beneficial mutations for adapting to the new optimum first. Consequently, a less assortative genotype in the second species would benefit by increased interbreeding, resulting in weakening of the species boundary, counterbalanced by breaking up of favourable ab and AB combinations.

This simple model shows the interplay between species boundaries and contemporary evolution. The dynamics of adaptation are influenced by the species boundary. Either both populations adapt independently to the change in environment or the populations adapt in concert by sharing environmental genes through recombination. In turn, the strength of species boundaries is determined by the pattern of selection pressures operating over time. Real environments constitute multiple dimensions varying at multiple spatial and temporal scales. Characterization of the pattern of that variation and its impacts on organisms is needed in order to quantify forces operating on species boundaries.

There are further complications to these ideas. The above model assumes that genes behind adaptation in each population are compatible. If incompatible mutations arose in each population (Maheshwari and Barbash, 2011), this would strengthen selection against gene exchange across species boundaries. In addition, other responses are possible under the above selection regimes aside from changing strength of assortative mating. For example, fluctuating environments plus constant selection on the ab and AB genotypes could select for increased linkage of locus 1 and 2 (see chapter 6), so that beneficial interbreeding for adapting to environmental change would not disrupt adaptation to the two patch types. Genetic assumptions of the models could themselves evolve in response to fluctuating selection.

7.4 Empirical evidence for permeable species boundaries and selection upon them

Section 7.3 considers scenarios when the strength of species boundaries is under selection. The alternative is that reproductive isolation and species boundaries are a by-product of selection and isolation, and not under direct selection themselves. It might be inevitable that many species—either through long-enough period of isolation or adaptation to divergent-enough niches—no longer exchange genes. So, what is the evidence for leaky species boundaries and does selection play a role in maintaining the capacity for leaks?

Mallet (2008) reviewed cases of races with permeable boundaries allowing around 1 per cent gene flow (described in section 2.6). Harrison and Larson (2014) further reviewed cases of hybrid zones, and argued that most species boundaries are semipermeable. The frequency of such scenarios is unclear, however, since geneticists might be more interested to work on such cases than species with strict boundaries: species pairs that display no hybridization would not have been included in either review. Whole-genome sequencing within and between species for whole clades or biotas will allow more systematic evaluation of patterns of gene flow between species. Such work could test whether particular types of genes are more likely to cross species boundaries, for example those coding for β-traits associated with shared environmental features rather than α-traits behind species coexistence.

One compelling case of introgression facilitating joint adaptation to a new environment in a pair of species is the spread of insecticide resistance in *Anopheles colluzzii* and *A. gambiae sensu stricto*, previously known as the M and S molecular forms of *A. gambiae* (Clarkson et al., 2014). Although morphologically indistinguishable, the two forms are adapted to different larval habitats (the former in large rice paddies, the latter in small puddles) and distinct at a few large islands of genomic divergence. Pre-zygotic isolation is maintained through swarming behaviour (Diabate et al., 2009), but no postzygotic isolation is detectable in the laboratory. Hybrid individuals are found occasionally in the wild. In response to spraying of insecticides to control the mosquitoes, which are the main vector of malaria in sub-Saharan Africa, mutations arose in the S form in the voltage-gated sodium channel gene (*Vgsc*). The *Vgsc* protein is targeted by insecticides, and the mutations reduce their binding affinity. The beneficial mutations introgressed from the S form into the M form and spread rapidly (Fig. 7.4). This homogenized one island of genomic divergence between the two forms, but other differences remained and there was no subsequent loss of reproductive isolation. Observed hybridization rates remained low throughout this period—but possibly marginally higher during the early stages of increase in the resistance mutation (Fig. 7.4B). Other introgressed genes were removed by purifying selection (Hanemaaijer et al., 2018).

Relaxed species boundaries might therefore be favoured at times of uniform selection pressures affecting multiple species. It has been observed, for example, in Darwin's finches that selection between species and morphs within a species can be disruptive in drought years– that is, driving the population apart – but uniform in wet years (Hendry et al., 2009). It would be interesting to explore similar scenarios with experimental evolution, to demonstrate that the degree of gene flow can respond to

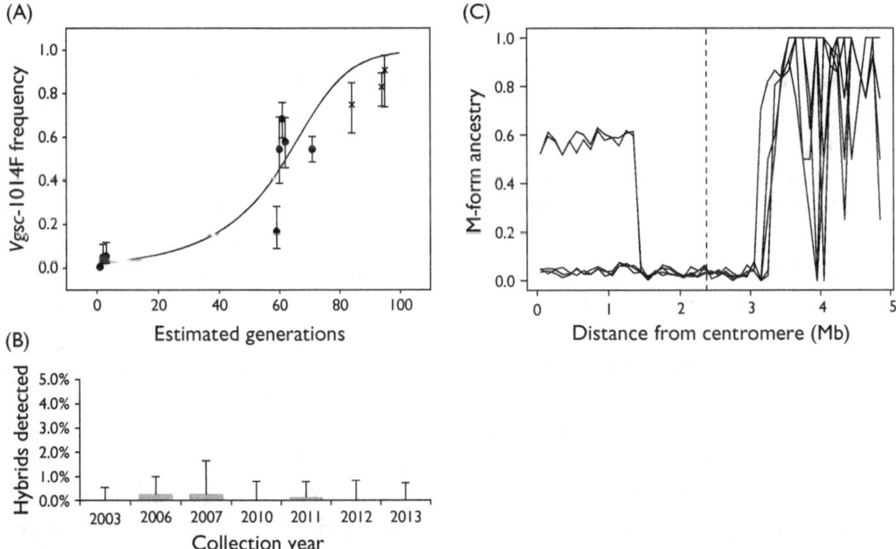

Fig. 7.4 Spread of insecticide resistance by introgression between M and S forms of *Anopheles gambiae* in Ghana. (A) Mutation in the voltage-gated sodium channel gene that originated within the S form entered the M form by hybridization and subsequently spread rapidly due to strong selection from insecticide control measures. (B) Frequency of observed hybrids in samples of hundreds of individuals remained below 1 per cent during this period. (C) Spread of resistant allele reduced divergence between M and S forms within the so-called 2L genomic island of divergence: where previously M-form individuals were divergent from S forms across this region, the five sampled individuals had a large chunk of DNA from S forms around the *Vgsc* gene (at dashed line). In two individuals, further recombination is restoring M ancestry towards centromere (left) end of region. (Simplified with permission from Clarkson et al. (2014).)

selection. While not readily feasible in larger plants and animals, this could be investigated by experiments allowing assortative mating to evolve in yeast or algae.

Finally, has genome structure evolved to allow varied levels of gene flow among different loci? There are several genomic features in eukaryotes that potentially protect some genes from homogenizing effects of gene flow while permitting gene flow at other loci. Chromosomal inversions protect sets of genes from recombination between populations, while wider genomic background is free to recombine, and these are widely implicated in speciation (Rieseberg, 2001; Hooper and Price, 2017). Sex chromosomes also restrict recombination and tend to be highly differentiated between species, although many other processes might explain their special status for speciation (Payseur et al., 2018). Genes on mitochondrial and plastid genomes are also largely protected from recombination. Organelle DNA seems to cross species boundaries more easily than nuclear genes and appears not to be under divergent selection between species (see chapters 3 and 4), hence it does not seem to generally serve to protect local adaptation, although it is implicated in generating incompatibilities. More generally, any regions of low recombination across chromosomes protect against introgression between species (Martin et al. 2019).

7.5 Genetic interactions between bacterial species

Bacteria possess a wide array of mechanisms for transferring genetic material between cells. Some of them involve agents capable of their own selfish behaviour, such as plasmids and viruses, whereas others involve mechanisms fully under bacterial control, such as transformation via homologous recombination. These can be viewed in a similar light to that discussed in section 7.4: linkage groups transmitted clonally or in single blocks serve to protect coadapted suites of genes, whereas some genes are transferred promiscuously. There may be genetic constraints on what types of genes behave in different ways. Some traits require functioning sets of genes, and shuffling combinations of those would likely result in failure of the entire pathway. Other traits might be easier to plug in and out of a genomic background, bringing a single useful function without disrupting other cellular processes. Prokaryotic genomes as a whole are more modular, with genes contributing to a given pathway found in the same or neighbouring operons (Yin et al., 2010). This is largely driven by selection for efficient transcription and regulation, but also potentially facilitates HGT, as transfer of whole operons is common (in addition to transfer of individual genes within operons; Omelchenko et al., 2003). Many plasmids encode secreted molecules that detoxify the environment, but those molecules do not interfere with other proteins in the cell.

So again, we can ask whether the range of mechanisms are a haphazard outcome of multiple competing forces, or whether they contain design features optimizing a combination of vertically and horizontally transmitted features. Furthermore, for a given mechanism, what are the consequences of genetic transfer for adaptation? For transfer involving an agent such as a plasmid, then, transfer offers costs and benefits that vary between the donor, the agent, and the recipient. Who, if anyone, controls the transfer process? Plasmids are well known to transfer between bacterial species, but what is the typical range of plasmid transfer? Are most of them species-specific, clade-specific, or truly promiscuous?

7.6 Genetic interactions via plasmid exchange in bacteria

Bacterial plasmids have received a lot of attention as agents of gene transfer. Evolutionary studies have focused on coevolution between the host and plasmid genomes, and costs and benefits of plasmids to the host (Harrison and Brockhurst, 2012; Carroll and Wong, 2018). In principle, plasmids might be purely parasitic on host bacteria and behave as selfish elements maintained in a balance between drive and selection against them. In reality, plasmids typically contain genes that are beneficial to the host in particular environments, such as those carrying antibiotic resistance or heavy-metal tolerance. In non-selective environments, plasmids are often costly and reduce growth rates, because of the metabolic energy required for their replication and transfer, or genetic interference with host cellular processes. The cost can decline over time, as both the host and plasmid evolve compensating mutations and coadapt to each other (Harrison et al., 2015). In selective environments that favour genes on the plasmid, the plasmid should spread. If conditions persist for long enough, selection can favour integration of plasmid or constituent beneficial genes into the host chromosome, so that the beneficial traits become vertically inherited (Bergstrom et al., 2000).

The benefits of transfer differ between the plasmid, the donor cell initially carrying the plasmid, and the recipient cell that initially lacks a plasmid but receives one (Box 7.1). In simple models, there is always a short-term benefit for a plasmid to transfer itself into an empty host. Once the plasmid is fixed in the population there is no longer selection to maintain transfer—every potential host already contains the plasmid and the ability to transfer can be lost. A meta-analysis of 1730 plasmids found that half were not mobile: that is, they were vertically inherited as secondary chromosomes (Smillie et al., 2010). Maintenance of transfer ability longer term requires a continuing source of new recipient cells, either through dispersal from a selective into non-selective environment or through a change in conditions (Bergstrom et al., 2000). There is therefore a longer-term dynamic of short-term selection for spread and invasion of new host populations to maintain the element longer term, which is equivalent to other selfish elements (Goddard and Burt, 1999).

In contrast, the donor receives no advantage or disadvantage in transferring the plasmid in the simplest model: its own genotype is unchanged by the transaction (Box 7.1). Selection may favour loss of the plasmid in unselective environments, which occurs by segregational loss during cell division independently of transmission. The key effects are on the recipient: transfer enhances fitness in selective conditions and reduces fitness in non-selective conditions. One might predict therefore that control of transfer might rest more with the plasmid and recipient than with the donor. A meta-analysis of laboratory measures of transfer rates from the literature found that transfer rate depends on the identities of donors, as well as recipients and

Box 7.1 A simple model of transfer rate evolution in a bacteria-plasmid system.

Consider a single bacterium species and a single plasmid species growing in a chemostat (Stewart and Levin, 1977). The bacterium population includes cells with the plasmid—donors, at density N_p—and cells without the plasmid—recipients, at density N (Fig. 7.5). Cells grow on a single substrate that flows into the chemostat at concentration S_0 and dilution rate D. Both cells and substrate are also removed from the chemostat at the same dilution rate, D. The bacterial population grows at a per cell growth rate of k and k_p for recipient and donor cells, respectively, in proportion to substrate concentration in the chemostat, S. The difference between k and k_p reflects the cost or benefit of containing the plasmid. Plasmids are lost from a fraction of λ new donor cells as a result of segregational loss during cell division. Finally, plasmids transfer from donors to recipients at a rate of $\tau_N N_p N$, reflecting mass action encounters between donors and recipients with a transfer rate of τ_N.

Solving the model assuming a constant transfer rate shows that plasmids can persist when the transfer rate exceeds any cost plus segregational loss. Sheppard et al. (2019) used invasion analysis to assess the fitness of invading mutant hosts or plasmids into a population at equilibrium. A mutant plasmid can invade whenever $N^* > 0$, i.e. there are recipient cells present at equilibrium, and when it has a positive effect on transfer rate relative the rate of the current stable population. In other words, there is selection on the plasmid for transfer rate to increase as long as empty cells are present. In contrast, a mutant host affecting transfer rate can invade when $N_p > 0$, i.e. when plasmids are present, and when $(\tau_{NM} - \tau_N)(k_p - k) > 0$. This means that mutant host cell enabling a higher transfer rate from a non-mutant donor can invade when the plasmid is beneficial to the host (both terms in the second inequality are positive) and mutant hosts with lower transfer rate from non-mutant donors can invade when there is a cost to the plasmid (both terms negative). Selection therefore acts on recipients rather than donors, because the donor phenotype is unchanged by the transaction of donation. Other scenarios could lead to selection on donors (see main text).

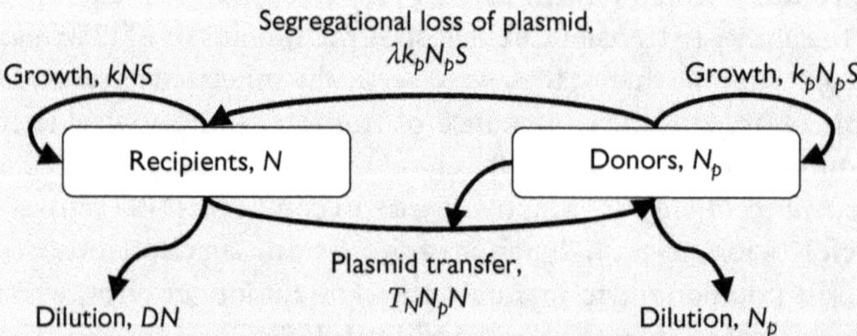

Fig. 7.5 Flow diagram showing terms resulting in changes in donor and recipient densities each generation. Mathematically, change in donor density is $dN_p/dt = k_p N_p S - DN_p + \tau_N N_p N - \lambda k_p N_p S$; change in recipient density is $dN/dt = kNS - DN - \tau_N N_p N + \lambda k_p N_p S$; and change in substrate concentration is $dS/dt = D(S_0 - S) - S(kN + k_p N_p)$.

plasmids (Sheppard et al., 2019). Molecular studies indicate that genes influencing transfer rate are found in the donor, recipient, and plasmid to varying degrees in different systems (Koraimann and Wagner, 2014). Genetic mechanisms for receiving plasmids in recipients that currently lack one show that plasmid systems constitute a specific mechanism to facilitate transfer upon the switch from non-selective to selective conditions—this is not just an incidental outcome of the spread of an autonomous, selfish element.

The above predictions depend on the diversity and nature of ecological interactions between the donor and recipient cells (Barraclough et al., 2012). The simple model assumes these are genetically equivalent otherwise, as would be the case for a plasmid spreading through a single host population. When donors and recipients differ, as the case of a transfer between species, this can change the costs and benefits of transfer. A donor under selective conditions is sharing beneficial genes with another cell that currently lacks them. If it has a competitive interaction with the recipient, for example two species sharing overlapping resources, then it is sharing a competitive advantage with another species that might break the conditions for coexistence. In extremis, this could lead to the donor species' extinction. If instead it has a facilitative interaction, for example the recipient produces a useful resource for the donor, then transferring the ability to resist antibiotics, for example, is beneficial to both. If the dominant interaction among co-occurring bacterial species is competition, as inferred from culture-based assays (Foster and Bell, 2012), it is expected that plasmid transfer between species should be favoured only by donors during unselective conditions, as it passes on the cost of carrying the plasmid to competitors (Dimitriu et al., 2016), but is favoured only by recipients during selective conditions. As in the simpler model, alternating conditions are needed to maintain plasmids in the system over longer periods.

By extension of these arguments, some kinds of traits might be more likely to transfer between species than others (Wiedenbeck and Cohan, 2011; Barraclough et al., 2012). Classic examples tend to involve β-niche traits that open up invasion of new environments, such as the acquisition of antibiotic resistance or heavy-metal tolerance. Only species with the trait can survive in those locations, and so there is strong selection for species that already possess the trait or can acquire it horizontally. These might also constitute traits that are easy to transfer in a modular block but are difficult to evolve by *de novo* mutation of bacterial genomes. In contrast, there is not the same selective filter for evolving α-niche traits that promote coexistence among species. These traits might both require and be feasible with substitutions across multiple loci in the bacterial genome to modulate the relative expression across different metabolic pathways involved in extracting energy from different resource molecules. Finally, plasmids often carry genes for secreted products, which in part might reflect traits that exert minimal costs from interference with cellular machinery, but also reflects public goods (McGinty et al., 2013). The donor and whole local community may benefit from sharing the gene and increasing the external concentration of a beneficial product, such as a chelating agent for toxic metal ions.

The nature of ecological interactions could also change selection on the plasmid. Normally transfer is beneficial for the plasmid, but it is possible to concoct a hypothetical scenario where it is not. For example, consider two species of bacteria: one slow growing that initially lacks the plasmid, and one faster growing species that has the plasmid. Genes in the plasmid and the slower growing species interact to produce a toxin that inhibits the growth of the faster-growing bacterium disproportionately more than the growth of the slower bacterium containing the plasmid. Now, transfer into the slow-growing species would reduce the fitness of the plasmid.

To summarize, plasmids carrying beneficial genes are vessels of horizontal transfer for adapting to a changing environment. Their existence and mode of action only makes sense when conditions favouring the genes they carry fluctuate over time or space (Bergstrom et al., 2000). From the host perspective, they are useful when entering new conditions, but the benefit is then lost relative to vertical transmission in static conditions. From the plasmid perspective, they can exhibit selfish behaviour and be maintained by drive, but only until they become fixed in the host population, upon which horizontal transfer no longer plays a role as all hosts are 'infected'. The costs and benefits depend on both the selective environment and the nature of interactions among the participants. The same arguments apply to other mechanisms of transfer such as by viruses or integrons. The sharing of antibiotic resistance elements among co-occurring and competing species is something of an evolutionary paradox. If bacteria communities are predominantly shaped by competition rather than facilitation, a mutant or colonist with a private, vertically transmitted resistance mechanism should be greatly favoured over a cell that shared its resistance mechanism via horizontal transfer with the wider community.

7.7 Evidence for gene transfer networks and adaptive species boundaries in bacteria

The increasing volume of whole-genome sequence data for bacteria permits broad surveys of horizontal transfer networks. Does horizontal transfer generally occur promiscuously independently of the degree of relatedness between participants, or does it tend to be more restricted within narrower taxonomic groups? Smillie et al. (2011) reconstructed gene transfer events affecting 10,700 mainly coding gene regions among a sample of 2235 bacterial genomes. Around one-quarter of events included a match to a mobile element such as a plasmid, which left a large proportion that gave no obvious signal of involving such an element. They found that transfer was more strongly predicted by ecological similarity and geographical proximity than by phylogenetic relatedness. This might reflect greater chance of transfer due to higher encounter rates or greater benefits because shared DNA is more likely to be beneficial when the environment is shared—a broader version of the niche-specific gene pool idea described in chapter 6.

Tamminen et al. (2012) performed a more focused analysis on the 2343 complete plasmid sequences available at the time of study. They reconstructed a network of gene sharing among plasmids (Fig. 7.6). Clustering in the plasmid gene-sharing

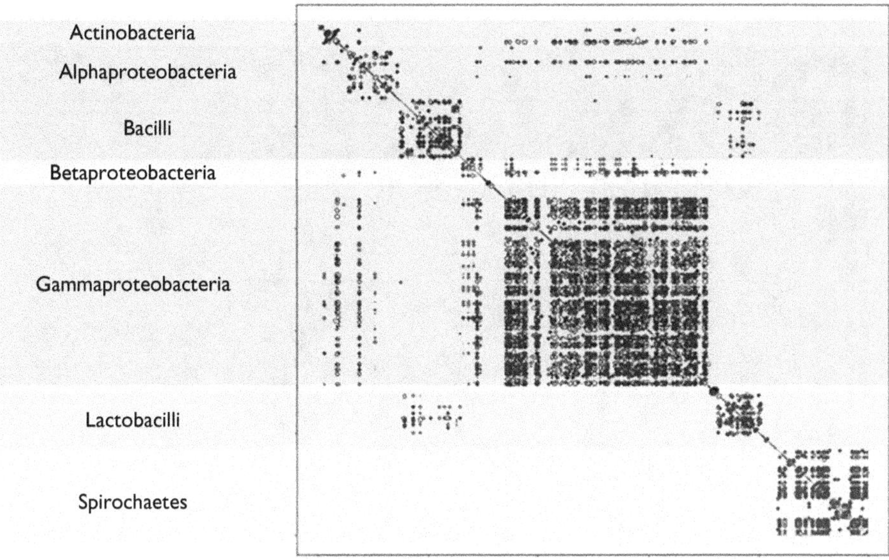

Fig. 7.6 Patterns of gene sharing among 2343 complete plasmid sequences in relation to phylogenetic affinity of host. Black dots in matrix indicate genes are shared, and sharing tends to cluster within major taxa with a few exceptions (e.g. bacilli plasmids share genes with lactobacilli plasmids and there is some sharing between Actinobacteria and Gammaproteobacteria plasmids). (Reprinted from Tamminen et al. (2012) with permission.)

network closely matched phylogenetic groupings of the host bacteria, consistent with there being barriers to sharing plasmids between distantly related bacterial clades. They did find evidence of wider sharing, however, for example in a trend for directional transfer of antibiotic resistance genes from Actinobacteria to Gammaproteobacteria. This matches the Smillie et al. (2011) results that antibiotic resistance genes, whose transfer has been promoted by the recent massive increase in antibiotic concentrations due to human use, are less constrained in transfer than other traits.

Popa et al. (2017) applied similar methods to reconstruct recent transfer mediated by bacteriophages, that is, transduction, among 3982 genomes. Transfer was mostly restricted to closely related donors and recipients (Fig. 7.7). About half of phages received genes from more than one donor bacterium, but 53 per cent of those cases belonged to the same species and 31 per cent to the same genus. Passing on genes from a phage to a recipient was even more restricted: only 7 per cent of phages passed on genes to more than one recipient, and 78 per cent of them were restricted to one species. This difference might reflect differences in host specialization between lytic and lysogenic interactions of the phage (Popa et al., 2017). Alternatively, uptake into the phage genome might be less constrained than uptake into a recipient bacterium, which requires genes from the same species in order to be able to function and persist in that genomic background. Phage therefore constitutes a mechanism of horizontal transfer that operates mostly within the taxonomic rank of species rather than more

(A)

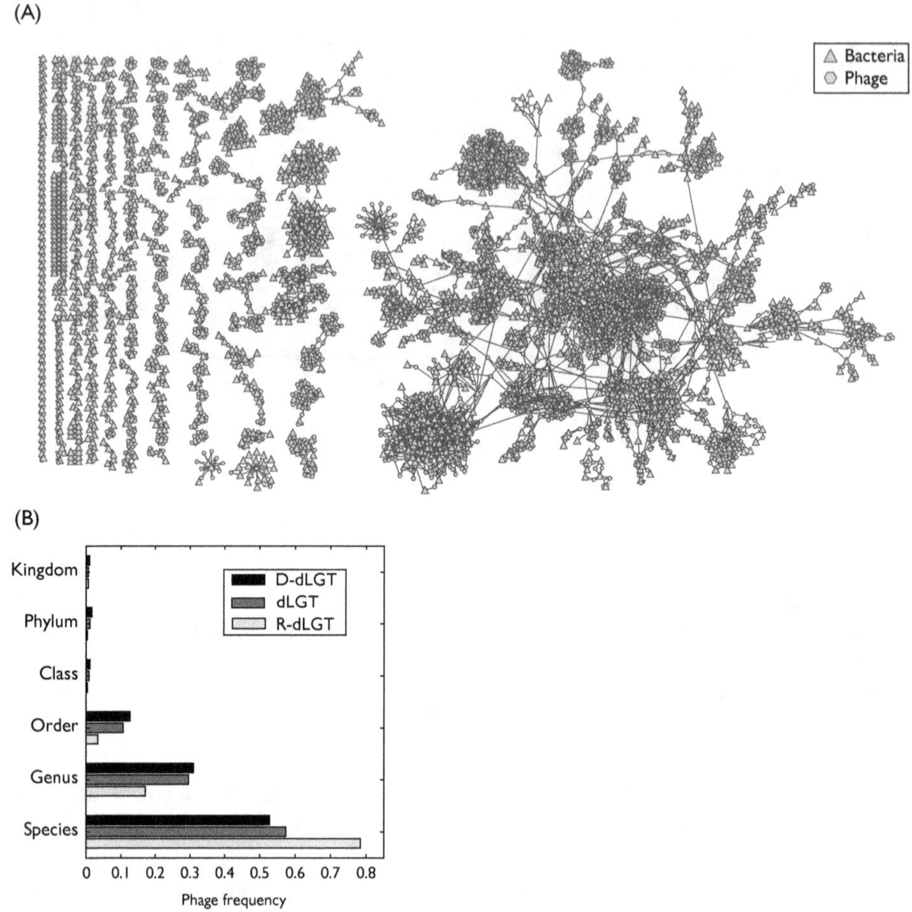

(B)

Fig. 7.7 Patterns of gene sharing in host–phage transduction networks. (A) Networks are shown with links indicating shared genes between hosts and phages. Taxonomic identities are not shown, but in constituent networks to right of plot, visible clusters tend to be found with shared taxonomic identities of hosts. Those on the left tend to link multiple hosts. (B) Frequency of occasions in which a given phage only has links within a given taxonomic level: in most cases, phages are connected by gene transfer only with members of the same species. dLGT indicates all links of lateral gene transfer. R-dLGT indicates just those with a bacterium (recipient) receiving a gene from a phage. D-dLGT indicates just those with a phage receiving a gene from a bacterium (donor). Phages are less taxonomically selective in their uptake of genes than in their donation of genes. (Reprinted from Popa et al. (2017) under creative commons license.)

broadly. In contrast to the Smillie et al. (2011) conclusions, phylogenetic relatedness was a better predictor of transfer than ecological similarity.

These studies are starting to reveal the structure of gene transfer networks in bacteria. It will be very instructive to reconstruct similar networks for multicellular eukaryotes once a denser sample of whole genomes become available. The shape of these networks will determine the extent of genetic interactions as species adapt to new conditions. A more finely resolved sampling is required to determine exactly the scale of transfer networks, and whether these typically occur within clusters of closely related strains or at higher taxonomic levels. From existing evidence, it seems that there are differences in specificity among different agents of transfer in bacteria. We should therefore observe different kinds of genes being transferred by different agents and different consequences and roles for adapting to different types of environmental change. We still need to bridge longer and broader-scale patterns with shorter and narrower taxonomic scales (see chapter 9). The extent to which bacterial interactions with plasmids and phage represent optimized mechanisms to promote horizontal transfer remains open. Although common in bacteria relative to animals and plants, horizontal transfer might still be rare enough that selection mainly operates at the point of arrival of the new variant: that is, on whether it persists or is lost, rather than strongly shaping the mechanisms of uptake.

7.8 Conclusions

It has long been recognized that the transfer of genes across species boundaries is sometimes a creative force in promoting both adaptation and speciation, in contrast to the more basic effect of eroding genetic diversity between species. A key question is therefore whether species boundaries are under selection themselves in order to provide an 'optimal' structure for adapting to the environment, or whether they emerge purely as a by-product of other processes. Even in some systems associated with widespread transfer of genes, such as phage-mediated transduction in bacteria, most transfer events seem to be restricted within species. Further elaboration of the mechanisms behind restriction, perhaps bridging long-term genomic reconstructions with selection experiments, will help to answer this key question.

8

Species interactions and contemporary evolution

8.1 Introduction

As a result of the cumulative effects of processes outlined in earlier chapters, species are a ubiquitous feature of life. There are complications in terms of blurred boundaries, ongoing gene flow, and multiple stages of completion, but every habitat on the planet is home to many species living together in ecological communities. A study of a deep goldmine in South Africa was noteworthy for reporting an ecosystem that appeared to be dominated by a single species—a chemoautotrophic bacterium called *Desulforudis audaxviator* (Chivian et al., 2008). But subsequent work revealed that, sure enough, other species are present even in this extreme environment far from our surface world, including nematodes that eat the bacteria (Borgonie et al., 2011).

In a single ecosystem at local scales, species are mostly unambiguously separate—it is easier to discriminate species in a local community than in a survey of a clade across a wide geographical area (Bergsten et al., 2012). Furthermore, species are ecological units as well as evolutionary units. Except in the theoretical case of strict ecological neutrality (see chapters 2.5.2 and 10.5), species exhibit differences in resource use, habitat preferences, and the pathogens and predators that target them. The central question for this chapter is what are the consequences of species diversity and ecological interactions among co-occurring species for ongoing evolution of each species (Barraclough, 2015)? Do species adapt to environmental changes independently, or is evolution shaped by the network of interactions? To answer these questions first requires an outline of the determinants of evolutionary rates and trajectories in the absence of species interactions. I then describe the kinds of ways that species interactions can influence evolution, before describing a model of adaptation in a guild of competing species and experiments in a tractable model system, bacterial communities from tree-holes of beech trees.

8.2 Evolution in a changing environment: single species

Much evolutionary theory takes a focal species approach and assumes that evolution can be understood on a species-by-species basis, without explicitly incorporating

The Evolutionary Biology of Species. Timothy G. Barraclough, Oxford University Press (2019).
© Timothy G. Barraclough 2019. DOI: 10.1093/oso/9780198749745.001.0001

species interactions. As a result, there is abundant theory and evidence on the determinants of evolutionary rates and trajectories in single populations adapting to new conditions (Lande, 1979; Burger and Lynch, 1995; Barraclough, 2015). Consider a single species growing on a defined resource in a physical environment represented by a single axis (e.g. temperature). A trait of the species determines its ability to use the resource and a second trait determines the optimum environment for growth of the organism (Fig. 7.1). Both traits exhibit heritable variation and therefore can evolve.

In a static environment, the population evolves towards optimum resource use and to align its optimum temperature for growth with the ambient temperature. If the environment changes, there is selection to track the change. The rate of evolution for a trait encoded by multiple genes is determined by generation time, genetic variation for the trait, genetic covariance with other traits, and the strength of selection pressure acting on the trait (Fig. 8.1; Lande, 1979). Genetic variation for the trait itself depends on mutation, the effective population size, and the level of recombination among genes encoding the trait. Genetic covariance reflects the genetic architecture—that is, whether some genes also influence other traits, and, if so, in a positive or negative fashion (Siren et al., 2017). For example, in a scenario of changing environment but stabilizing selection on resources, a positive covariance between the resource trait and optimum environmental conditions would reduce the response to selection (Hansen and Houle, 2008).

Current genetic architecture depends on past selection acting on those traits (Hansen, 2006). For example, if resources and the environment tend to fluctuate independently over multigenerational timescales, selection will act to reduce the

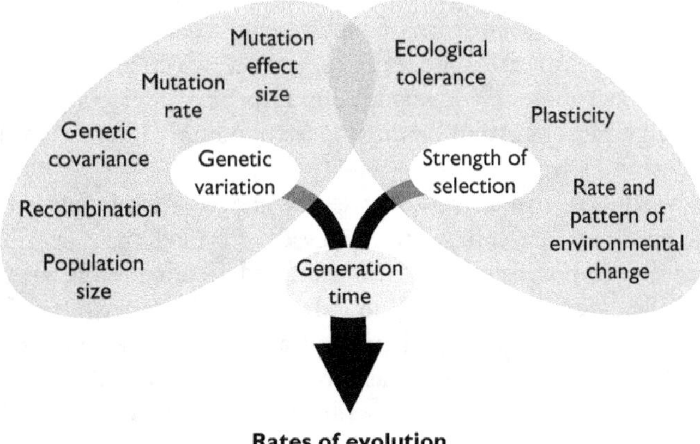

Fig. 8.1 Rate and direction of evolution depends on interaction between genetic variation and strength of selection pressure, which in turn depend on several components. Many models of evolution specify per generation rates of change, but in comparing multiple species, generation time is expected to play a major role determining not only rate of response but also type of selection experienced in a fluctuating environment. Theory and evidence for the role of each component in determining evolutionary rates is reviewed in Barraclough (2015).

correlation between traits, assuming it is mechanistically feasible for separate genes to encode the two traits (Steiner, 2012). Environmental fluctuations over shorter time-scales might select for plastic expression of the traits instead (Chevin et al., 2010). Genetic variation and covariance among traits therefore determine 'evolvability' or the intrinsic potential for the species to evolve, and these in turn depend on the past history of selection (Draghi and Whitlock, 2012) as well as mechanistic constraints (i.e. some traits are necessarily encoded by overlapping sets of genes).

The second aspect controlling evolution is the selection pressure acting on the trait. This depends in part on the magnitude of change in the environment: the rate of evolution should increase with the rate of change in the environment up to a threshold above which the population cannot keep pace with the change and will go extinct (Burger and Lynch, 1995). In addition, it depends on the effect of environmental change on fitness (Wade and Kalisz, 1990; Hunter et al., 2018). Organisms with broader environmental tolerance or greater phenotypic plasticity will experience weaker selection from the same environmental change than organisms with narrow environmental preferences and low plasticity. The type of environmental change interacts with the types of genetic variation present: the response to selection is greatest when new mutations have a large selective advantage, which depends on types of mutations, their effect on the phenotype, and the mapping from environment via phenotype to fitness (Dittmar et al., 2016).

If all of these aspects could be quantified, the rate and trajectory of evolution during periods of environmental change could be predicted, assuming no interactions among species. In practice this is challenging, in part because of the multidimensional space of possible genetic responses, lack of predictive knowledge of how genetic changes affect phenotypes and fitness, and because some of the above quantities (e.g. mutation rate, genetic covariances, environmental tolerances) themselves evolve. Also, the above discussion concerns adaptation to changes in external resource availability and environment. Some traits experience selection and evolve even in a constant external environment. For example, traits for sexual selection and antagonism are typified by high rates of evolution due to coevolutionary interactions between the sexes (Connallon and Clark, 2014).

Relatively few studies have tested how well the above determinants actually predict rates of evolution. There are comparative analyses of evolutionary rates and selection compiled from contemporary studies (Kinnison and Hendry, 2001; Kingsolver et al., 2012). For example, Pitchers et al. (2014) found evidence for faster rates of evolution of sexual than non-sexual morphological traits in animals (but not plants), but neither genetic covariance among traits nor measured strengths of selection explained the observed variation well, despite themselves varying widely among species. Phylogenetic comparative analyses have also looked for correlates of rates of evolution along branches reflecting millions of years of change, but the amount of variation explained was again low (Chira et al., 2018). Just as with the theoretical determinants of speciation rates discussed in chapter 5, it is difficult to obtain systematic data for all parameters of interest—instead, proxies are used, such as life history and range size. The lack of knowledge of how phenotypes depend on specific underlying genes might also limit the power of current approaches (Brookfield, 2016). Joint surveys of phenotypic and genomic change should help to alleviate this issue. Finally, measures of evolutionary rates for a whole community of organisms living together would be useful to supplement compilations of single-species studies and clade-level surveys. How do a suite of organisms with very different characteristics adapt to the same changes in abiotic environment?

8.3 Evolution in a changing environment: two species

Now consider two species inhabiting the same unstructured region. In order to coexist stably, a mechanism for coexistence is required. A simple mechanism is to assume that the species use different limiting resources. If we also assume initially that resource use is fixed and does not itself evolve over the timescales of responses to environmental change, then numbers of individuals in each species are independent (assuming that they do not share predators or interact in other ways). In this scenario, the species are expected to adapt independently to environmental change (Fig. 7.1B). Quantifying genetic potential and environmental pressure as outlined in section 8.2 would be sufficient to predict evolution. Predictions change, however, if the species do interact, by one of four main processes:

Alterred population density–Interacting species affect each other's density. For instance, competition for partly overlapping resources will lower the density of one or both species relative to a single species on its own. A reduction in census population size shortens the time to deterministic extinction in new and adverse conditions, limiting the time available to adapt to the change. Furthermore, reduced effective population size and genetic variance will slow the rate of adaptation if population size limits the rate of evolution (Osmond and de Mazancourt, 2013). Competition might therefore tip the balance from persistence to extinction as the environment changes (Fig. 7.1D). A reduction in population sizes caused by competition is 'easy' to take account of, however, and would not require specific modelling of interactions to predict evolution—it simply alters the measures of population size used in the single-species theory outlined in section 8.2 (Barraclough, 2015).

Resource use changes–More interesting effects occur if the resource axis that underlies coexistence changes. For example, Johansson (2008) modelled two species partitioning resources on a single linear niche axis (e.g. two consumers partitioning prey size). Under a scenario of increasing prey size, both consumer species evolved to track the change. The species that was initially adapted to eat smaller prey dwindled to extinction, however, because it was outcompeted by the species pre-adapted to eat larger prey (Fig. 8.2). This occurred even when the genetic variance of both species was sufficient for both species to have tracked the change in prey size in the absence of competition. The effect was amplified by a decline in the rate of evolution of the species adapted to the smaller prey (a linear effect of population size in the model used). But in this case the evolutionary response was not predictable purely from single species quantities outlined above but depended on the nature of species interactions.

Directionality–Outcomes vary with other types of species interaction. For example, with a predator and prey species, the predator is reliant on the prey adapting to the

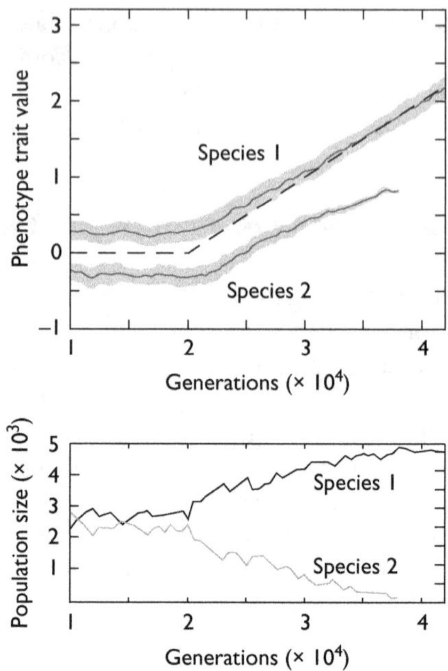

Fig. 8.2 Competition impairs rate of adaptation to a changing environment in a model of two species. In constant environment, species partition a Gaussian distribution of resources around a single optimum (dashed line, top panel); shaded areas show distribution of phenotypic trait values in each species. When the environment changes to increase the optimum value, the species with highest trait value has an advantage and evolves to keep pace with the change, whereas the species with lowest trait value lags behind the change in environment and, in this case, dwindles to extinction. Both species shared the same genetic variation and hence genetic potential to adapt to change. (Modified from Johansson (2008) with permission.)

new conditions, whereas the prey benefits from the predator failing to adapt. This leads to directionality that lower trophic levels must adapt before higher levels are able to. If lower trophic levels tended to have faster generation times and larger population sizes as well (e.g. phytoplankton in the sea), this would increase the chances of sustaining the whole system through a period of climate change. A similar 'evolutionary cascade' might occur with cross-feeding relationships in bacteria: the consumers of input resources must adapt before users of downstream resources (Fig. 8.3). With obligate mutualists, both partners must adapt simultaneously to new conditions, hence at the rate of the species with the lowest genetic potential.

Reciprocal coevolution–The above scenarios all involve ecological interactions that are either fixed or mediated via changing resource levels or densities of the species. A final type of evolutionary interaction is the classic scenario of reciprocal coevolution where matching traits in each species directly influence the strength of interaction (Slatkin and Smith, 1979; Kopp and Gavrilets, 2006). For example, traits in a parasite determine its ability to infect its host and traits in the host determine its ability to resist infection. Now, there can be an arms race with escalation in virulence and

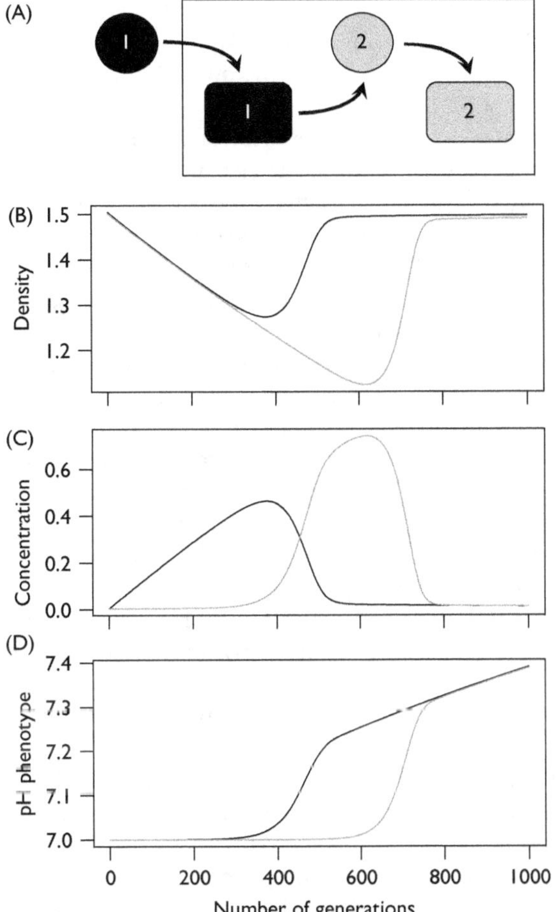

Fig. 8.3 Evolutionary cascade in response to changing abiotic environment. (A) Consider two bacterial species (rounded rectangles) with facilitative interaction: species 1 degrades an input resource (substrate 1, circle), and as a result produces a waste product (substrate 2) that is used by species 2. Both species are adapted to a shared physical environment (e.g. pH of their growth medium) but differentially adapted to metabolizing the two resources. There is genetic variation for traits determining optimum pH for growth and resource use, which affects the growth rate of each species. If the environment changes through a shift in pH from 7.0 to 8.0 at time zero, then densities of both species initially decline (B), leading to an accumulation of unused substrate 1 (C). It is only after the first species adapts to the new pH (D—dark grey line showing pH phenotype), and thereby begins generating substrate 2 again as waste, that the second species is able to grow sufficiently to adapt to the new pH conditions (B and D—light grey line). Depending on relative ease of evolving changes towards a new optimum pH versus resource use, an alternative possible outcome is that species 1 adapts to the new pH, species 2 is unable to adapt and goes extinct, and then species 1 diversifies to use both resource types. Similar scenarios could be envisaged for a predator–prey trophic interaction.

resistance, or cyclical evolution as the species fluctuate between alternate, frequency-dependent states (Sasaki and Godfray, 1999; Abrams, 2000). These dynamics occur even in the absence of external changes in environment. Other examples are the evolution of antimicrobial production and resistance between two competing microbe species or traits determining the relative rewards to a plant and its pollinator. In competitive interactions, an arms race can also occur with an escalation in competitive ability (Abrams and Matsuda, 1994), or alternatively species might diverge in resource use to reduce the strength of negative interactions between them (Case and Taper, 2000).

There are therefore three types of relevant traits for consideration: α-niche traits mediating coexistence of species on a defined set of resources, β-niche traits determining growth within the shared physical environment within a single region, and biotic niche traits that directly target interactions with other species (which can be independent from resource acquisition). The only scenario in which the presence of the other species does not influence evolution is when species do not interact and the environmental change is orthogonal to resource axes underlying coexistence (i.e. α-niche traits) and to biotic niche traits. All other cases generate eco-evolutionary interactions: the way each species evolves depends on how the other species evolves, which in turn alters ecological interactions (Hendry, 2017).

Furthermore, selection on one type of traits could have knock-on effects on the other types of traits (Northfield and Ives, 2013). Even if resource availability itself does not change, it is possible that adapting to changing physical conditions could trigger evolutionary shifts in resource use. For example, if one species fails to adapt to new conditions, it would free up resources that the other species could evolve to exploit, by expanding its resource use or by diverging into ecotypes. Similarly, while selection on biotic niche traits is not directly dependent on a change in abiotic environment, it is possible that environmental change could affect densities or costs and benefits of the interactions and stimulate further coevolution as a by-product.

Given that species diversity is ubiquitous across ecosystems and communities, and assuming that species interactions are also ubiquitous (at least over evolutionary timescales—they might not be measurable ecologically if species currently partition resources (Connell, 1983)), these effects should be a universal feature of evolutionary dynamics. With the possible exception of evolution in response to intraspecific competition, such as the case of sexual conflict mentioned in section 8.2, we cannot hope to understand and predict evolution of focal species without taking into account species interactions and responses of co-occurring species (Barraclough, 2015; terHorst et al., 2018).

8.4 Evolutionary dynamics in a guild of species: a theoretical model

Having outlined how species interactions affect evolution in the simple case of two species, theoretical models are needed to explore dynamics in more diverse communities. de Mazancourt et al. (2008) devised such a model to investigate the general effects of competitive interactions on a guild of co-occurring species. The model considered a set of species occupying spatially distinct habitat patches within a focal region. Each patch was defined by a single environmental variable, temperature, which varied among patches (Fig. 8.4A). Propagules dispersed among patches according to the fraction that entered a global, mixed pool versus remained locally (Fig. 8.4B). The growth of organisms within a patch, and subsequent production of propagules, depended on a single phenotypic trait, optimum temperature. A close match between the trait and the temperature in the patch led to maximum growth, which then declined according to a Gaussian function as the match between phenotype and environment declined (Fig. 8.4C). Input of variation by mutation and by recombination during sexual reproduction allowed the phenotype to evolve in response to selection. The model was envisaged to represent plant species adapted to habitat patches with different microclimates, for example locally endemic plant species in the Cape region of South Africa.

The region was populated initially with a set of species, one for each patch, and each of them optimally adapted to the local conditions in that patch. This situation is stable as long as the dispersal rate is low enough that plants tend to find themselves in the same conditions as their parents, and as long as the width of environmental tolerance is narrow relative to the difference in temperature between different patches. Otherwise plants evolve towards a more generalist phenotype that permits moderate growth under several patch conditions. Since the aim was to investigate the effects of diversity and species interactions on evolution, conditions were chosen that permitted initial diversity. The mechanism of coexistence is therefore specialization on spatially distinct niches. Species with similar optimum growth temperatures compete, whereas those with different phenotypes interact weakly.

In order to investigate the effects of interspecific competition on adaptation to changing conditions, the stable starting community was then exposed to a period of regional environmental change (Fig. 8.5A). The environment in each patch changed linearly over 50 generations to new conditions drawn again from a uniform distribution, but with higher limits than at the start. The envelope of temperatures therefore shifted to an increased distribution, equivalent to regional warming on average, but some patches experienced cooling by chance. The question then is how the community as a whole adapted to the changes in terms of evolution of the component species and ecological changes in abundances and distribution (i.e. whether they stayed in their original patch or shifted to a different patch following change). In the context of mechanisms outlined in section 8.3, this scenario represents a case where the environment changes in a parallel direction to the niche axis being partitioned.

The outcome depended on the amount of dispersal among patches. With zero dispersal, species are limited to their original patch and either adapt to the new temperature (Fig. 8.5A) or—if the rate of change in that patch is too rapid for them to keep up—go extinct. With intermediate dispersal, low enough to sustain local adaptation initially but high enough for offspring to occasionally disperse to different patches, the dynamics shift towards ecological responses (Fig. 8.5B). Species tend not to evolve and instead shift to patches with new conditions similar to their initial

(A)

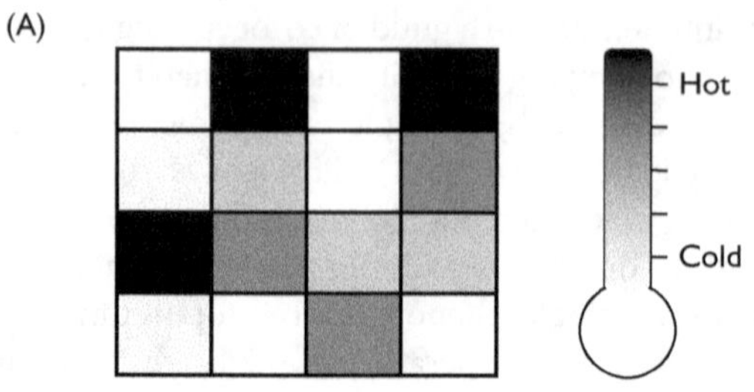

(B)

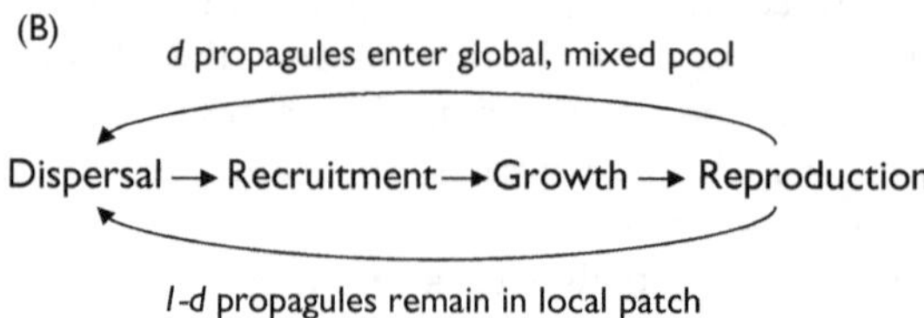

(C)

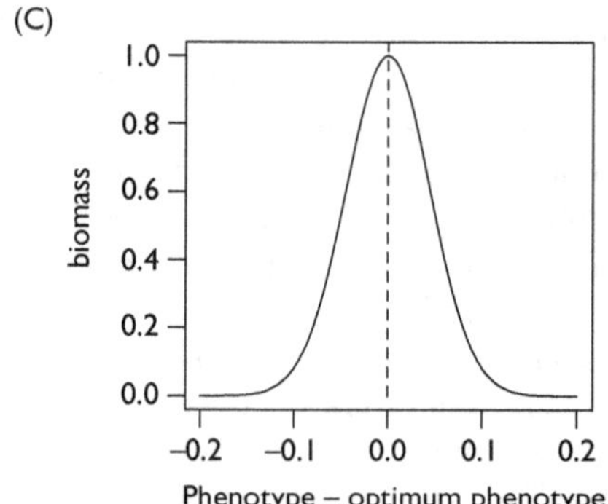

Fig. 8.4 Schematic of the model of a guild of species adapting to environmental change by de Mazancourt et al. (2008). (A) The environment comprises a set of patches with different average temperatures (A). Organisms complete the following life-cycle: (B) propagules either disperse within a globally mixed pool or remain in their local patch; recruitment is via a lottery up to a fixed carrying capacity; growth occurs in proportion to the match between temperature of patch and a single phenotypic trait, x, which represents the organism's optimum temperature for growth (C); reproduction is by random mating. Number of propagules produced is proportional to biomass accrued during growth. Individuals that are better adapted to their local patch produce more propagules for seeding the next generation.

optimum phenotype. Comparison of simulations with just one patch occupied initially versus all patches occupied showed that this resulted from competition. In the absence of competitors, species tended to diversify into ecotypes occupying a wide range of patches (Fig. 8.5C), but this was constrained by the presence of other species.

De Mazancourt et al. (2008) therefore concluded that competition tends to inhibit evolution of component species. Furthermore, there was a directionality to responses.

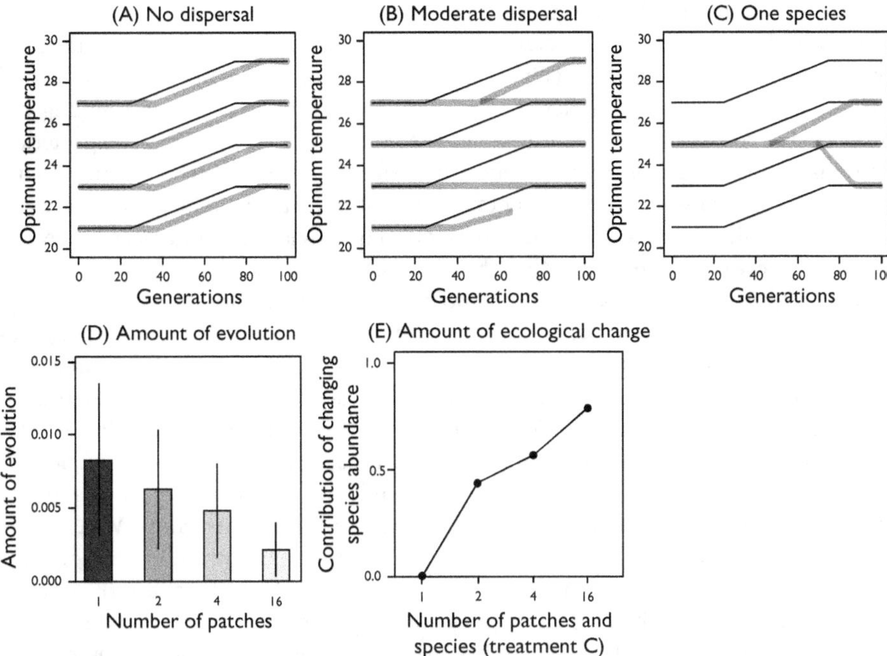

Fig. 8.5 Summary of results of the de Mazancourt et al. (2008) model. (A) With no dispersal among patches, species adapt locally to changing temperature as long as mutation rate is sufficient to keep pace with change in temperature: black lines indicate temperature change in each of four patches; thick grey lines, change in phenotype of the four species adapted to each starting patch. (B) With intermediate dispersal, species are initially specialized on separate patches, but when the environment changes, species tend to disperse and colonize final patches with similar temperatures to their starting patch. New patches warmer than the initial envelope are colonized by the most warm-adapted species, and the species adapted to the coolest patch originally is driven extinct by competition from other species. (C) With just one species present initially, the species diversifies to occupy empty patches. (D) Across repeated simulations with random initial and final patch temperatures, average amount of evolution per species declines with increasing numbers of species. (E) Concurrently, total amount of change in phenotypic trait distributions that results from ecological changes in species abundances increases. (D and E redrawn from de Mazancourt et al. (2008) with permission.)

Species initially adapted to cooler patches tended to go extinct, because there were no similarly cool patches after change and competition prevented them from adapting to warmer patches—even if they could have adapted to warming in the absence of other species. Obvious parallels can be drawn with global warming, and the prediction that high latitudes will be colonized by pre-adapted species from warmer climes. In contrast, evolution is required to adapt to the warmest patches after change, and these patches tended to be colonized by the species from the warmest patches initially. Finally, species initially adapted to the intermediate temperature range exhibited niche conservatism and responded by dispersing to new patches with favourable conditions.

Across simulations with varying numbers of species, the amount of evolution per species declined with more species (Fig. 8.5D), whereas the relative importance of ecological responses increased (Fig. 8.5E). These findings have implications beyond the details of the specific model. Any scenario with standing variation among co-occurring species that is relevant to a change in environment would be expected to demonstrate the same processes. Almost by definition (see chapter 2), genetic variation in ecologically and environmentally relevant traits tends to be greater between species than within species. Selection therefore acts on relative abundances of species, resulting in ecological sorting, rather than on the frequency of genotypes within species, which encompasses less variation and hence a slower response.

An explanation for persistence of standing variation in the focal trait among co-occurring species is needed in order for this mechanism to work. For changes in physical environment, this is facilitated by open systems connected by dispersal, so that species adapted to different local conditions can persist, ready to colonize new areas as conditions change. But equivalent models based on resources in a single area would work: each patch in the above model could be a resource on a linear axis, for example prey size, and the distribution of prey sizes available then shifts. When these criteria are met, diversity should tend to inhibit evolution because competition constrains the ecological opportunity for other species. This begs the question of when species evolve, especially in open, hyper-diverse systems. Within the model, evolution occurs in patches with new conditions outside the range exploited by current species (i.e. new marginal habitat). Multidimensional environments might offer even greater chance of completely new sets of conditions arising. Other possible solutions are that occasional events create new ecological opportunities (e.g. extinction occurs independently of environmental change) or evolution predominates in low-diversity systems (e.g. oceanic islands).

The central conclusion that diversity inhibits evolution is at odds with the intuition, expressed in metaphors such as the Tangled Bank and Red Queen, that diversity should stimulate evolution (Voje et al., 2015). One possible cause for the discrepancy is that the model considers a guild of competing species rather than trophic interactions: perhaps antagonistic or mutualistic interactions are required for diversity to stimulate evolution. Guimarães et al. (2011) modelled species interacting in a mutualistic network with traits affected by coevolution and background evolution independent of interactions. Allowing coevolution led to cascading effects of evolution through the network and sped up evolutionary rates overall.

Similarly, the de Mazancourt et al. (2008) model does not include biotic niche traits or the potential for arms races. Species do not adapt to the traits of other species directly but to availability of suitable habitat. Species evolve a change in optimum temperature, but two species with the same optimum temperature compete equally. Biotic niche traits could be added, such as allocation to traits that enhance competitive ability (e.g. root mass) or through the production of allelopathic compounds to harm competing species directly (Lawrence and Barraclough, 2016). My hunch is that resource competition is more direct and immediate, and that a change of external environment would still convey a large advantage to species most pre-adapted to new conditions. Even if another species initially had the upper hand in an allelopathic interaction, its density would fall in the new environment and hence the weapon would become less effective. Several models have demonstrated how competitive or antagonistic interactions can stimulate Red Queen coevolution, but most of these focus on diversification phenomena (such as stasis and punctuated changes) over macroevolutionary timescales (Voje et al., 2015). More work is needed to consider how these processes might operate during recent and ongoing changes (Melian et al., 2011; Moya-Larano et al., 2012).

8.5 Evolutionary dynamics in microcosm experiments: tree-hole bacteria

The de Mazancourt et al. (2008) model predicts that species interactions in diverse communities inhibit evolution, whereas classic coevolutionary theory predicts that diversity of biotic interactions increases evolution. Genetic studies in plants and animals have measured selection caused by interactions such as competition, herbivory, and pollination, either in specialist reciprocal interactions or via so-called diffuse coevolution with generalist herbivores or pollinators (Strauss et al., 2005). Ideally, experiments are needed that track evolution over multiple generations to test these ideas, however, which is difficult with animals and plants.

Lawrence et al. (2012) turned to experimental evolution with bacterial communities. The experimental design was fairly simple (Fig. 8.6). Five distinct species of bacteria belonging to different genera were isolated from rainwater pools formed by the roots of beech trees, called tree-holes (Bell et al., 2005; Bell, 2010). The species were subjected to regular serial transfers into fresh medium every few days, for a period of 8 weeks, corresponding to a standard design for experimental evolution (Buckling et al., 2009). The resources for growth were dissolved compounds in the growth medium, which was beech tea made by boiling autumn leaf fall. The abiotic environment encompassed both the resources provided and standard laboratory conditions: warmer than the wild, more uniform because cultures were shaken, and with continuous growth because of the regular transfer as opposed to intermittent arrival of new resources followed by periods of famine in real tree-holes. The experiment was repeated for two treatments: each species on its own in monoculture versus all five species together in a community (polyculture). Ancestral isolates and the evolved isolates of each species re-isolated from the end communities were frozen at −80°C for later assays (Fig. 8.6). One species failed to survive during the experiments and so was discarded from later assays, leaving four species for comparison.

The first set of assays measured how well the isolates had adapted to the abiotic conditions, namely growth on beech tea and under laboratory conditions. As expected, monoculture isolates of each species grew faster on average than ancestral isolates consistent with adaptation to laboratory conditions (in all except species A, which showed no significant increase (Fig. 8.7A)). But surprisingly, polyculture isolates of three of the surviving species grew far worse on the beech tea medium in the absence

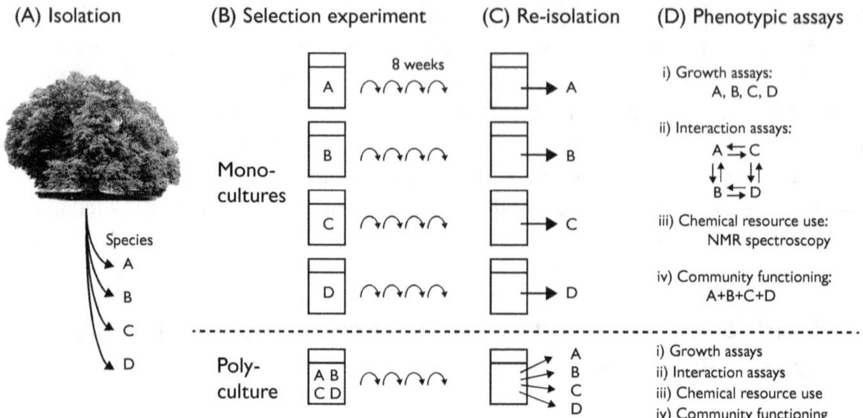

Fig. 8.6 Experimental design for the Lawrence et al. (2012) study of coevolution in a bacterial community microcosm. (A) Five species were isolated from pools formed by roots of a beech tree—one did not survive the full course of the experiment, hence only four surviving species are shown here. Each species belonged to a different bacterial family. (B) Monocultures of each species and a polyculture containing all of them were set up and cultured for 8 weeks with serial transfer twice per week. (C) At the end, each species was re-isolated from the monoculture and community cultures and frozen together with the ancestral isolates of each. (D) Various phenotypic assays were performed to measure changes in the isolates: (i) growth in isolation on the beech tea medium used during the experiment; (ii) growth on beech tea already used by one of the other species to assay ecological interactions; (iii) use and production of chemical metabolites in the medium by each species; and (iv) respiration rates of a reassembled community as a measure of whole-community functioning. Assays were repeated for the ancestral, monoculture, and polyculture isolates of each species in turn.

of other species than did the ancestral isolates (Fig. 8.7A). This indicated that by adapting to the biotic environment of co-occurring species, they had all but lost their ability to grow alone in laboratory conditions. In other words, there was a trade-off between adapting to the presence of other species versus in their absence.

The next set of assays sought to investigate further how species adapted to each treatment, by measuring the strength and type of interaction among species. By growing each isolate on filtered media previously used by another isolate, we quantified negative versus positive interactions. If the second species grows worse on used media than on unused beech tea, this indicates a negative interaction, for example because they grow on the same resources. If instead the second species grows better on the media previously used by the first species, this indicates a positive interaction. For example, the second species might metabolize waste products of the first species for growth, a well-known phenomenon called cross-feeding.

Ancestral isolates tended to interact negatively, consistent with their use of overlapping sets of resources in the medium (Fig. 8.7B). Monoculture isolates evolved in a way that, if anything, strengthened the negative interactions among them. But most strikingly, polyculture isolates evolved towards a greater number of positive

(A) Growth assays

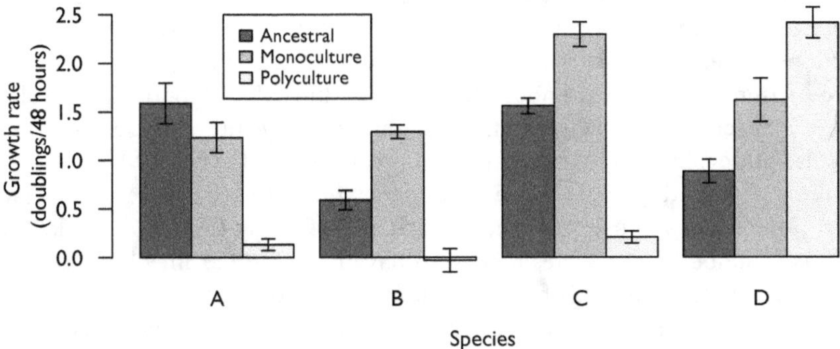

(B) Interaction assays

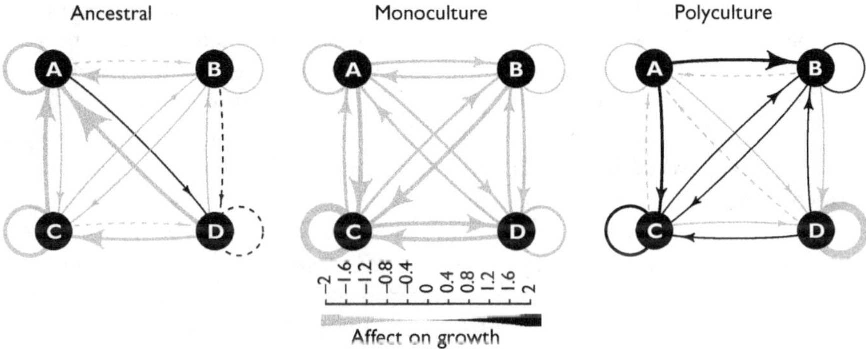

(C) Chemical resource use

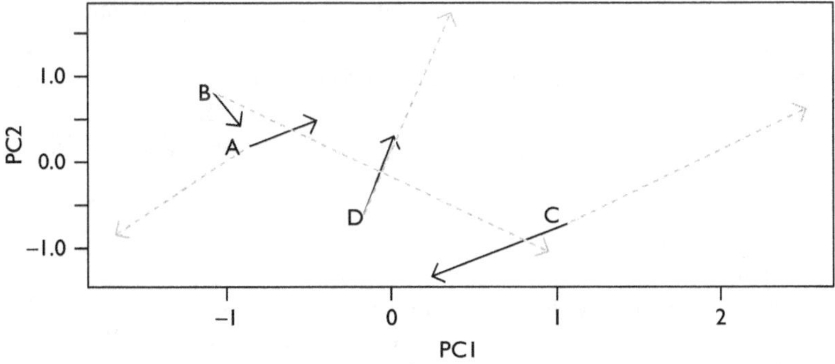

Fig. 8.7 Results from the Lawrence et al. (2012) experiment outlined in Fig. 8.6. (A) Growth of isolates that evolved in monoculture was, on average, significantly greater than growth of ancestral isolates on beech tea medium. In contrast, species A to C evolved in polyculture so that their growth in isolation on beech tea was reduced, whereas species D evolved an even higher growth rate in polyculture than monoculture treatment. (B) Interactions were assayed by growing each isolate on filter-sterilized tea already used by another isolate. Grey solid arrows

interactions among the species. Three species in particular grew better on tea used previously by another polyculture species than on unused tea. Further assays measuring the chemical composition of used and unused tea showed that the changes in interactions related to phenotypic changes in resource use. Changes in metabolic use and production were more pronounced for polyculture than monoculture isolates (Fig. 8.7C). Specifically, metabolic differences among polyculture isolates increased, consistent with evolution of niche partitioning. Species had evolved to use more of compounds produced more as waste products by other species: there was a match between production and use, as expected with cross-feeding (Lawrence et al. 2012). In contrast, monoculture isolates tended to have more similar metabolic resource use, consistent with the finding of more negative interactions in the earlier assays. The genetic basis of changes was not tested directly, but assays regrew isolates in common garden conditions for at least three generations; hence any contribution of plasticity was heritable over multiple doublings.

Taken together, these results showed that species interactions altered the trajectory of evolution in the community treatments. Species adapted to different sets of resources and indeed to dependency on the waste products of other species, establishing a cross-feeding community. As a consequence, their ability to grow in isolation under the laboratory conditions was diminished. Phenotypic changes relative to ancestors as measured by both growth assays and resource use measurements were greater in the community than in isolation.

An interesting question is why ancestral isolates had overlapping resource use when they co-occur in the wild, since they were isolated from the same water sample. Competition is a general pattern for random pairs of tree-hole isolates (Foster and Bell, 2012). Why don't they evolve niche partitioning there as well? One explanation is that fluctuating resource availability favours generalist strategies in real tree-holes, but specialism is favoured with simpler resource conditions. Or in an open system with widespread dispersal, species interactions might lack the constancy of the laboratory set-up to generate consistent selection for specialization. Finally, perhaps the many hundreds of species present in tree-holes constrain specific coevolutionary interactions that are feasible in lower diversity systems. The Lawrence et al. (2012) results match a classic scenario of adaptive divergence in an empty environment,

Fig. 8.7 Continued

indicate the second species grew significantly worse than on unused beech tea; black solid arrows indicate the second species grew significantly better. Thicker arrows indicate a proportionately greater strength of interaction. (C) Metabolite use and production was measured by comparing nuclear magnetic resonance (NMR) spectra of unused tea and filter-sterilized tea following growth of a species. Summarizing the chemical profiles with principal coordinate (PC) analysis, the solid arrow shows change in resource use/production in monoculture isolates from ancestor, and dashed arrow shows change in polyculture isolates: the latter diverged into four corners of the plot, indicating chemical resource niche partitioning. (Redrawn from Lawrence et al. (2012) under creative commons license.)

except starting with four colonizing species that partition resources rather than a single ancestor that diversifies.

Subsequent work addressed these questions. Fiegna et al. (2015a) investigated the effects of increasing diversity by constructing communities with 1, 2, 3, 6, and 12 species from a pool of 12 isolates (Fig. 8.8A). Communities were grown in standard beech tea for 2 weeks before transfer to three parallel environments for 5 weeks: standard beech tea, acidified beech tea (pH 5), and spruce tea (which contains different resources). A split partition design was used such that each species was represented in two different community compositions at each level of diversity. This design partitions effects into those due to species composition versus species richness and allows statistical estimation of whether changes in the total yield of each community result from ecological sorting, shared evolutionary trends across all richness levels (i.e. the response observed in monocultures), or evolution of species interactions (that varied depending on diversity level and which other species were present). The analyses estimated that 18 per cent of changes in yields of whole communities resulted from additive effects of evolution observed in monocultures, 14 per cent were due to evolution of less negative interactions at higher richness levels, and 47 per cent were due to changes in the strength of species interactions in particular species compositions (Fig. 8.8B, see also Rivett et al., 2016). Most changes were observed in the acidic beech tea treatment. Composition was therefore more important than richness in determining outcomes. Growth assays confirmed that isolates had reduced ability to grow independently after a period coevolving with other species (Fiegna et al., 2015b), again indicating a trade-off between adapting to the presence versus absence of other species. This effect was saturated with increasing diversity levels, however, (and

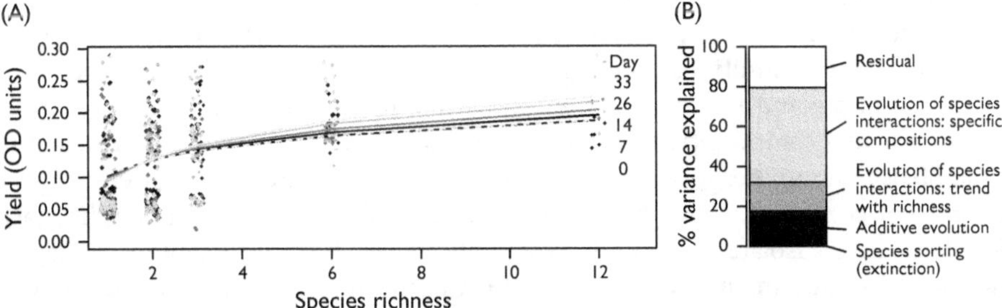

Fig. 8.8 Inferring evolution of species interactions from changes in growth yields of bacterial communities cultured for over 5 weeks following a shift to new culture conditions. (A) The shape of how community yields vary with species richness changed over time following a shift to acidic beech tea (pH 5). Curves became steeper over time (i.e. greater increase in yield over time in more diverse communities). (B) For these data, together with two further treatments for a shift into spruce tea and control beech tea, statistical analysis of the shape of changes in communities versus monocultures allowed partitioning of variation into five main components as shown: species extinction explained a negligible amount of variation among communities; additive evolution represents changes predicted from adding changes observed in monocultures; the final component of explained variation indicates that species interactions evolved to be more positive. (Redrawn from Fiegna et al. (2015a) under creative commons license.)

reversed for growth rates in beech and spruce tea), which is potentially consistent with the idea that higher diversity constrains coevolutionary interactions compared to few-species systems.

Real tree-holes contain many more species. Scheuerl et al. (2019) devised a microcosm method for tracking the evolution of focal species in different whole-community backgrounds. Species were inoculated into a dialysis bag inserted into the well of a 96-well plate that contained an inoculum from a whole tree-hole community. Metabolites and signalling molecules are able to diffuse across the membrane, allowing chemically mediated interactions, but the focal species is kept physically separated from the other species, allowing ease of isolation. Cultures were grown for 24 weeks with regular replenishment of beech tea medium, and growth assays compared the relative growth of ancestral versus final isolates of the focal species. Factorial combinations of 22 species and 9 communities were cultured to investigate the important intrinsic properties of the focal species versus properties of the community in determining responses. Overall, changes in growth phenotype were explained equally by species (13 per cent), community (12 per cent), but especially their combination (38 per cent). Specifically, species evolved higher growth rates relative to the ancestor when the community was dominated by relatively few species, and when the focal species was rare in the community and was initially poorly adapted to new conditions. Complex communities constrained the evolution of focal species.

Taking these ideas into the wild is hard, however, because of the challenge of tracking the evolution of constituent species in diverse systems with hundreds or thousands of species. Lawrence et al. (2016) investigated the effects of immigration by performing selection experiments on whole tree-hole communities seeded into glass bottles exposed to ambient versus warmed conditions in the field. The immigration treatment compared responses in open versus closed bottles. The whole community 'adapted' to local conditions over time—that is, warmed communities grew better in warm conditions and ambient communities better in ambient conditions. The growth rate of warmed communities was reduced by immigration, whereas the growth rate of ambient communities was increased—perhaps because source inocula were adapted to ambient conditions (having arrived by dispersal from the surrounding habitat). Random species isolates from each treatment did not show consistent local adaptation to their experimental conditions, which again implicated species interactions rather than parallel evolution of each species in the overall community response. The key limitation, however, was that tracking the same species over time was not feasible. Prospects for tracking evolution in whole communities in the wild is discussed in chapter 9.

Although few controlled experiments have been performed for communities of multicellular organisms over multiple generations, similar results have been found from a long-term field study on plant biodiversity–ecosystem functioning relationships. Zuppinger-Dingley et al. (2014) tested for trait evolution of 12 plant species over 8 years of the Jena Experiment, which maintained plots with different levels of species richness of the focal species. Plant species grown from the final period displayed greater differences in height and specific leaf area, indicative of ecomorphological differentiation compared to ancestors, and increased community yield, consistent with greater complementarity in their contribution to whole-community functioning. More multigeneration experiments are now needed to examine evolution with alternative interaction types: for example, in food web and host–parasite mesocosms (Gomez and Buckling, 2011; Betts et al., 2018).

8.6 Conclusions

The theory and examples described in this chapter illustrate that species interactions are a key component for understanding evolutionary responses to new conditions. They also provide some initial qualitative predictions for different types of dynamics (Box 8.1). More theory and experiments in model communities are needed to explore the full range of responses and understand when different outcomes should arise. In particular, the work described here emphasizes adaptation to the physical environment and resource availability, whereas many studies emphasize coevolution of biotic niche traits (Betts et al., 2018). The relative importance of these various types of evolutionary interactions remains an open question. Biotic niche traits involved in arms races exhibit some of the highest rates of divergence between closely related species, yet much coevolution is diffuse rather than strong reciprocal effects between pairs of species (Strauss et al., 2005). Few studies have attempted an inventory of all sources of selection acting on species (c.f. Collins and Gardner, 2009) and investigated the interactions among responses to those different sources.

Another open question is understanding how species interactions evolve. The similarity of phenotypes in wild tree-holes implies that species converge in resource use (at least those that can be readily isolated and cultured in the laboratory), whereas the same isolates diverged in resource use in the laboratory. Ecological theory identifies two kinds of mechanism for coexistence. Niche partitioning to reduce competition is

---

Box 8.1 Summary of qualitative predictions.

- Species adapt to environmental change in parallel and independently when there are no species interactions among them, because of either niche partitioning or geographical isolation.
- If species interact ecologically, such that the density of one species affects the density of other species, the way each species adapts to environmental change is interdependent.
- Changes in physical environment can trigger further evolution in resource use and biotic niche traits as a result of changing species densities. Evolution of biotic niche traits can also affect evolution to the physical environment if there are trade-offs between adapting to different sources of selection.
- The relative role of evolution versus ecological sorting in responding to environmental change decreases as diversity increases, and in systems that are more open to immigration and therefore contain a larger global pool of species that might be pre-adapted to new conditions.

a stabilizing mechanism, as it reduces the interactions among species through ecological differences. In contrast, species might converge ecologically, leading to nearly-neutral coexistence—the timescale of ecological replacement slows to the timescale of drift—which is called an equalizing mechanism (Chesson, 2000). Although neutral theory is viewed as an abstraction, because of well-documented ecological differences among species, it remains possible that equalizing mechanisms apply in some communities, especially coupled with the ability to survive unfavourable conditions. Under what conditions do we expect niche partitioning versus convergence to evolve, or versus an arms race in competitive ability, perhaps leading to extinction of losers in that race?

These examples illustrate that the dynamics are important not only for predicting evolution but also for explaining coexistence of species over evolutionary timescales. Species must not only coexist ecologically in a static environment. As environments change and species adapt to those changes, they must also adapt in a way that maintains species diversity. The timescale of diversification to recover lost diversity is longer than the timescale of extinction of species failing to adapt to environmental change or being outcompeted by their compatriots. A fluctuating environment with alternating increases and decreases over multiple generations could lead to erosion of diversity and eventual persistence of a single generalist species (Tilman and Lehman, 2001). To maintain species diversity, either species must adapt independently to environmental change or survive adverse periods for multiple generations (e.g. in seed, resting egg, or spore banks (Chesson, 2000)).

The potential for evolutionary responses among a set of species increases the scope for species interactions. Species might not interact at present because of niche partitioning or geographical isolation, but under longer-term changes the way that each species adapts to those changes depends on the responses of the others. Species might therefore evolve independently over the short term, but there remains evolutionary interdependence as habitat types and environmental conditions wax and wane over longer timescales. The effects of these connections on broad-scale diversity and diversification are addressed in chapter 10.

As a final point connecting to earlier chapters, responses to current change should also depend on the history of species origins. For example, genetic architecture of selected traits could depend on whether species diverged in the presence of gene flow (even if there is no current gene flow) and whether species experienced correlated or uncorrelated selection on those traits in the past. Such factors will influence whether species display correlated or uncorrelated genetic potential for adapting to a given environmental change. How species evolve will be shaped by the history of their divergence and past interactions between them. The genetic basis of phenotypic changes was undetermined in the above studies, in part due to constraints when working with multiple, non-model species, but this is becoming feasible due to improvements in genome sequencing technology (see chapter 9). Future work will be able to bring together inference on past evolution, genetic architecture, and observed genetic change during multigeneration experiments, integrated with ecological forces considered here.

9

Predicting evolution in diverse communities

9.1 Introduction

Chapters 7 and 8 showed how interactions among co-occurring species affect their evolutionary trajectories, using simplified theory and experimental microcosms. The big challenge is to take these ideas into the real world, not just to understand observed evolution but also to predict how diverse systems will change in the future. Humans depend on many species for food, natural products, and ecosystem services such as nutrient cycling and clean water supply. Other species such as pathogens and parasites have detrimental effects. These beneficial and adverse effects depend both on the abundance and distribution of species and on their functional traits (i.e. phenotypes). When conditions change, for example through climate change, the invasion of a new species, or a management action, the ecological and evolutionary responses of species will determine the new distribution of traits, and hence the functioning of ecosystems in the new conditions (Barraclough, 2015; Alberti et al., 2017; Cadotte et al., 2017). Managing natural systems therefore requires prediction of how assemblages of species, and the phenotypic traits behind functions that affect human well-being, will respond to a given change.

Needless to say, this is a hard task. As outlined in chapter 8, many processes affect the evolutionary responses of a set of species, encompassing genetic and extrinsic factors, which are compounded by the complexity of real species interaction networks. Characterizing the present-day environment and its effects on the fitness of multiple species is daunting enough, without attempting to forecast all the effects of future environmental change. Direct effects of a management intervention such as clearing a forest might be relatively easy to predict, but many indirect effects are hard to predict, such as secondary changes in local microclimate or local people's behaviour. Predictions of future climate comes with a large degree of uncertainty of the full set of processes that together determine global climate, in no small part because of unknown feedbacks with living systems themselves. In short, the natural world is hideously complex to understand mechanistically, with a large essentially stochastic element.

Although prediction is hard, it is not futile. While it is unlikely that we will ever be able to predict every aspect of how a diverse community responds to environmental

The Evolutionary Biology of Species. Timothy G. Barraclough, Oxford University Press (2019).
© Timothy G. Barraclough 2019. DOI: 10.1093/oso/9780198749745.001.0001

change, certain aspects of responses can be predicted (Evans et al., 2012; Holmes, 2013; Lassig et al., 2017). Do species evolve to have stronger or weaker interactions? Does the community become more or less stable? Will particular aspects of functioning increase or decrease? Which types of species are likely to be lost versus able to adapt to a given change? How important is gene exchange between species? Such emergent properties might be predictable even if the detailed dynamics of every species are not. To do this requires robust theory for guiding predictions—simple enough to be generalizable but encompassing the minimum set of mechanisms needed to reflect the real world—and validation by tracking evolution of whole communities in response to experimental change. It further requires quantification of uncertainty and the time horizon for useful forecasts (Petchey et al., 2015). After first describing some motivating cases where a multispecies evolutionary approach is needed, I outline theory for predicting evolutionary dynamics of microbial communities, methods for tracking evolution in whole communities, and ways to connect evolution of component species to ecosystem functioning.

9.2 Motivating examples

Many evolutionary problems are tackled from a single-species perspective. But every time a species adapts to new conditions or evolves resistance to human control, it does so in a milieu of hundreds of other species. Three examples illustrate the importance of a multispecies perspective.

9.2.1 Managing marine food webs adapting to climate change and fishing

Evolutionary responses to climate change are widely documented, in addition to demographic responses, range shifts, and phenotypic plasticity (Merila and Hendry, 2014). Furthermore, it is recognized that phenological mismatches among interacting species can arise as climate changes (Renner and Zohner, 2018). If species have different thermal responses, the relative strength of interaction will change in warmer climates (Bestion et al., 2018). For example, consider the food web of marine organisms affected by warming and acidification in the North Sea (Fig. 9.1). Metabolic theory can be used to predict the outcome of warmer temperatures (O'Connor et al., 2009): species tolerance curves predict some species will grow faster, others more slowly, and this will alter the strength of interactions across the network and the risk of species loss. Equivalent theory on pH tolerance could be used to predict the joint effects of warming and acidification. But these changes also impose selection on each species (Moya-Larano et al., 2012). Can we predict how species will adapt to warming and acidification and the effects on food web properties, such as the biomass of larger fish?

Thinking of an alternative impact on the same system, can we predict how the food web will respond to a particular fishing strategy? Early approaches to fisheries management focused on populations of each species separately. More recently, multispecies strategies have been developed (Thorpe et al., 2016). For example, body size

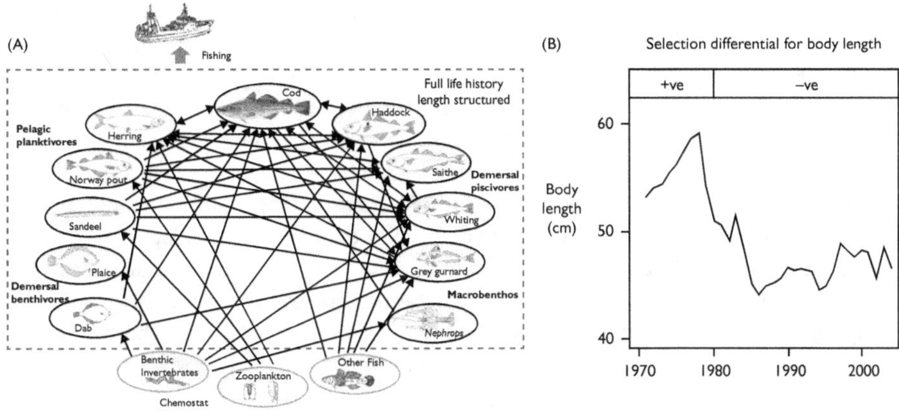

Fig. 9.1 Body size is a simple trait with an important role in structuring marine food webs. (A) Graphical representation of the length-structured model of the North Sea food web by Speirs et al. (2016) (reprinted under creative commons license). Model predicts food web responses to size-structured fishing regimes, but similar models could explore responses to climate change, mediated by mechanistic links between body size, metabolic rate, and temperature. (B) Cod body size declined in the North Atlantic over several decades due to a switch in selection differential on body size in 1980. Body length average for 6-year-old fish and selection differential for 4-year-olds are shown (redrawn using data from the Gulf of St Lawrence from Swain et al. 2007). What data would be needed to predict body size evolution of the whole ecosystem under different fishing and climate conditions?

versus abundance distributions are used as a simple, one-dimensional axis (Blanchard et al., 2012) for considering both food web interactions (larger fish eat smaller fish, total biomass across trophic levels depends on energy flow and efficiency) and fishing practices (size-selectively of nets). Models then predict the response of the whole food web to fishing (Speirs et al., 2016, Fig. 9.1A).

So far, the approach assumes fixed traits of species, but these could be integrated with evolutionary models of body size to predict interdependent ecological and evolutionary responses. Classical work showed that size-selective fishing led to a decrease in body sizes of North Atlantic cod over a few decades (Fig. 9.1B). It has since been confirmed that the decrease resulted from evolution, not just demographic changes in age structure of the population (Swain et al., 2007). Integrating evolution into food web predictions would require quantification of evolutionary genetic and phenotypic parameters for the constituent species, including plankton, as well as the whole fish community. This is unfeasible using gold-standard methods applied to genetically tractable model organisms or long-term studies of focal species.

The ramping up of genome sequencing technology opens up the possibility for whole-genome sequencing of the entire community and resequencing of population samples. So far, genome sequencing of a focal species, cod, has revealed high frequency of polymorphic tandem repeats that are heterozygous in 4 per cent of coding and 8 per cent of promoter regions, which has been argued to offer variability for rapid evolution (although not linked specifically to body size changes, Torresen et al., 2017). Further work revealed the presence of an inversion containing genes linked to survival at low salinities that protects local adaptation between connected open-ocean and fjord-dwelling populations (Barth et al., 2017). At present, genome data are mostly used descriptively and for explaining how current adaptations came about. But in a future awash with genomic data, there is scope for new methods for quantitative prediction; for example, can we predict the genetic potential of each species to adapt to warming, acidification, trophic interactions, and fishing pressures from its genome structure and variation? We cannot do this at present—obtaining a genome sequence is far from being able to predict how organisms will evolve in response to selection—but we should strive to make this feasible in the not too distant future (chapter 11.5).

9.2.2 Improving food security through the control of crop pests and pathogens

Nearly 35 per cent of global food production is lost annually to pre-harvest pests and disease (Popp et al., 2013). Tackling this problem is vital in order to meet rising food demands while protecting natural areas from increased exploitation. With 25 per cent of global primary productivity appropriated by humans (Krausmann et al., 2013), there are strong selection pressures for organisms to adapt to exploiting these resources. Pesticides are the front-line defence against crop pests, but their efficacy is reduced by evolution of resistance among insects, fungi, and other problem taxa (Gould et al., 2018). Also, they can disrupt other organisms, such as pollinators and natural enemies of crop pests, leading to unanticipated negative effects on crop yields. Can we predict the responses of the entire farmland ecosystem to pesticide application? Which species will evolve resistance? What is the optimal level or pattern of application to reduce the pest species, taking into account effects on interacting species?

Another method with interesting coevolutionary implications is biocontrol, which uses organisms that parasitize or consume crop pests to control them. These have the advantage that self-sustaining populations can establish and maintain low pest abundances by coevolving with the pests. Evolution of the biocontrol agent or pest (Tomasetto et al., 2017) or unexpected ecological interactions with other species can impair efficacy, however. For instance, the biocontrol agent can become a pest itself. The design of biocontrol programmes would therefore benefit greatly from the ability to predict ecological and evolutionary responses in the biocontrol agent, pest, and the wider farmland ecosystem of interacting species (Roderick et al., 2012).

Species diversity is again inherent to these problems. A diverse array of potential pest organisms resides in the environment, and controlling one pest simply opens up an empty niche for another to occupy. For example, the fungal genus *Fusarium* encompasses a large array of strains associated with disease of crop plants. Regular outbreaks emerge on particular crops—coffee, bananas, lettuce—and species such as *F. oxysporum* are known to infect a broad taxonomic range of plant species (Gwynne et al., 1997). The strains causing disease on a particular crop, for example Panama disease of bananas, can have multiple independent origins within the species complex (O'Donnell et al., 1998). Genome sequencing revealed that pathogenicity can be acquired through the transfer of mobile chromosomes that carry effector genes for pathogenicity between unrelated strains (Fig. 9.2; Ma et al., 2010; van Dam et al., 2017). Can we now apply genomic methods and phenotyping across the whole clade to develop predictive models of the emergence and spread of new variants, and then use this information to identify risk factors and manage pathogen loads? This would require characterization of the network of transfer of elements carrying effector genes, as well as understanding ecological adaptations of each species, for example to abiotic conditions.

9.2.3 Antibiotics and alternative treatments for disease-causing bacteria

A familiar example of evolution in action is the emergence of antibiotic resistance in bacteria. Roughly 10 years after the introduction of each new antibiotic for medical use, resistant strains of major bacteria causing disease such as *Streptococcus pneumoniae* and *Staphylococcus aureus* have spread (Clatworthy et al., 2007). Selection experiments on single species in the laboratory have tested the determinants of the spread of resistance (MacLean et al., 2010), including the effects of dispersal among metapopulations with different antibiotic exposure (Perron et al., 2007), the effect of combination drug therapies (Roemhild et al., 2015), and how resistance persists through amelioration of costs in antibiotic-free environments (San Millan et al., 2014). Such experiments also provide early warning for types of genetic change that will evolve.

While the focus is often on controlling a particular strain that causes human disease, the evolution of resistance in reality always occurs in a context with many other species (Fig. 9.3). The human body is home to many thousands of bacterial species. For example, *S. pneumoniae* is a major cause of pneumonia, but many other inhabitants of the nasopharyngeal tract, such as *Haemophilus influenza* and *Staphylococcus aureus*, also cause the disease. Vaccination programmes show that these bacteria compete for space and resources—reducing numbers of one strain by vaccination leads to its place being taken by an alternative strain or species, which consequently becomes a more important cause of disease (Hausdorff and Hanage, 2016).

Similarly, in cystic fibrosis (a genetic disease that causes a build-up of mucus in the lungs), serious lung infection is associated with the transition from a diverse lung microbiome to one monopolized (typically) by *Pseudomonas aeruginosa*. The bacterium adapts genetically during this time, including the acquisition of multidrug resistance in response to antibiotic treatment (Marvig et al., 2015; Winstanley et al., 2016). Other organisms are associated with later stages of infection, however, including anaerobes, and successful treatment of *P. aeruginosa* infections has led to the emergence of alternative pathogens (Surette, 2014), similarly to the outcome of vaccination for *S. pneumoniae*. Also, coevolution with *S. aureus* led to genetic changes in *P. aeruginosa* that make it more resistant to antibiotics (Tognon et al., 2017). These discoveries raise the potential for treatments aiming to manage the lung microbiome, for example with probiotic

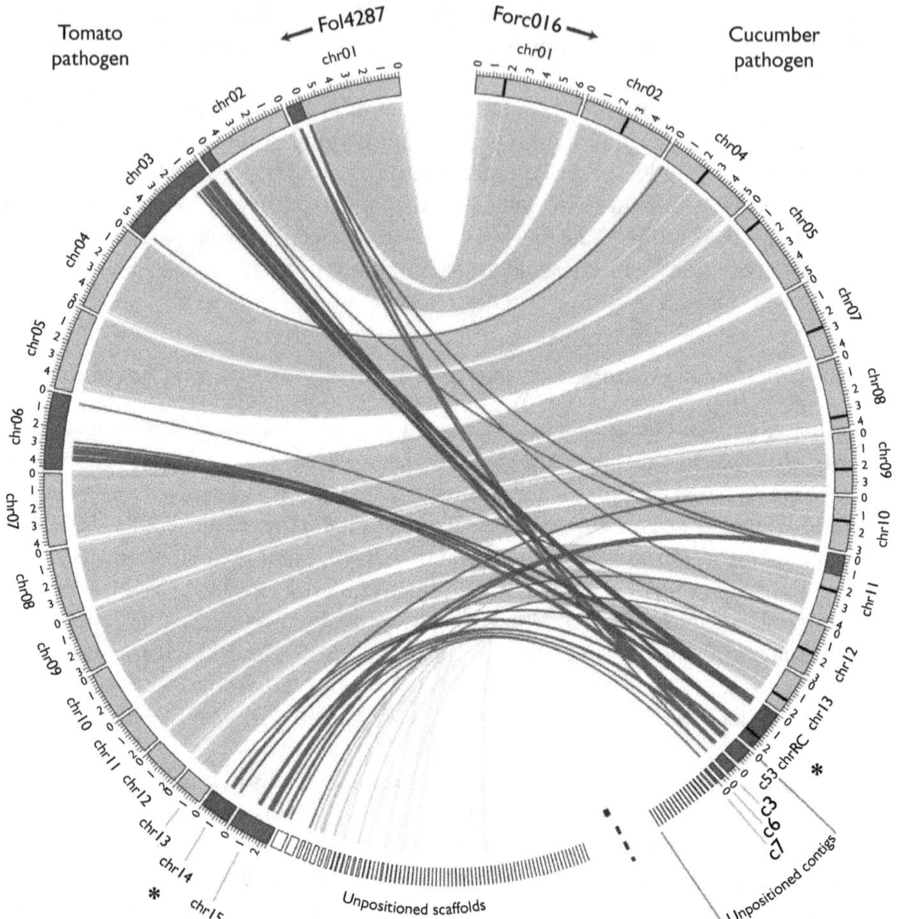

Fig. 9.2 Genomic basis of acquisition of pathogenicity via horizontal transfer in *Fusarium oxysporum*. Right part of circle shows the genome of a cucumber pathogen, *F. oxysporum* forma specialis *radicis-cucumerinum* strain Forc016 (forma specialis just means that the strain was isolated from the plant species with the following Latin name). The genome is mapped to the reference genome for a tomato pathogen, *F. oxysporum* f. sp. *lycopersici* strain Fol4287. The cucumber pathogen has 11 core chromosomes (light grey) that share conserved gene content and order with the reference genome (light grey curved connecting lines). It also contains one pathogenicity chromosome (called chrRC, dark shading, asterisk) that has fragmentary homology with multiple accessory chromosomes in the reference genome (dark shading), and contains a high density of candidate pathogen effector genes that are absent in the reference genome. The tomato pathogen genome has its own pathogenicity chromosome (chr14, asterisk) that shares little homology. All of the accessory chromosomes have a high density of repeat regions and low gene density. Experimental transfer of the mobile chromosome to a non-pathogenic strain confirmed transfer of ability to infect cucumbers. (Modified from van Dam et al. (2017) under creative commons license.)

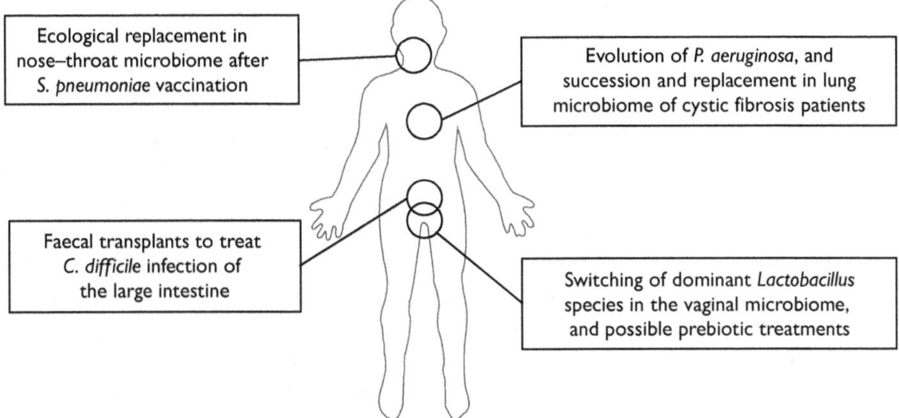

Fig. 9.3 Summary of examples where bacterial diversity is important for treatment of human disease. References supporting each example are in the text.

application of species associated with healthier lung function (Huang and LiPuma, 2016). The changes in body conditions or immune system that lead to commensal organisms becoming disease agents are not fully understood, but these occur in a community of interacting species (O'Brien and Fothergill, 2017). Understanding the ecological and evolutionary succession leading to adverse colonization of the lung could help to mitigate disease progression.

Another important bacterial community for human health is the vaginal microbiome. A survey of North American women revealed a relatively discrete set of community types (Ravel et al., 2011): the microbiome was dominated by either one of four different *Lactobacillus* species or a more diverse community. The last type included species traditionally associated with an unhealthy state, such as *Prevotella* and *Gardnerella*. Vaginal pH varies markedly across these community types, being much lower in *Lactobacillus*-dominated communities than the diverse community-type. A high pH is associated with higher fertility but lower protection from infection, whereas low pH is protective against disease, such as sexually transmitted infections like *Chlamydia*, but is associated with lower fertility because it is spermicidal. The community in a single individual can shift between types, for example from *L. crispatus* to *L. iners* dominated, but two *Lactobacillus* species cannot coexist both at high frequency for long (France et al., 2016). Reverse ecology from genome sequences inferred that the former thrives in sugar-rich environments, whereas the latter survives low-resource conditions by liberating resources from human cells. Intriguingly, predominance of *Lactobacillus* is unique to the human vaginal microbiome, being largely absent in other primates, for reasons yet to be determined (Miller et al., 2016). Prebiotics (providing resources for bacteria with beneficial properties) and probiotics (supplementing bacterial populations) are being explored as therapeutic treatments to inhibit the growth of pathogens (Al-Ghazzewi and Tester, 2016). How do species evolve during changes in community type? At present, there is no information, but this would make an excellent and tractable system for evolutionary studies in vitro or in vivo.

By far the most diverse body region is the large intestine, which harbours several kilos of bacteria, encompassing 10^{14} cells and between 500 and 1000 species. Collectively, these bacteria encode 150 times as many enzymes as the human genome (Qin et al., 2010) and play a major role in human physiology (Flint et al., 2012; Nicholson et al., 2012). The bacteria grow anaerobically on undigested polymers arriving in the colon, including polysaccharides such as indigestible starch, pectin, and inulin, which are collectively called fibre, and undigested proteins. The polymers are broken down by bacteria into numerous metabolites, of which the short-chain fatty acids acetate, butyrate, and propionate play a key role in human physiology (Zhao et al., 2018). These compounds provide an energy source (especially in traditional diets), contribute to healthy functioning of particular organs (e.g. propionate in the liver, butyrate for colon cells), and feed into hormonal pathways that regulate appetite and many other aspects of human health (Sleeth et al., 2010). No single bacterium is able to break down input material into these vital compounds; it depends on the collective actions of multiple species. The bacteria interact through competition for resources, space on the gut wall, indirectly via phage, and through cross-feeding (Robinson et al., 2010). For example, *Ruminococcus bromii* breaks down starch into acetate, which is converted to butyrate by *Eubacterium rectale* (Kovatcheva-Datchary et al., 2009).

Healthy gut functioning is strongly connected to the functioning of the bacterial ecosystem. Dysbiosis, represented by low bacterial diversity and an excess of disease-causing species such as *Clostridium difficile*, is associated with ailments such as irritable bowel syndrome and diarrhoea (Brandt et al., 2012). Impaired butyrate production has been associated with increased risk of colon cancer. Considerable efforts are therefore placed on designing interventions for this system. Prebiotics are food supplements that stimulate growth of beneficial bacteria, such as *Bifidobacterium* species that metabolize fibre and facilitate butyrate production (Delzenne et al., 2011). Probiotics introduce bacteria themselves, a successful treatment to redress dysbiosis following disease. More extreme is the use of faecal transplants, which represent whole-community samples from another person that greatly enhance recovery from *C. difficile* infection (Brandt et al., 2012). This raises the prospect that mechanistic knowledge of the community could be used to enhance functioning even before extreme problems arise, for example to reduce expansion of *C. difficile*.

These examples clearly demonstrate that bacterial disease in humans occurs within a microbial-community context. But the ecological and evolutionary mechanisms behind changes leading to disease states remain poorly understood. Sometimes evolution is documented, as in the cystic fibrosis example, but the selection pressures and effects of ecological interactions in reaching that end-point are not fully known. In others, such as managing *C. difficile* in the gut, we have little idea of how the focal species interacts with other species, or how much species adapt genetically to changing conditions. It is known that species differentially tolerate and adapt to antibiotic treatment—with *C. difficile* emerging as a winner—but what aspect of its genetics versus its ecological niche and species interactions determine that? Can we predict when it will expand to high numbers and when antibiotic treatment versus alternatives (e.g. multispecies probiotics) would work?

Beyond treating disease, we might wish to alter diet or use supplements to enhance production of beneficial metabolites and thereby health. But how will microbiome species respond to interventions? Will they adapt to use new resources and produce new metabolites? Gut conditions seem perfect for evolution—census population sizes are huge (on average ~ 10^{11}) and regular flow provides the resources for rapid generation times. Yet sheer diversity might act against evolution through mechanisms described in chapter 8. Whenever resources or conditions change, there might always be another species (perhaps at low density initially or in resting spores) to expand and take advantage of the opportunity, just as new crop pest species arise as climate and weather patterns change.

9.3 A model for predicting evolution in communities

All of the motivating examples would benefit from a theoretical and experimental framework for predicting evolutionary outcomes for sets of interacting species (Ghoul and Mitri, 2016). Is that possible without drowning in an overabundance of variables and parameters?

A critical part is deciding how to model species interactions. Traditionally, dynamic models of species interactions use a Lotka–Volterra approach by specifying a matrix of pairwise interactions (Clark and Neuhauser, 2018). This is problematic because interaction coefficients scale with the square of the number of species, they are impractical to measure directly at large scales, and they do not constitute a trait in a single species that can be assumed to evolve. A more tractable approach is to model the evolution of traits that determine species interactions. It is possible that interactions are determined by idiosyncratic traits depending on which species are interacting. This does not help much in terms of scale. A simplifying approach is to assume that a single axis can be used to represent ecological interactions among species, and that each species position on this axis is determined by a single trait. All other aspects of species identity and ecology are assumed to be negligible relative to this main structuring axis. An example would be the use of body size as a structuring trait for marine ecosystems (see section 9.2.1). Knowing the body size of an organism provides information on which species it can eat, or be eaten by, and its metabolic rate and growth rate. Clearly this is a gross simplification of species interactions in the sea, but perhaps it is sufficient to understand and predict major changes.

The next component is evolution. Evolutionary models of body size and food web structure have tended to incorporate assembly via speciation (a longer-term phenomenon) and invasion of mutant phenotypes—more like the colonization of new species—rather than evolution by genetic change along lineages that interact with one another (Loeuille and Loreau, 2005). Instead, to model body size changes as they might occur in a marine food web exposed to change, a gradual model of evolution of mean phenotype within each species could be considered. For this, an expression is needed to link mean fitness to the genetically determined body size of each organism: for example, the gradient of per capita growth rate of a species with respect to body size around its current mean body size. The response then depends on the product of the selection gradient and the genetic variance–covariance matrix for the trait (see section 8.2; Lande, 1979). Many details of evolution are simplified by modelling phenotypic changes in this way: the aim is to simplify as much as possible, while still considering the main structuring forces.

Perhaps a reason for the lack of development of such models is the difficulty of tracking evolutionary responses of multiple species over multiple generations, and consequently lack of datasets requiring such a model. Bacteria offer more feasible options because multiple generations occur over a period of weeks and new methodology is making tracking evolution viable (see section 9.5). I therefore focus on describing an approach for modelling evolution of bacteria communities, motivated by the case of gut bacteria described in section 9.2.

Barraclough (2019) introduced a model for evolving bacteria communities motivated by the case of gut bacteria described in section 9.2 (Fig. 9.4). Based on classical models of growth in chemostats (Stewart and Levin, 1973), the model assumes that resources flow into the system and that resource, metabolites, and cells flow out (Kettle et al., 2015). Bacteria grow by metabolizing input resources or substrates derived from the metabolism of input resources. The breakdown pathway is a predetermined feature of the system, but species vary in which substrates in the pathway they are able to metabolize. Species interactions result entirely from patterns of resource use: species growing on the same resource compete by scramble competition, whereas a species growing on a derived substrate is facilitated by the species that produces that substrate as waste (i.e. cross-feeding).

The trait underlying resource use is the level of allocation to different metabolic enzymes. Each species produces the same total amount of enzyme, but cellular enzyme can be allocated to different metabolites (Berkhout et al., 2013). There is therefore a linear trade-off between using one versus multiple resources: a generalist species produces proportionately less of a given enzyme than a specialist could. This assumption can be altered to allow non-linear trade-offs: for example, that there is an additional cost to being a generalist, leading to a further decrease in enzyme production relative to specialists. Enzyme allocation is the trait under selection: species evolve to use more of abundant resources and less of scarce resources. Consequently, there are eco-evolutionary feedbacks (Hendry, 2017): ecological interactions affect selection on the trait (via their effect on resource levels), and in turn ecological interactions are altered by evolution of the enzyme allocation among species.

Considering first the case of two species and a linear trade-off, coexistence requires differential growth on different substrates. A classical result is that the species that lowers the concentration of a limiting resource most—which was called R^* by Tilman (1982) when applied to plant ecology—will outcompete other species solely dependent on that resource. However, with two input resources, two species can coexist as long as one species metabolizes the first resource more effectively, and the second species metabolizes the second resource more effectively (Stewart and Levin, 1973). Allowing for evolution, species therefore evolve to converge in resource use towards more abundant resources while maintaining partial specialization to partition the resources (Fig. 9.5A). The evolutionary end-point depends on the starting conditions, because

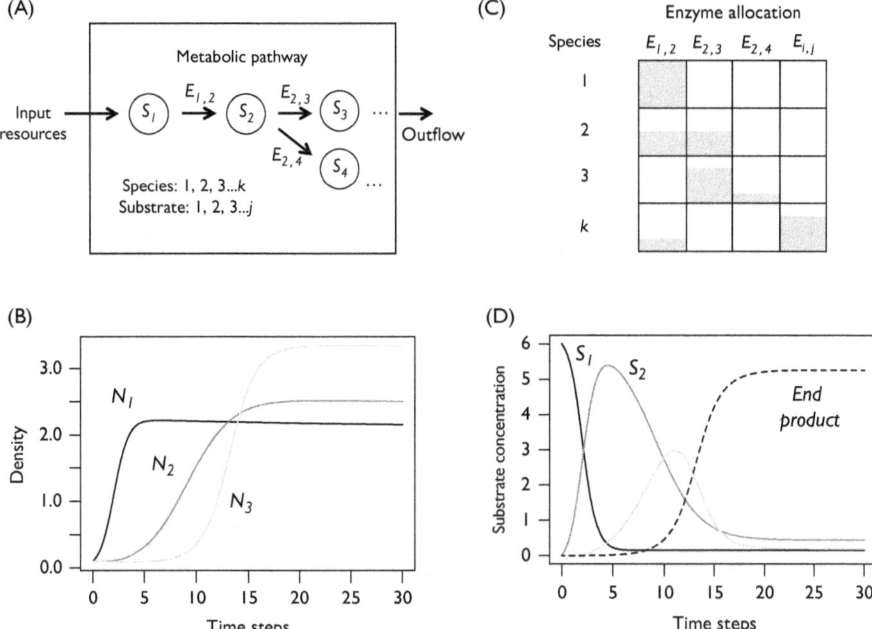

Fig. 9.4 Schematic of a chemostat-style model of gut bacteria metabolizing input resources into a series of derived metabolites. (A) Indigestible materials arrive in the colon, which for simplicity is treated as a single mixed chemostat. Substrates and cells flow out. (B) Bacteria grow by metabolizing chemical substrates, according to growth equations described in Box 7.1, and produce waste products that other bacteria can grow on. (C) Each bacterium has a fixed amount of metabolic enzyme per cell, but species vary in how much of that quota is allocated to each type of enzyme (defined as an enzyme metabolizing substrate i into substrate j), as shown by the allocation matrix. Species that overlap in metabolite use (e.g. species 1 and 2) compete for shared substrates, whereas species that use waste of another species exhibit a cross-feeding, facilitative interaction (e.g. species 3 uses substrate 2 produced by species 1). Enzyme allocation evolves in order to maximize the per cell growth rate of each species. (D) Substrates are used up, and waste products produced as bacteria grow.

the range of solutions permitting stable coexistence is very broad. Similar predictions apply for substrates connected by a metabolic pathway. For example, with two substrates for growth, the second of which is produced by metabolism of the former, both species evolve to use both of the resources, but with one species marginally more specialized on the former and one species more specialized on the latter.

The predictions change if there is a stronger, concave trade-off between specialism and generalism. If generalists grow proportionately worse than specialists, the two species evolve towards specialist phenotypes devoting most of their enzyme to growth on one substrate (Fig. 9.5B). Niche partitioning and character displacement are classical results for eukaryotic organisms (Case and Taper, 2000), and it seems intuitive that strong trade-offs would exist. For example, a generalist needs additional machinery

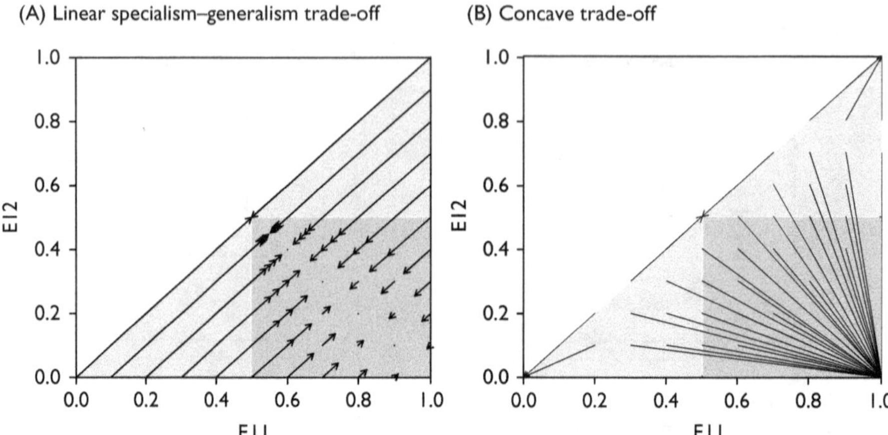

Fig. 9.5 Evolution of resource use in a system of two species partitioning the use of two separate input resources. (A) Linear trade-off between specialism and generalism, such that a generalist allocating 50 per cent of its enzyme to two reactions grows at 50 per cent the rate on each as corresponding specialists. Each arrow shows evolution in enzyme allocation during a single run, starting with different initial allocation of enzymes by the two species. Starting value of E11 (proportion of enzyme allocated to substrate 1 by species 1) is greater than or equal to starting value of E12 (proportion of enzyme allocated to substrate 1 by species 2). (Because there are only two substrates, proportion of enzyme devoted to substrate 2 is 1–E11 and 1–E12, respectively.) Input substrate concentrations and energy rewards from their metabolism are assumed to be the same for both substrates. Species converge in resource use but retain partial specialization. (B) A concave trade-off, such that the generalist allocating 50 per cent of its enzyme to two reactions grows < 50 per cent the rate as corresponding specialists would do. Most starting values evolve to the lower right corner, indicating divergence, so that species 1 is specialist on substrate 1 and species 2 is specialist on substrate 2. Dark grey region indicates the region of ecological coexistence in the absence of evolution; lower grey triangle, species 1 persists alone; upper grey triangle, species 2 persists alone. (From Barraclough 2019.)

to ensure expression of the two metabolic enzymes, and must replicate more DNA than a streamlined specialist, which adds a metabolic cost (Lynch and Marinov, 2015). Examples of evolutionary diversification with cross-feeding support the assumptions of the trade-off: *E. coli* grown on glucose diverges into two ecotypes, one that metabolizes glucose and one that metabolizes glucose and acetate, rather than a single genotype metabolizing both resources (Herron and Doebeli, 2013).

These predictions can be extended to multispecies communities growing on multiple input and derived substrates, as mirrors resource supply in the human gut (Fig. 9.6). With a linear trade-off, species converge towards similar enzyme profiles matching the profitability of different resources passing through the system. With a stronger trade-off, a series of specialists perform each step. The model can be used to investigate the interplay between ecological determinants of responses, such as the starting resource use of each species, versus evolutionary determinants, such as the

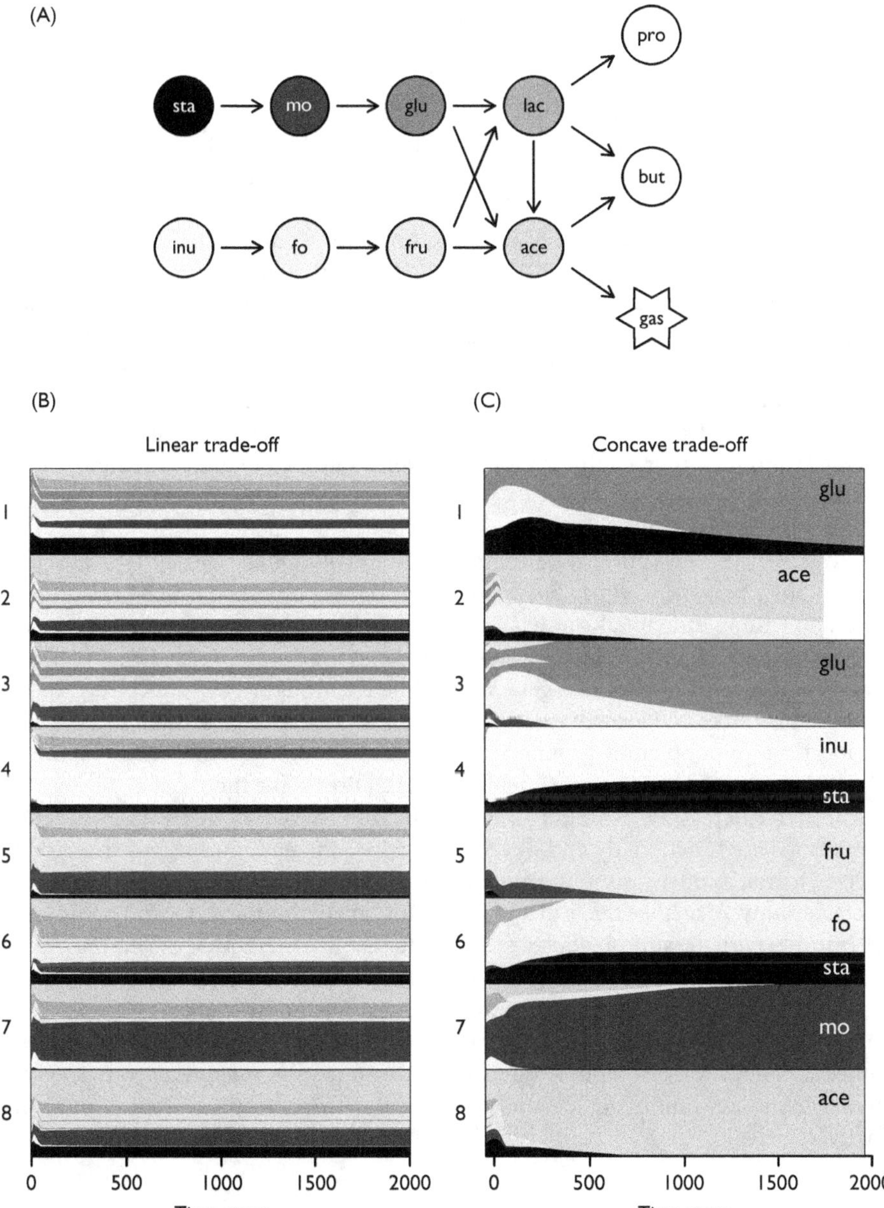

Fig. 9.6 Evolution of resource use in a simplified model of gut microbiome. (A) The model considers two input resources: starch (sta), which is metabolized via malto-oligosaccharides (mo) and glucose (glu) into lactate (lac) and acetate (ace), and inulin (inu), metabolized via fructo-oligosaccharides (fo) and fructose (fru) into the same derived metabolites. The other two short-chain fatty acids, propionate (pro) and butyrate (but), plus gas, are produced from

evolutionary rate of each species and genetic correlations among different traits (Barraclough, 2019).

Clearly, many other features will be needed in order to attempt prediction in real communities. Evolving β-niche traits such as pH tolerance of each species could be added in addition to resource use, where pH is modified by metabolic activity of each species as well as by external environment and host secretions in the example of gut microbiomes. Biotic niche traits such as production of antimicrobials or allelopathic chemicals could also be added. For example, for a simple case of two interacting species, Lawrence and Barraclough (2016) modelled allocation between metabolic enzyme versus production of a costly antimicrobial between two competing species. It was generally more beneficial to invest in acquiring resources than to expend metabolic energy on antimicrobials. Additional interacting agents could be included such as plasmids that transfer metabolic or other traits among species, or bacteriophage, which would lead to additional indirect interactions among the bacterial species (Koskella and Brockhurst, 2014). A lot more work is needed to extract general understanding of these models.

There are further challenges to extending evolutionary models to multispecies perspectives. One is the appropriate units of time. Evolutionary theory typically uses generations as a unit of time. For comparing multiple species, which may vary in generation time, individual models can be scaled by generation time to represent evolution in units of time that are comparable across species (e.g. days, weeks, or years). But in many organisms, generation time itself might vary depending on the access to resources and growth rates. Deciding how to map from growth rates to generation time to population genetic processes of mutation and selection, especially when models are linked to experimental data, requires some thought.

Another interesting problem is how selection connects to traits and genes. Some traits are functionally and developmentally coherent entities, but similarly to 'species', a 'trait' is often a simplification of a multidimensional entity, chosen to permit tractable study. Also, the effects of a given change in environment depend on how the organism experiences those effects (see chapter 8). But predicting evolution is hard without being able to separately measure environment and effects—in how many cases can we predict an optimum in a new environment other than retrospectively observing that a population evolved towards that trait value? In a multispecies context, much more work is needed to understand the mapping from external environment conditions, to selection on the organisms, to phenotypic traits, and finally to genes.

Fig. 9.6 Continued

lactate and acetate. The model was started with eight species present with random allocation of possible enzymes. (B) Evolution in enzyme allocation with linear trade-off over 2000 time-steps: species evolve limited specialization and all of them retain a degree of generalization on all resources. (C) Evolution with concave trade-off leads to much greater specialization of each species onto one or two substrates. (From Barraclough (2019).)

9.4 Evolution and ecosystem functioning

One of the main motivations for being able to predict ecological and evolutionary responses in whole communities facing environmental change is to manage ecosystem functioning. Productivity, nutrient cycling, decomposition rates, and metabolite production in the gut are all predictable in theory from the set of species and their traits found in the local ecosystem (Kettle et al., 2015). If the abundance of different species changes or their functional traits change via evolution, this will change overall functioning of the ecosystem.

One interesting question is whether evolution of constituent species tends to enhance aspects of functioning of the whole community (Barraclough, 2015). From one perspective, there is no reason to expect that evolution will necessarily improve functioning of the whole community or ecosystem. Selection acts on genotypes to enhance individual fitness, not the performance of the whole community (Bell, 2007). The best strategy for rate of resource acquisition for particular species might be detrimental across the wider community, as evident through 'selfish' strategies that reduce benefits across the entire system.

Alternatively, evolution might tend to improve functioning of communities. A defined set of resources are available for metabolism, and selection should act to increase the utilization of those resources—underutilized resources provide an opportunity for a species to exploit. Irrespective of which species use particular resources, selection should increase the rate or efficiency of metabolism of the resources (depending on whether resources are regularly replenished or in limiting supply, respectively), which will translate into high measures of overall functioning such as respiration rate, nutrient cycling, or decomposition. A set of species that initially use resources relatively inefficiently provides the opportunity for spread of new genotypes with enhanced resource use, which collectively should increase functioning for the whole community.

The model outlined in section 9.3 can be used to investigate when evolution enhances or impairs measures of overall functioning. Under the assumption of a linear trade-off among species, a community of generalist species tends to yield a higher metabolic rate for a decomposition pathway than a community of specialists (i.e. lower starch and higher butyrate concentration; Fig. 9.7). This is because performing the whole pathway requires intermediate steps that do not yield enough energy to sustain a population to still be performed adequately. With specialists, those steps are not sustained, as the specialist performing them dies out. With evolving generalists, they are not sustained either, as the generalists allocate enzymes away from less-profitable steps. Therefore, evolution reduces functioning on average—even though the collective metabolic rate and productivity of the whole system would be higher if species did devote enzyme to performing those steps.

In contrast, with a concave trade-off and a cost of generalism, overall functioning is enhanced by resource specialization of different species, which is also optimal for individual selection on those species. Now evolution tends to enhance functioning of the whole pathway relative to baselines for non-evolving communities (i.e. lower

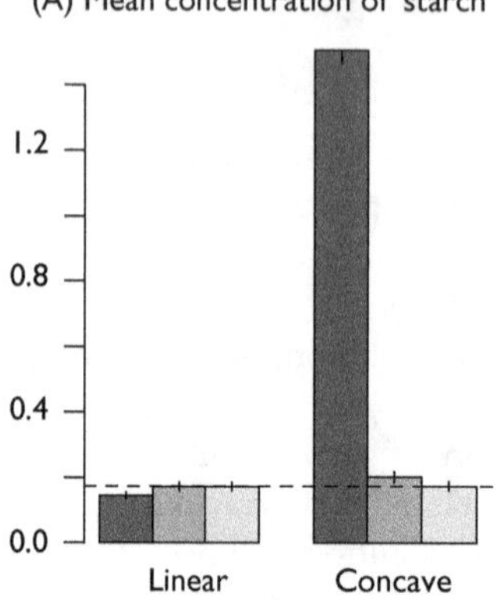

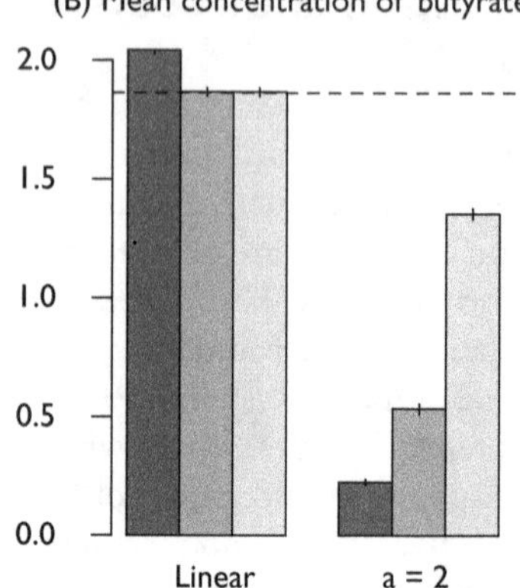

Fig. 9.7 Average metrics of the model of gut functioning outlined in Figs 9.4 and 9.6. (A) Average concentration of starch. Higher values indicate less-efficient degradation. (B) Average concentration of butyrate—the short-chain fatty acid with known benefits for colon health. Values summarize averages from repeated trials starting with different initial random allocation of enzymes across species. Dark grey indicates non-evolving generalists; mid-grey, evolving communities that start as generalists; light grey, evolving communities that start as specialists. Standard errors are shown. Dashed line indicates non-evolving community of specialists. (From Barraclough 2019.)

starch and higher butyrate concentrations; Fig. 9.7). The system still does not evolve to the theoretical optimum level of functioning, however. Furthermore, there are many aspects of functioning of interest, and some are reduced by faster overall rate of metabolic pathway—for example, if the focus is on production of an intermediate metabolite, evolution might reduce standing levels of that metabolite by increasing its rate of degradation. The answer depends on which metrics of functioning are of primary interest.

Experimental evolution with microbial communities tends to support the finding that overall ecosystem functioning increases with evolution. Lawrence et al. (2012) found that niche partitioning and cross-feeding of coevolving species (see chapter 8) led to increased overall respiration rate for the microcosms compared to isolates that evolved in monocultures. Similarly, Fiegna et al. (2015a) found that functioning measured as overall yield increased as microcosm communities adapted to new environments, again inferred to result from the evolution of less-negative interactions among constituent species, consistent with niche partitioning.

The Jena long-term experiment on flowering plant communities described in section 8.5 also found evidence for increasing functioning due to evolution and specifically niche partitioning. Reassembled communities of evolved species displayed higher combined biomass yield than communities assembled with naïve representatives of the same species (van Moorsel et al., 2018). The effect was only pronounced, however, for low-diversity plots with two or four species, which matches the predictions and observations described in chapter 8 that coevolutionary effects might decline in more diverse systems. Nonetheless, these experiments show that evolution is important for ecosystem functioning over timescales relevant for management even of eukaryote communities—responses to land-use practices and climate change need to consider evolutionary interactions within diverse communities. Other eukaryotic systems might be open to tracking multigenerational evolution, such as time-series of plankton revived from resting eggs from lake sediments (Decaestecker et al., 2007) and ancient environmental DNA from sediment cores (Balint et al., 2018), or by analysing genetic or phenotypic changes in historical collections in museums over past decades or centuries.

Most real communities are open to colonization from outside, and this will introduce additional complications. Similar arguments could be made for the effects of colonization—will it tend to enhance or impair community functioning (Strayer, 2012)? It could be argued that colonists that are able to persist and establish necessarily must use resources more effectively than existing species, and so overall rates of resource use must necessarily increase. However, there are numerous examples of invasive species degrading ecosystem functioning—for example, invasive plant species disrupting hydrology by blocking streams—again indicating that there are many more aspects of ecosystem functioning than simply derive from resource use.

One interesting aspect of predicting functioning is that the effects of evolution depend on the starting properties of species. In experiments, starting properties are often arbitrary and rather unrealistic—species start away from adaptive optima and are not already pre-adapted to the growing conditions. In the plant biodiversity experiments, arbitrary seeds are used and the plants are not initially coadapted to just the set of species found in their plots. Akin to the understanding of speciation described in chapter 5, the starting conditions and responses are part of an ongoing dynamic, and studying involves imposing an arbitrary 'start' and 'end'. In order to make general predictions of the outcome of evolution in communities faced with environmental change, or the effects of colonizing species, we need to predict the properties of species at the start, as well as how they will respond to the environmental change of interest.

The responses of species to current or future changes will be further shaped by the history of how the species originated and changes they experienced in the past. For example, genetic architecture of related species varies depending on whether they evolved in the presence of gene flow between differentially adapted populations (see chapter 7), and whether they experienced correlated or uncorrelated environmental fluctuations in the past (see chapter 8). Even if the species do not currently interact, the genetic potential of each species to adapt to a given change will be shaped by the history of their divergence and past interactions between them.

9.5 Tracking evolution in whole communities

The main limitation to opening up this area at present is a shortage of suitable data. New methods are needed to track evolution in whole communities in order to test the kinds of predictions outlined in sections 9.3 and 9.4 and to validate the assumptions of the models. Understanding of the diversity of microbial communities has been revolutionized by applications of DNA sequencing methods (Thompson et al., 2017). Species composition can now be measured readily at high temporal and spatial resolution by sequencing marker genes such as 16S. The initial applications of this technology focused on observation of natural variation (Ravel et al., 2011; Huttenhower et al., 2012), but combining this method with experimental perturbation of whole communities is now needed to gain more mechanistic and dynamic understanding. For example, replicates of a whole microbial community can be exposed to a change in environmental conditions such as a shift in resource availability or an increase in temperature. The ecological responses of component species can then be readily tracked over time. Metabolic functioning can also be measured easily through high-throughput assays of productivity or respiration rates, and metabolomic measures of the concentrations of multiple metabolites (Frost et al., 2014).

The hard part is to measure whether surviving species evolved to adapt to new conditions. A traditional approach would be to isolate species from the starting and final communities and perform single species assays or perhaps in factorial combination with the starting and final communities to determine phenotypic responses. This is laborious to perform manually in communities with many hundreds of species, and only works if the majority of species can be cultured in isolation. Single-celled isolation methods and the use of genomics to design culture conditions might increase throughput enough to make this a practical approach (Browne et al., 2016). An alternative is to label focal species. For example, fluorescently labelled *E. coli* has been used to investigate adaptation to life in a mouse gut. Transposable elements played a large role in causing changes to metabolism and the balance of aerobic and anaerobic respiration in the same underlying genes across replicate experiments (Barroso-Batista et al., 2014).

An obvious alternative is to use metagenome sequencing to detect genetic changes in constituent species between the start and the end of an observation period. For example, Minot et al. (2013) tracked the gut metagenome of a single adult individual over 30 months and tracked eco-evolutionary changes in bacteria and their phage. As well as inferring changes in frequencies of different types, they took advantage of the bacterial clustered regularly interspaced short palindromic repeats (CRISPR) system of defence to infer coevolution. DNA sequences from invading viruses are incorporated as spacers into the bacterial genome, which then allows CRISPR spacer RNAs to target destruction of the invaders. Spacers were found in bacterial genomes that matched to viral genomes recovered from the metagenome. One case was observed of a virus gaining a base mutation in the target region, and the mutant version increased in frequency relative to the original (Fig. 9.8), which was putatively interpreted as Red Queen dynamics of the phage evading the defence system.

Subsequent studies have reconstructed genetic evolution in bacterial species over time. Zhao et al. (2017) sequenced genomes of *Bacteroides fragilis* from seven individuals over 2 years. They uncovered parallel changes in cell-envelope biosynthesis—which could be evolving in response to phage attack or the host immune system—and

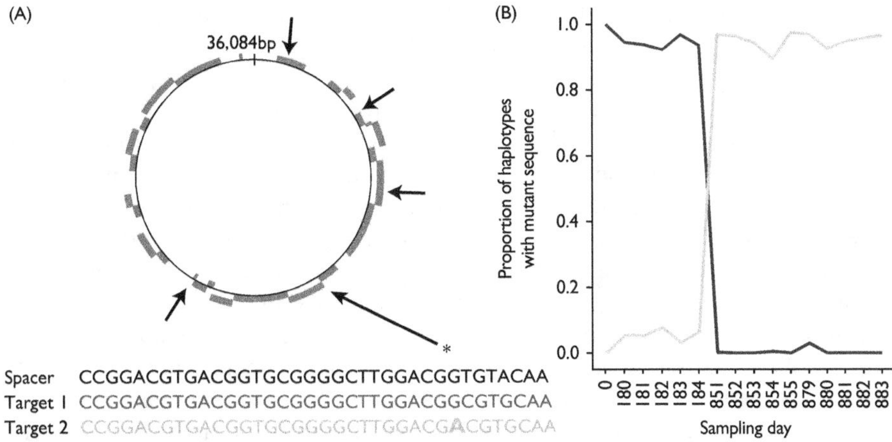

Fig. 9.8 Putative example of bacteriophage evolution in human gut microbiome. Bacterial CRISPR spacer regions could be matched to target virus genomes assembled from metagenomes sampled during a 2.5-year survey of one individual. (A) CRISPR target sites mapped onto a reconstructed virus genome. (B) A mutant form of virus (grey line) arose later with a single base-pair mutation in a target region. It eventually replaced the earlier one (black line). (Redrawn from Minot et al. (2013) with permission.)

in polysaccharide use—which might reflect adaptation to fluctuating levels of different resources (section 9.3). Similarly, Garud et al. (2017) tracked genotypic dynamics in 30 common species across Human Microbiome Project samples at multiple times from the same individual. They detected selective sweeps of single nucleotide variants or single gene gains or losses that were interpreted as within-host evolution rather than replacement by existing variants. Some events involved recombination. These dynamics predominated over 6-month timescales, whereas species replacement seemed to be the dominant force of change over longer, decadal timescales. Bendall et al. (2016) found similar results tracking bacteria species in a lake over 9 years.

The challenge is to connect changes to particular phenotypic effects. For some traits such as resource use it might be possible to infer that a particular SNP leads to upregulation or downregulation of a given metabolic enzyme, but experimental demonstration of the effect of each genetic change is not feasible across the whole community. One alternative way to infer phenotypic evolution is to perform whole-community assays to measure growth rates of constituent species together in the initial and final environmental conditions. This is hard to interpret because the growth of each species will differ in part on its interactions with other species, which also differs between the start and the end. One possibility would be to use time-series and perturbation data to fit a model of species growth and interactions under the starting conditions (c.f. Venturelli et al., 2018). That model could then be used to predict interactions among surviving species and see whether changes in resource use or kinetic parameters are necessary to fit time-series data collected at the later time-point.

An experimental approach should be beneficial as well in understanding determinants of evolution in complex systems. Johnson et al. (2019) made a preliminary attempt to tie together these approaches to investigate adaptation of human gut communities to altered resource use. Three communities seeded from faecal samples from three healthy volunteers were cultured in chemostats and exposed to a shift from a typical mixed diet to a diet consisting solely of inulin as a carbon source. Inulin is widely used as a prebiotic to stimulate the growth of beneficial bacteria and production of beneficial metabolites, and it was thought that selecting communities on this resource might produce useful probiotic communities to use in concert with inulin—a so-called synbiotic approach. The composition of each community was tracked over time using 16S sequencing, metabolite production was tracked using nuclear magnetic resonance spectroscopy, and whole metagenomics sequencing was used to search for genetic changes in the predominant species. The shift to inulin as the sole carbon source led to a major shift in the species composition of each community and decline in species richness (Fig. 9.9A): two communities became dominated by *Megasphaera elsdenii*, whereas the other was dominated by *Acidaminococcus*, both species associated with growing on lactate as a metabolite, which indeed increased in concentration in the chemostats. Other species such as *Bifidobacterium* species that metabolize inulin recovered over longer timescales of several weeks, as did the balance of metabolites. Metagenome sequencing revealed single nucleotide changes in *Megasphaera* between the start and end of the inulin treatment, including some in metabolic enzymes of potentially relevant pathways (Fig. 9.9B).

Similar studies are needed for eukaryotic communities. A key challenge there is how to track evolution over a timescale of hundreds to thousands of generations—too long for contemporary observation in most animals and plants, but too short to resolve from the fossil record or infer from phylogenetic information. Imagine the rewards when the challenges for measuring evolution across multiple species in real communities are solved. A new scale of evolutionary inquiry would open up. One avenue would be comparative investigation of contemporary evolution—why do some species evolve more or less than others, which species are able to survive periods of change, and how do interaction networks shape evolutionary responses?

Another direction would be to audit the sources of selection and responses of populations over an extended period of time. What proportion of selection pressures faced by each population derives from the surrounding community of co-occurring species versus the physical environment? The answer will determine what the typical evolutionary dynamics are for that species and across all the species in the community. The classical Red Queen scenario assumes that the dominant selection pressures come from coevolution with other species—even in a static physical environment, there would be continual selection in order to keep pace with competitors and antagonists (Voje et al., 2015). Alternatively, the so-called Court Jester scenario proposes that biotic interactions come and go, and the main selection derives from changes in physical environment, which occur continually and at varying spatial and temporal scales (Ezard et al., 2011). This distinction has been considered for fossil records. It is equally germane for contemporary communities, yet remains unknown for any

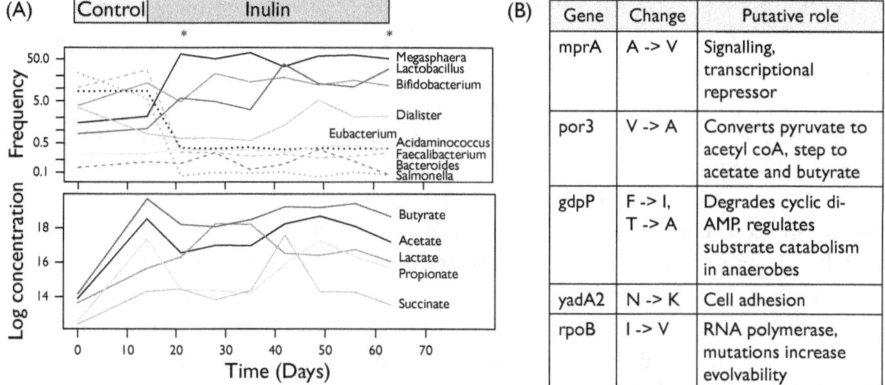

Fig. 9.9 Experimental evolution of whole human gut communities in vitro. (A) Cultures were initiated from faecal samples from a healthy volunteer, and cultured on a control diet indicative of resources typically reaching the large intestine over 2 weeks, followed by a shift to medium with inulin as the only carbon source. Relative frequencies of a selection of bacterial taxa and concentrations of a selection of key metabolites are shown. (B) Metagenome sequencing shortly after switch to inulin and at end of experiment (sampling times indicated by *) revealed changes in amino acid sequence (using standard one-letter codes) in several genes of possible relevance to experimental conditions. Drawn with data from Johnson et al. (2019).

community. Model communities are needed in the same way that model species have provided great insights into genetic mechanisms of evolution. There might also be clues from retrospective analysis of the genetic architecture of traits, in terms of which types of traits display architecture conducive to rapid evolution.

9.6 Do species matter?

Within the framework outlined in the earlier sections, species are the functional and evolutionary units for ecosystem functioning and responses to environmental change. But is this correct? Functional redundancy is a common feature of ecological communities. The exact species membership might change, but aspects of overall functioning remain robust because multiple species are able to perform overlapping aspects of function. Do species really matter for predicting the functioning of whole ecosystems? Perhaps aspects of overall functioning derive as an automatic consequence of there being a set of organisms that strive to use available resources, irrespective of how functional traits are packaged into species. This possibility is of particular relevance for microbial communities, where various approaches to investigate functioning by sampling genes and metabolites irrespective of which species they belong to. For example, metagenomics looks at the profile of enzymes present in the community and thereby attempts to predict functioning (Qin et al., 2010). Similarly, flux-balance analysis applied to whole communities seeks to predict aspects of functioning based on simplifying assumptions of overall optimality for the community (Shoaie et al., 2013). Does this work? Can we predict functioning and dynamic responses from the soup of genes (i.e. functional traits) present in a community? Or is the packaging of traits into species important?

The model in section 9.3 can be used to investigate this question (Barraclough, 2019). By holding fixed the composition of enzymes and their kinetic parameters, but varying how they were allocated to species, the effects of species as receptacles of functional traits could be investigated. Despite quite wide functional redundancy in these models, the packaging of genes into species did affect overall functioning (Fig. 9.7). This was because the ecosystem-level concentration of different enzymes is constrained by the growth of individual cells, rather than being free to adjust to optimal levels of functioning. Packaging introduces constraints into the growth of species. Allowing evolution of trait composition still led to an effect of species packaging—further constraints such as whether particular trait combinations are evolving.

In a community context, species are both functional and evolutionary units of diversity. Arguably this is the clearest context where species are relevant and important units—free from the ambiguities of hierarchical levels of diversity that afflict species delimitation and speciation (chapters 2 to 6), these are units affecting how living systems respond to environmental change.

9.7 Conclusions

Predicting evolution in the wild requires knowledge of the effects of interactions in diverse communities on evolutionary outcomes. Furthermore, these dynamics are relevant for understanding how ecosystem functioning relevant to humans change over management-relevant timescales of weeks (in bacteria) to decades (in animals and plants). Very little theory or evidence is currently available to make such predictions, in part because it is hard to track evolution over multiple generations across all the species in a diverse assemblage, but the necessary approaches are starting to emerge. The time is ripe to solve these problems and incorporate evolutionary dynamics into multispecies and ecosystem studies.

10

How does species richness accumulate over time?

10.1 Introduction

Why are there more species of animals and plants in the tropics than in equivalent areas of higher latitudes? Why are there so many species of beetles, but so few species of flying frogs? Why are there more species of benthic marine organisms than pelagic? Is the number of species on earth still increasing over time or has a plateau been reached? These questions illustrate that species are the fundamental units for considering broad spatial scales and longer timescales in the evolution of biodiversity. Studies of the fossil record often use higher taxa such as genera or families as units, but this is because of practical necessity—if species-level data were available they would be used in preference. Furthermore, even if data were available to resolve the number of populations over comparable spatial and temporal scales, I wager most people would still use species in preference as the evolutionary unit of diversity.

This chapter considers the evolutionary processes behind the proliferation and demise of species diversity within lineages and geographical regions. The focus is no longer on evolutionary dynamics within species, but on the dynamics of numbers of species over geological timescales. There has been an explosion of interest in this field over the last few decades, fuelled by the expansion of phylogenetic methods and data (e.g. Bininda-Emonds et al., 2007; Jetz et al., 2012; Rabosky et al., 2018). At the same time, statistical models of species birth and death, developed initially for analyses of fossil datasets, have been adapted to infer evolutionary processes from phylogenetic reconstructions of extant taxa (Nee et al., 1994; Morlon, 2014; Harmon, 2018). These approaches expanded the scope for inferring evolutionary processes behind diversity patterns greatly, because many extant clades (including, sadly, all the taxa that I have worked on) lack a resolved fossil record. Yet caution is needed in interpretation, since these methods infer processes purely from the shape of relationships among extant species, without any direct evidence of extinction (Rabosky, 2010). I discuss these concepts under three headings—rates, opportunity, and turnover—which lead into discussion of alternative models and concepts for large-scale and long-term diversity patterns.

The Evolutionary Biology of Species. Timothy G. Barraclough, Oxford University Press (2019).
© Timothy G. Barraclough 2019. DOI: 10.1093/oso/9780198749745.001.0001

10.2 Diversification rates

10.2.1 Background theory

Species richness refers to the number of species within a defined sampling unit such as a clade or a geographical region. The theory for changes in species richness within a clade developed by analogy to theory for changes in the number of individuals within a population. Simply, the number of species, N, at time $t+1$, is the number of species at time t plus new species resulting from speciation events and minus species lost through extinction:

$$N_{t+1} = N_t + (b - d)N_t$$

where stochastic models are used, and b and d are the average per lineage speciation rate and extinction rate, respectively.

The temporal dynamics of species richness and shape of phylogenetic relationships among species depend on how speciation and extinction rate are patterned over time and across the clade. The simplest model is that both rates are constant over time and across species, called the constant birth–death model. In this case, assuming $b > d$ and the clade avoids chance extinction when it contains only a few species, the clade grows exponentially without limit, at a net diversification rate of $b-d$. The expected variation in species richness among clades of equivalent age is geometrically distributed (Fig. 10.1; Nee et al., 1992). This means that considerable variation in species richness is expected among lineages even if there is no inherent difference among them in their propensity to diversify, because of chance and the compounding effects of exponential growth. If significant variation is observed relative to the null expectation, for example a clade has more species than expected, such as rodents and bats in Fig. 10.1, then in this framework it could result from higher speciation rate or lower extinction rate than other clades.

A range of more biologically realistic models can be specified, limited only by the imagination of the researcher (Morlon, 2014). For example, a sub-clade might acquire a different speciation rate either because it encounters new conditions or because it evolves new traits that make speciation more likely, which would lead to expansion of that clade relative to the other lineages. This creates imbalance in the distribution of species among sub-clades relative to the constant birth–death model and also boosts the net diversification rate at the time the change occurred. Alternatively, speciation rates might decline or extinction rate might increase as the clade grows, if there are density-dependent effects as the clade occupies available geographical or ecological space (Phillimore and Price, 2008; Rabosky, 2013). This is often called a niche-filling model. Density-dependent speciation rates have been demonstrated in planktonic foraminifera, one of the few clades to have a fossil record that is resolved sufficiently to infer ancestor–descendent relationships at the species level (i.e. phylogenetic relationships of those species) (Ezard et al., 2011). In this model, equivalent to a logistic density-dependent model for population growth, variation in species richness among clades might now result from different 'carrying capacities' in the number of species,

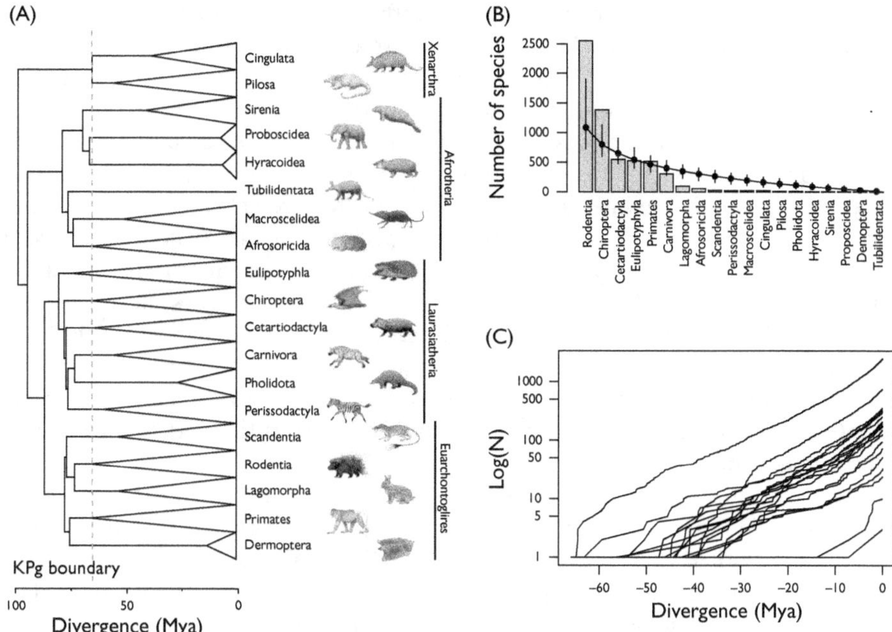

Fig. 10.1 Example of application of constant birth–death models to diversification of mammal orders. (A) Reconstructed and dated phylogenetic tree of the orders of placental mammals (reprinted with modifications and permission from Foley et al. (2016)). To a close approximation, all 19 living orders were represented by a single common ancestor at the Cretaceous Paleogene (KPg) boundary 66 Mya (dashed line). (B) Bars indicate number of species in orders of mammals ranked by their species richness (data from Burgin et al. (2018).) Dots and line indicates the median number of species expected for ranked orders under an equal rates null model: equivalent to apportioning total number of species among orders according to a broken stick model. Ninety-five per cent intervals are also shown. Rodents and bats have significantly more species than expected from uniform branching rates across orders, and there are many orders with lower species richness than expected, notably in Afrotheria. (C) Log of number of lineages over time reconstructed from phylogeny of extant species for one random trial of 19 lineages diversifying over 66 Myr according to a constant birth–death model. Simulations used the sim.bdtree function in the R package geiger (Harmon et al., 2008), a net diversification rate of 0.187 per species per million years, and an extinction rate of 0.1 per species per million years (De Vos et al., 2015). Final numbers vary greatly across lineages, despite equal average rates.

based on the geographical area available to them or the amount of ecological resources their lifestyle provides access to. This can lead to fundamentally different patterns in tree shape from a model in which clades are growing or shrinking based on diversification rates but without limits on the numbers of species. The effects of limits and species turnover over time are explored further in sections 10.3 and 10.4.

One complication for inference is the reconstruction process. A phylogeny reconstructed from molecular data represents the relative timing of branching events leading to extant species, with lineages leading to extinct species 'pruned' from the tree

(Nee et al., 1994). Under the constant birth–death model, this introduces a 'pull of the present', whereby recent branching events are less affected by extinction than deeper branches. Plotting the logarithm of the number of lineages in the tree against time towards the present results in an upturn towards the present, from a slope of $b-d$ towards a slope of b at the present (Fig. 10.1C). There is a perceptual bias towards interpreting reconstructed trees in terms of 'growth' of number of species, because inevitably the number of lineages increases towards the present (Ricklefs, 2007). The reconstructed phylogeny of a clade in steady decline from an earlier hey-day still 'grows' from root to tips. In principle, extinction rate can be estimated from its effects on the shape of the tree (Nee et al., 1994), but so many other processes affect the shape of trees in similar ways that this is a challenging task (Barraclough and Nee, 2001; Rabosky, 2010). For instance, the signal of recent up-turn indicative of background extinction could also result from a recent increase in speciation rates, a recent decrease in extinction rates, or a mass extinction event occurring at the start of the up-turn (Barraclough and Vogler, 2002).

The ability to fit complex diversification models has increased massively over recent years, but great care is needed in evaluating those models. Even when based on multiple markers, phylogenetic trees contain many blips and wrinkles in reconstructed dates due to processes such as saturation of markers, uncertainty in fossil or other priors, and other errors. With likelihood and Bayesian approaches, a model can yield a major improvement in likelihood even if just one aberrant data point is explained better. They tell you which of a set of focal models is best supported by the data, but they may be less robust to unaccounted-for errors or processes than old-fashioned frequentist approaches requiring replication to support significance. In some cases, for example in estimating extinction rates, it is debated whether there is sufficient signal in the available data to discriminate processes as claimed (Rabosky, 2010).

Despite the challenges, birth–death theory has proved extremely useful in providing a logical framework for analysing diversity patterns. The focus has been on statistical inference and providing methods to test the kinds of questions posed in this section, rather than on conceptual or mechanistic theory (Morlon, 2014). The models do not make predictions about what sorts of diversity patterns should be observed or when speciation or extinction rates should vary, but they allow alternative hypotheses on these quantities to be tested. The hypotheses themselves have typically come from verbal theory often itself motivated by informal empirical evidence.

10.2.2 Causes of variation in diversification rates

Under the framework outlined in section 10.2.1, one explanation for observed variation in numbers of species among clades is that lineages vary in characteristics that affect the probabilities of speciation or extinction (i.e. net diversification rates). For example, observations that sexual selection by female choice promotes reproductive isolation led to the hypothesis that strong sexual selection might increase speciation rates and diversity. Barraclough et al. (1995) tested this hypothesis by comparing a set of sister clades of passerine birds that differed in their degree of sexual selection. They used the proportion of species in each clade that were sexually dichromatic as a surrogate for strength of sexual selection by female choice. There was a statistically significant correlation between the proportion of sexually dichromatic species and the total species richness of the clade, consistent with the hypothesis. The relationship was weak, however, and explained only a small amount of variation in species richness. Subsequent work confirmed a trend of a weak but significant relationship (but not across all birds (Phillimore et al., 2006)).

Another hypothesis of this kind is the idea that faster rates of molecular evolution might speed up rates of evolution of reproductive isolation and ecological divergence, and hence speed up diversification. Early studies found a correlation between substitution rates and net diversification in flowering plants, which was consistent with the hypothesis (Barraclough et al., 1996; Barraclough and Savolainen, 2001). Two additional explanations were possible, however. The conditions generated by diversification such as subdivision into smaller population sizes might result in faster accumulation of nearly neutral mutations, rather than faster molecular evolution driving faster diversification. Or there could be confounding variables correlated independently with the two focal variables that give the appearance of a relationship where no mechanistic connection actually exists.

Davies et al. (2004b) expanded the investigation to consider the effects of environmental energy on both molecular rates and diversification. It has long been suggested that higher levels of environmental energy at low latitudes might explain high diversity there (by permitting a higher number of species to coexist). A more recent hypothesis was that organisms at low latitudes evolve more rapidly because higher energy inputs speed up molecular rates (for example, due to faster metabolism at higher temperatures (Gillooly et al., 2005)), which in turn speeds up diversification rates.

By compiling data on present-day average temperature and other measures of energy load across the geographical ranges of flowering plant families, Davies et al. (2004c) first confirmed that measures of environmental energy indeed correlate with species richness across flowering plant families. They then investigated whether this relationship was mediated via rates of molecular evolution or instead via effects on biomass (i.e. supporting a greater amount of life). The results indicated that molecular rates and diversification rates were independently correlated with different measures of environmental energy (Fig. 10.2)—diversity was more closely predicted by temperature, and molecular rates by latitude, out of a suite of environmental measures. The conclusion was that the earlier discovery of a correlation between molecular rates and diversification rates was due to confounding but separate effects of energy on each variable: energy does speed up molecular evolution and enhance diversification but apparently via separate mechanisms. The story is further complicated because the degree of climate change experienced since the last glacial maximum also predicts the species richness of flowering plant families, in a way that covaries with current climate (Jansson and Davies, 2008).

Clearly, the exact direction of causation remains uncertain in observational studies of this kind, but these examples serve to illustrate some issues with identifying causes

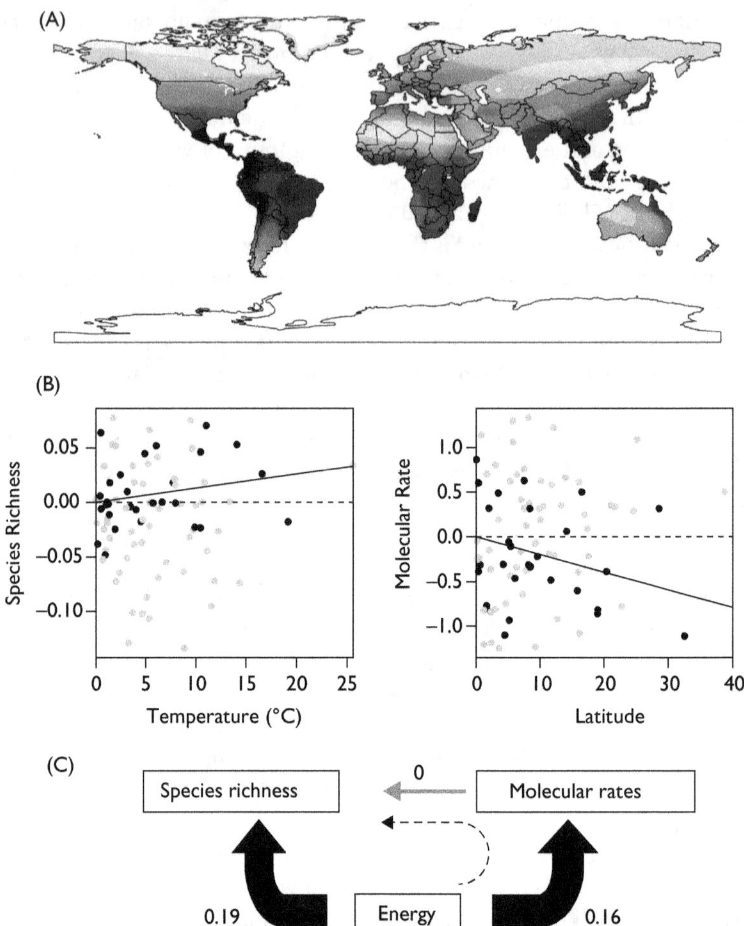

Fig. 10.2 Interplay between flowering plant species richness, environmental energy, and rates of molecular evolution. (A) Estimated global distribution of flowering plant species richness based on overlaying range maps for flowering plant families. (B) Independent contrasts in species richness and molecular rates across 86 pairs of sister families plotted against best-fit surrogate measures of environmental energy for each response variable. Richness contrast is log (richness sister clade 1) − log(richness sister family 2), molecular rate contrast is (molecular branch length sister clade 1) − (molecular branch length sister clade 2), and environmental contrasts are (mean value for range of sister clade 1) − (mean value over range of sister clade 2). Black lines equal regression lines through origin (the standard method for contrasts). Black dots are more phylogenetically robust comparisons that received higher weights. (C) Summary of linear model results. Energy has independent effects on molecular evolution and species richness, rather than mediating its effect on species richness via molecular rates. (Reprinted from Davies et al. (2004b, 2004c) with permission.)

of variation in diversification. Because of limited data at such scales, surrogates are used rather than a direct measure of the real quantity of interest, such as the strength of intersexual selection or biomass. It can be hard to separate causal factors from confounding variables. No phylogenetic resolution was available at the time within tribes or families that served as clades, and so it was not possible to try teasing apart speciation or extinction. More recent methods allow models to be fitted that estimate different speciation and extinction rates depending on the value of a trait, from resolved phylogenetic trees within clades (Maddison et al., 2007). There is ongoing study, however, into the robustness of these methods (Davis et al., 2013; Rabosky and Goldberg, 2015), in part because alternative processes can have overlapping or contrasting effects on patterns of trait distributions and branching rates. Comparison of repeated sister clades remains a robust experimental design for exploring correlates and circumvents some problems of reconstructing trait shifts in whole-tree studies.

A more general issue is whether rates are the right measure of diversification. Age is an important variable to consider in explaining species richness, and analyses need to take it into account (Ricklefs, 2007). Yet average rates can be misleading metrics for processes that are not time-homogenous. Family sizes do not correlate with their age (Magallon and Sanderson, 2001)—there is no general tendency for species richness to correlate with time (Stadler et al., 2014). Similarly, in the Cape flora of South Africa, more recently radiating clades do not tend to contain fewer species, because they exhibit faster net diversification rates (Linder, 2005). These observations do not fit with even rates of species proliferation over time, and instead might indicate growth towards an equilibrium limit in the number of species. In such cases, knowledge of how limits are set and why they vary among clades may be just as important for explaining variation in species richness among clades as how quickly they grow towards the limit.

To conclude, the impression from studies of correlates of diversification is that traits that correlate with diversification rates can be found, but that the amount of variation explained is fairly small. This could be because of the reliance on surrogates from available data rather than using primary causal factors of interest—the most challenging missing variable being good estimates of dispersal capability (see chapter 5). Alternatively, diversity patterns do not result from a race among a set of lineages with trait values controlling the accelerator for speciation and extinction rates. Instead, they depend on geographical and ecological opportunity, which in turn depends on the interaction between traits and the environment.

10.3 Geographical and ecological opportunity

A classic idea bringing opportunity to theories of diversification was that of key innovations—defined as new traits that triggered increased diversification either by opening up a new ecological zone for exploitation, or enabling finer partitioning of an existing zone (Hunter, 1998). These could be unique events such as the origin of powered flight in bats, or repeated events such as shifts to feeding on flowering plants in insects (Mitter et al., 1988). With unique events, inference is limited to pinpointing the shift and whether it coincides with particular trait changes of interest. For example, Sanderson and Donoghue (1994) demonstrated that the shift to increased diversification in flowering plants did not encompass all of the angiosperms because an early-diverging lineage displayed low rates similar to the outgroups. They argued that this ruled out synapomorphies of all angiosperms as key innovations. With repeated events, it is possible to test for statistical correlations as described in section 10.2.2: angiosperm-feeding insects are significantly more diverse than their non-angiosperm feeding sister clades (Mitter et al., 1988).

Understanding of these processes has evolved over time. Subsequent work with a phylogenetic tree of flowering plants resolved to the level of families indicated that there have been multiple shifts in diversification occurring repeatedly at all levels of the tree (Davies et al., 2004a). Accumulation of species richness has not resulted from a few shifts, but instead there are high- and low-diversity clades scattered across the tree (Fig. 10.3). This pattern is also found in birds (Ricklefs, 2003; Jetz et al., 2012) and, to a lesser extent, in insects, where four holometabolous orders (Hymenoptera, Coleoptera, Diptera, and Lepidoptera) have shifted to significantly higher net diversification rate (Condamine et al., 2016), again with patchy high- and low-diversity clades within them. Assuming traits are phylogenetically heritable, this pattern implies that characteristics typical of a higher clade such as flowering plants are associated with high diversification in some circumstances but low diversification in others.

Multiple sources of evidence indicate that geographical or ecological opportunity is a key determinant of diversification. First, the size of geographical area occupied by

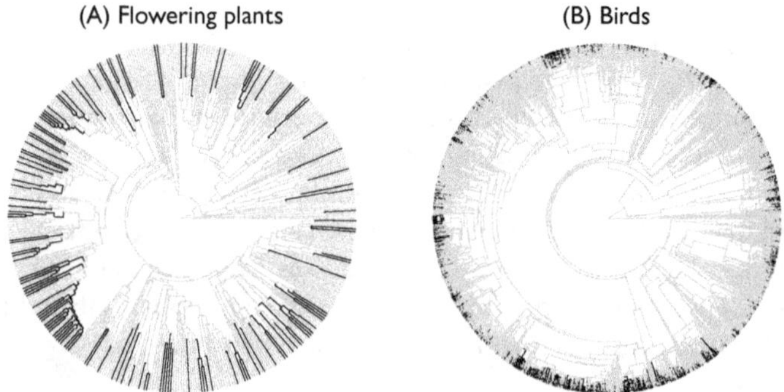

Fig. 10.3 Supertrees of (A) all families of flowering plants (Harris and Davies, 2016) and (B) all species of birds (Jetz et al., 2012). Dark branches indicate those with estimated net diversification rates in the top quartile of those observed across the tree. Diversification rates for flowering plant families and leading up to bird species were calculated as described in the source papers. Internal branches of the tree were coloured according to average ancestral values reconstructed by maximum likelihood. Branch lengths are proportional to reconstructed time: in the flowering plant tree, distance from centre to tips is 184 Myr, and in the bird tree, centre to tips is 128.5 Myr. (Trees were redrawn using data from the above sources.)

a clade is the strongest correlate of species richness in many taxa. Families of plants and birds found in large areas tend to have more species than clades restricted to small areas, such as islands (Ricklefs, 2003; Davies et al., 2005). How should we factor this into analyses of phylogenetic diversity patterns? On the one hand, a large geographical range could be seen as a consequence of successful diversification and spread, rather than a cause. Yet, area also poses a constraint. A lineage that finds itself in Madagascar is unlikely to accumulate as many species as a lineage able to colonize the whole of Africa. Models of tornado incidence, for example, use probabilities per unit time and space. Should we similarly estimate diversification in a spatially explicit way—per species per million years per square kilometre? For example, the flowering plant genus *Protea* radiated at similar rates in the Cape region with a Mediterranean-type climate as it did in the summer rainfall of tropical Africa, but the former radiation occurred over a tiny area compared to the latter (Valente et al., 2010). What is the expected scaling of diversification with geographical area, on average? Species-area curves at continental scales have an exponent of ~1.0, which indicates linear scaling (Rosindell and Cornell, 2007).

Second, traits that explain most variation in diversification tend to be those associated with accessing new opportunities. A survey of diversification across birds confirmed a tiny effect of levels of sexual dichromatism on species richness, but found more variation explained by geographical area, surrogate measures of colonization ability, and measures of the degree of ecological specialism versus generalism, with generalist clades more diverse (Phillimore et al., 2006). More tests of these features are still needed, but they have in part been limited by lack of available data. As argued in chapter 5, dispersal is a key parameter for determining the spatial scale of diversification and yet we lack direct measures of dispersal distances (which need to separate short-range versus rare long-distance events) for most taxa.

Third, there is evidence from multiple sources that diversification is enhanced in low-diversity settings, that is, those with greater opportunity. Weir and Schluter (2007) estimated speciation and extinction rates across a latitudinal gradient in the Americas and found, contrary to common intuition, that speciation rates were higher at high latitudes than in the tropics. Their analysis focused on sister-species pairs, and hence on time periods since 9 Mya. One explanation is that recent speciation rates were high because lineages are expanding out into regions emptied of other species by recent glaciation (Schluter, 2016; Schluter and Pennell, 2017). More direct evidence comes from analyses of planktonic foraminifera with a highly resolved fossil record, where speciation and extinction rates can be separated reliably and modelled against various causal factors. Speciation rates increase when fewer species were present—that is, more opportunity and vacant niches—whereas extinction rates were more strongly determined by environmental fluctuations (Ezard et al., 2011, Fig. 5.10). Similarly, mass extinction events, which constitute spikes of high extinction rates followed by low diversity, are followed by high origination rates (Alroy, 2008).

One feature of diversification associated with opportunity is that different traits are likely to be associated with radiations depending on what opportunity has arisen. Dispersal ability will be a general trait promoting the chance of encountering new geographical areas, as perhaps would the ability to establish from low numbers. But ecological establishment in a new area is promoted by different traits depending on the environmental conditions (Davies et al., 2005). Similarly, the emergence of new habitats or ecological resources due to environmental change will favour diversification in organisms with different sets of traits depending on the conditions of the new environment. For example, succulents exhibit fast diversification in arid regions (Valente et al., 2014), and north temperate taxa were pre-adapted to undergo spectacular radiations at high altitudes upon the uplift of the Andes (Hughes and Eastwood, 2006), whereas climbers display rapid diversification in tropical forests (Gianoli, 2004).

There are now lots of methods available for pinpointing shifts in diversification rates on trees (Morlon, 2014). These can model density-dependent speciation and extinction rates to allow for limits on diversification, and infer filling of niche space from patterns of trait disparity. A key missing component, however, is what sets the phylogenetic range of limits on numbers of species (Barraclough, 2010; Stadler et al., 2014). Why is diversity limited at the level of that clade rather than a more inclusive or less inclusive level? Studies often choose a genus or family to focus on; why should those species compete with each other and fill available niche space to the exclusion of other clades? There is more about this in section 10.6.

10.4 Species turnover

The final component for interpreting diversity patterns, and one that is not easily identifiable in a phylogenetic framework, is species turnover. Tracing ancestral lineages of extant species into the past, the number of lineages inevitably declines back in time. For example, 6111 extant species of placental mammals are descended from around 19 lineages at the Cretaceous–Paleogene boundary 66 Mya (Fig. 10.1). That does not mean that there were only 19 species: over 200 mammal genera are known from the fossil record of that time (Paleobiology Database, accessed 14th December 2018). This is indicative of species turnover over time—species have a finite life span; they eventually speciate or go extinct, leading to turnover (Barraclough, 2010).

In principle, turnover could result from chance speciation and extinction in a static environment. A further driver of turnover, however, is environmental change. Environments fluctuate on many timescales. For example, since the last mass extinction there has been a steady trend for cooling and drying of the earth. As a result, lineages adapted to cooler and drier habitats experienced an increase in opportunity, whereas those adapted to warmer, wet habitats experienced a relative decline in opportunity (Sun et al., 2014). In theory, all species could adapt in parallel to these changes (see chapter 8) and diversity patterns would remain unaffected. But that is not what happens—clades vary in their degree of pre-adaptation to new conditions or evolvability, or by chance some clades encounter new conditions before others. The snapshot of the tree of life that connects present-day species reflects the waxing of clades in their boom time and the waning of formerly successful clades that are now in decline (Barraclough, 2010).

One example is angiosperm-feeding beetles, which are presently more diverse than beetles feeding on conifers (Farrell, 1998). This has been attributed to special ecological interactions with angiosperms such as plant–pollinator interactions driving coevolution. These processes may occur, but a simpler explanation is that most terrestrial plant biomass used to be conifers and that shifted over time to flowering plants (Barraclough et al., 1998a). As the area occupied by flowering plants grew and the area occupied by gymnosperms shrank, it is unsurprising that beetles feeding on flowering plants are more diverse than those feeding on less-abundant gymnosperms. Presumably in the past, there were relatively more species feeding on gymnosperms than on flowering plants, before the rise to prominence of the latter. Many further radiations of beetles show no association with herbivory (Hunt et al., 2007).

A clue towards the dynamics of species turnover is imbalance (Heard, 1992). It has been long recognized that phylogenetic trees are significantly imbalanced (Fig. 10.4), which has been interpreted as evidence that diversification rates vary across lineages. There is more to imbalance than just variation, however. A perfectly imbalanced tree entails that the next species to diversify is a descendant of the previous diversification event. Consider a model to generate such a tree. If the previous diversification event yielded species A and B, and next species A diversifies, species B is immediately consigned to not speciating itself, whereas a descendant of A will do. This is a strange model of diversification and implies progression—lineages that were previously successful (in terms of diversification) are more likely to yield lineages that will be successful in the future.

There are a few possible explanations for imbalance in phylogenetic trees. First, it could simply reflect numbers—if each extant species has the same probability of founding a future radiation, this is more likely to come from the currently successful lineage, resulting in a statistical trend towards imbalance. (Note this is subtly different from a constant birth–death model; it says that each species has the same per lineage probability of encountering new conditions that increase the diversification rate of its descendants compared to the rest of the clade.) Second, there might be long-term trends to improved competitive ability in successive radiations. Third, if environmental change is temporally autocorrelated, such that the environment varies over time but adjacent time points are more similar than distant time points, the most likely species to adapt to new opportunities during climate change are those thriving in recent past conditions: that is, from a recently diversified clade. Surviving lineages with low diversity near the base of the tree might be adapted to marginal or archaic conditions, still found in a restricted number of locations. Interestingly, similar dynamics have been observed with the phylogenetic patterning of influenza dynamics over observed time periods—epidemic lineages in subsequent years are significantly likely to be closely related but not identical to widespread strains in earlier years. In that case, coevolution to evade host immunity helps to pattern diversification, but the dynamics of turnover are still a result of competition among influenza strains in a changing environment (Bush et al., 1999).

Processes like these explain the patchy nature of diversification across large clades—ecological opportunities change as the environment changes, and a disparate

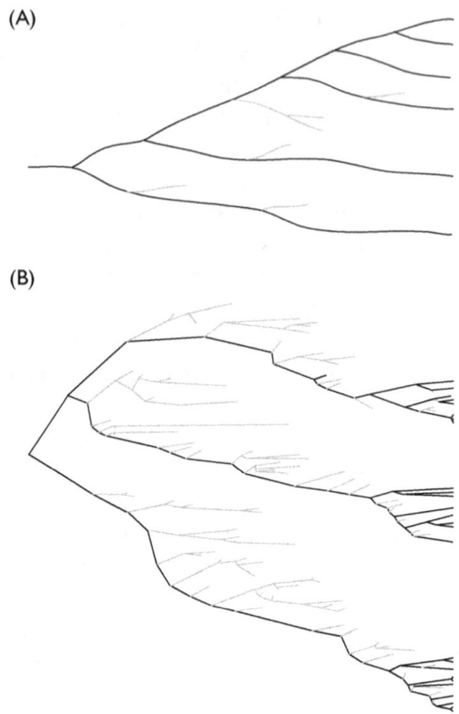

Fig. 10.4 Two phylogenetic patterns of interest for considering the role of opportunity and turnover in diversification. (A) Phylogenetic imbalance occurs when branching event $i+1$ in the tree is more likely to come from the lineage that contained branching event i, counting from root to tips according to relative age of nodes. In this example, the tree of surviving species (black branches) is perfectly imbalanced. (B) 'Broom-handle' pattern of three extant clades that separated from each other a long time ago, but extant species within each clade are very closely related to each other. This pattern is indicative of turnover occurring separately in each clade (grey branches, extinct species; black branches, relationships of extant species). The long stem branch leading to each clade does not reflect presence of a single species during that time, but instead that only a single lineage among the many species present in the past founded the later radiation in each clade.

set of lineages take advantage of new opportunities, although with a broad signature of imbalance indicative of progression. This is very different from the model of traits determining diversification rates outlined in section 10.2. New models are needed that incorporate competition and turnover more explicitly.

The alternative to turnover would be if clades initially radiate to fill geographical space or a new ecological opportunity, but once filled, the incumbent species persist indefinitely. This might be observed in a recent radiation, but even in so-called living fossils there is evidence for turnover. Horseshoe crabs are an ancient group of arachnids, with fossils dating back over 400 Myr (Lamsdell, 2016), and yet according to molecular dating the four extant species shared a single common ancestor 130 Mya

(Fig. 10.5), with further divergence since 40 Mya (Obst et al., 2012). Similarly, living species of cycads share ancestry dating back over 200 Myr. Nonetheless, the 354 extant species descended from just 10 lineages present 19 Mya, which happen to be classified as the 10 extant genera (Fig. 10.6; Humphreys et al., 2016).

This pattern of long stem branches and recent radiation of crown clades has been called the 'broom-handle' pattern in relation to plant clades occupying Mediterranean climate regions (Crisp et al., 2004). One explanation would be that a single species persisted for many tens of millions of years before encountering conditions that favoured its radiation. A more plausible explanation, however, is that there was standing diversity of the clade at all time points, but rapid species turnover, explaining why the most recent common ancestor of extant species is a lot more recent than the divergence of the clade from its nearest relatives (Fig. 10.4B). This pattern of clusters

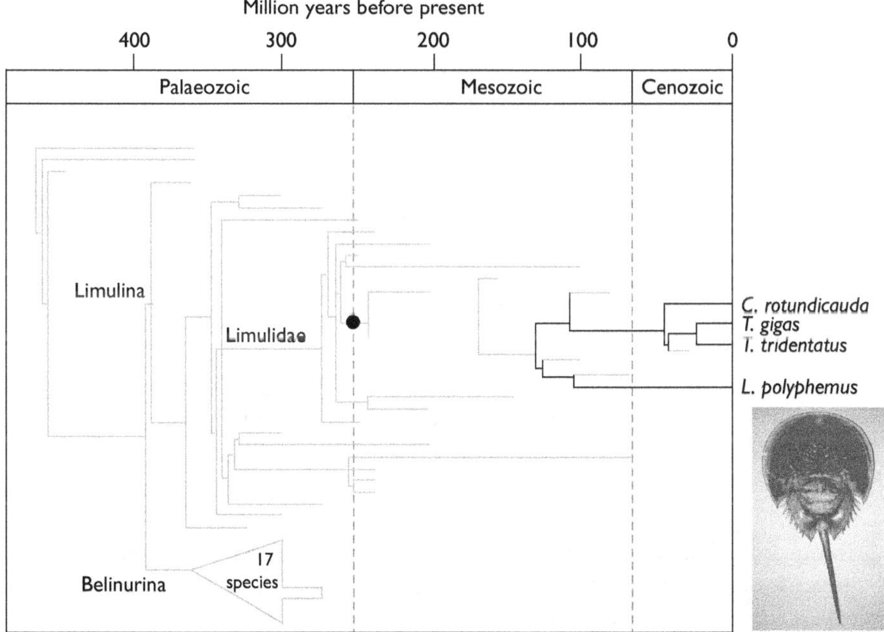

Fig. 10.5 Species turnover in horseshoe crabs (order: Xiphosura). The group diverged from its nearest living relatives over 400 Mya and many now extinct species left fossils between then and the present. The four extant species from genera *Limulus*, *Tachypleus*, and *Carcinoscorpius* share a much more recent common ancestor. Fossil branching and phylogeny are inferred from morphological analysis (redrawn from Lamsdell (2016) with permission). Branching events below node marked by black circle were redrawn to correspond to relationships and reconstructed ages from molecular dating by Obst et al. (2012). Divergence times inferred from morphological analysis of fossil and living species push the most recent common ancestor of living species back to around 260 Mya (Lamsdell, 2016). (Photograph of *L. polyphemus* by R.T. Barraclough.)

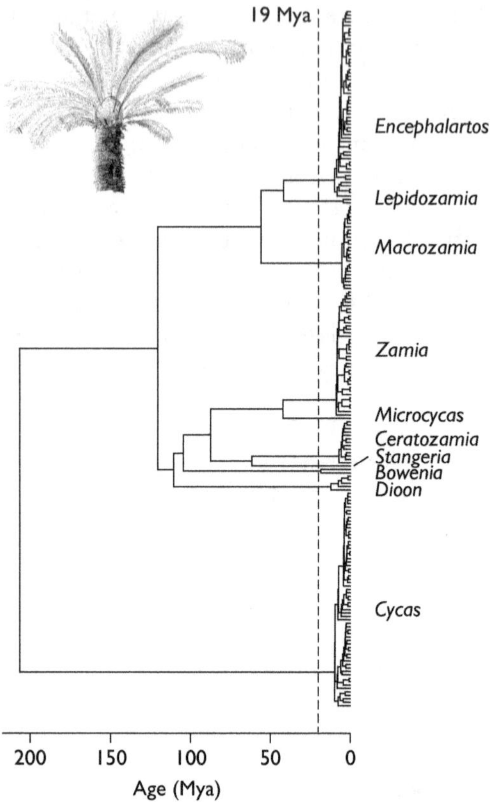

Fig. 10.6 Cycads originated over 200 Mya, but the 354 extant species originated from rapid radiations of just 10 lineages (which correspond to genera) in the last 19 Myr (dashed line). (Phylogenetic tree drawn using data from Humphreys et al. (2016). Drawing of cycad by C.E. Barraclough.)

of closely related species on a long stem branch mirrors the pattern described in chapter 3 for delimiting independently evolving species. That is because the same processes are at play, as discussed in section 10.5.

10.5 Competition-based models of diversification

A framework for modelling diversity patterns that incorporates the assumptions of limits and turnover arose when Hubbell (2001) extended his neutral theory of biodiversity to couple local and global diversity patterns via dispersal. The underlying assumption is that there is a fixed amount of geographical space and ecological resource within the region, defining a fixed total number of individuals, J_m, called the metacommunity size (Fig. 10.7). When an individual dies, it provides an opportunity

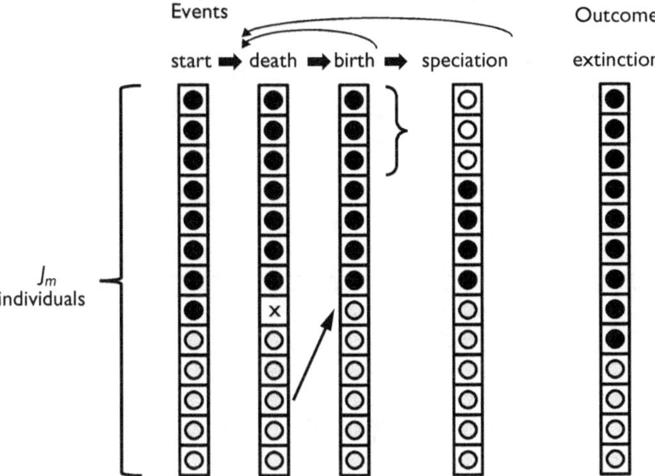

Fig. 10.7 Schematic of a neutral model of biodiversity applied to diversification. The simplest model assumes a fixed total metacommunity size of J_m individuals (circles). There are two species present initially, represented by shading of circles. When an individual dies, it opens up an empty 'slot' for birth and recruitment of a new individual, whose parent is chosen at random from the entire metacommunity, irrespective of species. An individual of the light grey species recruits into the empty 'slot'. This process loops repeatedly. There is a constant probability per individual per time step of a speciation event happening: speciation occurs by subdividing the species of the chosen individual into two randomly sized fragments according to a broken-stick model. Extinction is an emergent outcome of the model and occurs when, over time, abundance of a species dwindles to zero (i.e. last individual dies without issue). Here, the white species goes extinct.

for a juvenile to colonize: this can be visualized as a gap opening up in a forest canopy following the death of a tree, which allows a seedling to grow. Individuals therefore compete for available space and resources in a zero-sum game.

Neutrality comes in because it is assumed that all individuals have the same probability of colonizing a gap, even if they belong to separate species. There are no niche differences among species and all individuals compete on equal terms. The result is ecological drift: relative abundances of species fluctuate at random in the same way that frequencies of neutral alleles fluctuate within a population. Extinction therefore emerges mechanistically when the last individual of a species dies as a consequence of drift due to stochastic birth and death of individuals. In contrast, speciation must be specified as an additional process in the model. Without input of new species, the system will drift towards extinction of all but a single species after a sufficiently long period of time. With an input of new species, however, either by immigration from elsewhere or through speciation within the region, species richness is maintained at dynamic equilibrium and fluctuates around an average steady-state value (Fig. 10.8).

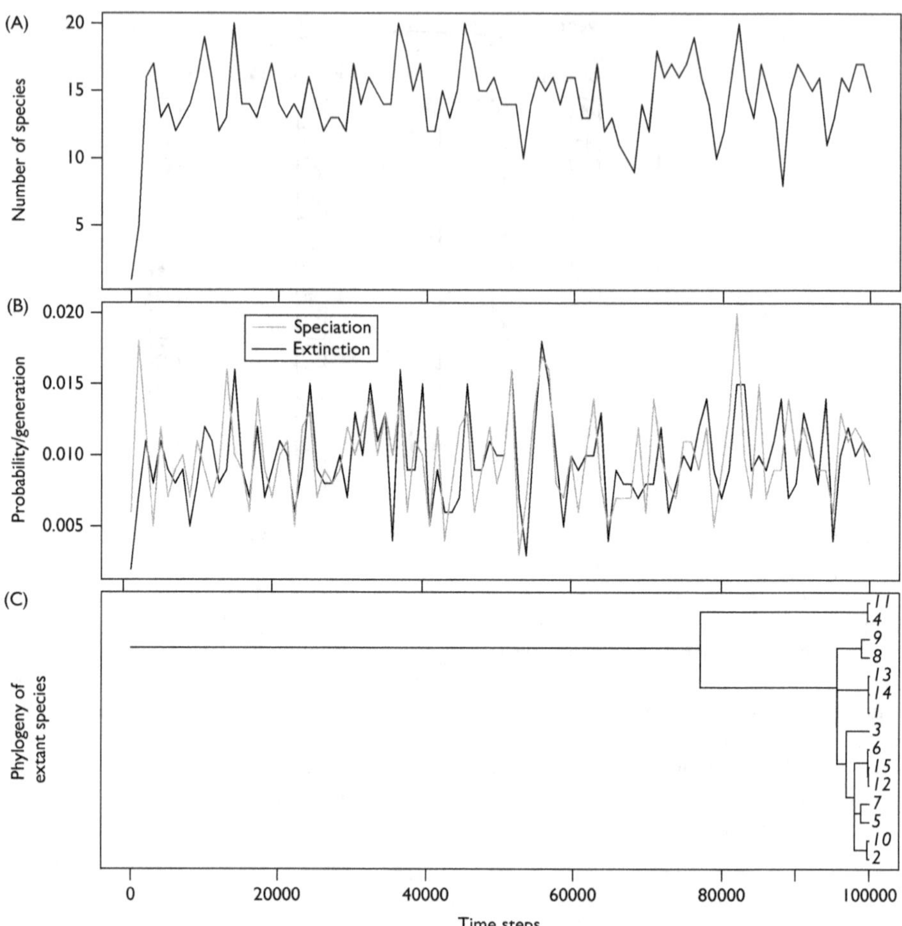

Fig. 10.8 Results of a neutral simulation of diversification using the model outlined in Fig. 10.7. A metacommunity with 10,000 individuals was run for 100,000 time steps with a speciation rate of 0.01 per individual per time step. (A) Number of species fluctuates around a steady-state value determined by metacommunity size and speciation mechanism (here a broken-stick model rather than point mutation). (B) Observed speciation and extinction rates per 1000 generations fluctuate: extinction rate (black lines) increases following periods of high speciation rate (grey lines) due to consequent increase in numbers of species with low abundance. (C) Phylogeny of extant species at end of simulation. Turnover leads to coalescence to a most recent common ancestor much more recently than the age of the metacommunity. (Run using simulation code from Humphreys and Barraclough (2014).)

The original version used a mutational model of speciation in which each individual has the same per generation probability that it would become a new species. Under this model, the expected steady-state species richness is a function of the speciation rate, v, and metacommunity size, J_m—analogous to population theory predicting allelic diversity as the product of mutation rate and effective population size (see chapter 3). The consequence of this assumption is that new species have low abundance and a very high risk of extinction. Although plausible in selfing plants experiencing auto- or allopolyploidy, for example, it is not generally assumed that new species arise as single individuals. An alternative model of speciation assuming fission of species ranges was considered (Fig. 10.7), which consequently increased the probability of newly formed species persisting. These alternatives were mostly raised in order to improve the fit of models to data. Yet they point to a question worthy of further study: namely how does species abundance influence speciation probabilities and vice versa (Pigot et al., 2010)?

From this relatively simple set of assumptions, the model predicts quantities such as steady-state diversity, the shape of phylogenetic trees, branching rates, and species abundances (Fig. 10.8). For example, the time to most recent common ancestor of all individuals in the metacommunity (Fig. 10.8C) is determined by the rate of turnover due to ecological drift and is proportional to metacommunity size. Again, this is analogous to coalescence theory applied to populations. Somewhat counterintuitively, the speciation process does not affect the ancestry of individuals in this model, which is solely based on random birth and death of individuals. Instead, it is a separate process 'layered' onto the pattern of ancestry, simply labelling whether particular branches constitute separate species from one another or not. The combination of these two separate processes generates diversification patterns, in the same way that the independent processes of coalescence and mutation generate patterns of genetic variation within populations (see chapter 3).

A major advance of neutral theory was to generate predictions for evolutionary dynamics of species richness under the assumption of limited opportunity and turnover. However, the quantitative predictions have not stood up to close scrutiny with empirical data. Predicted coalescence times due to ecological drift are much longer than the divergence times observed in real clades. Drift is not fast enough to explain observed branching patterns on phylogenetic trees (Nee, 2005; Ricklefs, 2006). Second, the speciation rate needed to explain observed standing levels of diversity, based on independent estimates of metacommunity size seems to be too high (Desjardins-Proulx and Gravel, 2012).

Various modifications have been made to rectify these discrepancies, but one general solution is to keep the assumption of competition for space and resources, but to relax the assumption of neutrality (Ricklefs, 2006). The problem of overlong coalescence times is solved if the environment experienced by the metacommunity changes over time. Individuals with new mutations conferring high reproduction or survival rate in the new environment will spread at the expense of their competitors, and as they increase in abundance will have higher probabilities of speciation and lower probabilities of extinction. Just as a succession of selective sweeps reduces genetic variation within a population, this mechanism would speed up turnover and lead to shorter coalescence times (Rosindell et al., 2015).

The problem of unfeasibly high speciation rates can be solved by relaxing the assumption of ecological equivalence. For example, suppose that there is environmental variation among patches in the metacommunity that promotes niche specialization. With appropriate trade-offs so that a single generalist species cannot perform as well in each patch type as a set of specialists (see chapter 9), diversity would be maintained through classic niche-based coexistence. This is analogous to frequency-dependent selection maintaining genetic variation within a population. In order to maintain species turnover in such a model, species in each niche must retain a certain probability of going extinct and freeing up opportunities for colonization. Drift in species abundances would be protected to some extent by niche partitioning, but additional causes of extinction such as local disturbance or disease outbreaks could maintain turnover. In neutral theory, the limit operates on abundance, whereas in niche-theory the limit might operate directly on number of species as reflecting the number of niches. The model by de Mazancourt et al. (2008) in section 8.4 represents a niche-based model with a limit on the number of individuals, but with separate niches available in terms of patches with different thermal conditions (Figs 8.4 and 8.5). A global increase in temperature in this model leads to extinction of cold-adapted species, and subsequent diversification of warm-adapted species to exploit the newly available warmer patches (i.e. species turnover).

There are clues from the fossil record to support the need for these modifications. Neutral simulations lead to extinction rates that lag behind speciation rates (Fig. 10.8), because a chance period of multiple speciation events is followed by elevated extinction risk due to the lower abundance of the resulting species. In contrast, analyses of the marine fossil record indicate that speciation rates lag behind extinction rates (Alroy, 2008). Extinction is caused by external, environmental events, which reduce diversity and create the opportunity for elevated speciation rates (Ezard et al., 2011).

Other scenarios can be imagined that retain the key features of these models but relax the unrealistic assumptions. Metacommunity size could itself vary over time—opportunities might emerge and be lost, leading to expansion and contraction of the metacommunity over time (Rosindell et al., 2015). This could still generate competition-based dynamics even if there is not a fixed limit to the number of individuals or species (Harmon and Harrison, 2015). Spatial scenarios can also be modelled, with spatial turnover of ecological equivalents across a larger region (Alzate et al., 2017). The process of species turnover might then occur over longer timescales than those predicted based on individual birth, death, and colonization dynamics. For example, the set of species within the Amazon Basin are not all interacting directly over short timescales, but as the environment changes over longer timescales, and ranges move around, the success of a given lineage is tied to the fate of all other lineages: if one lineage diversifies to fill all available geographical and ecological slots, others cannot.

More work is needed on how success of a lineage in terms of abundance maps to the number of species, and vice versa. The simplest hypothesis is that the two automatically go hand in hand. High abundance increases probability of speciation or reduces risk of extinction. Abundant lineages are more likely to encounter diversifying conditions than rarer ones, and perhaps more likely to respond because of their larger population size. In turn, high clade abundance is only achievable by diversifying to specialize on a range of different resources in different conditions, if trade-offs limit the performance of generalist species. Alternatively, high abundance need not equate with species richness: diversification might be dependent on traits that promote reproductive isolation and the coexistence of multiple differentially adapted species, yet that are not themselves targets of individual selection. The balance of evidence points to the former scenario being more widely applicable (see chapter 5).

To conclude, the assumption of competition in a zero-sum game, together with mechanisms connecting speciation and extinction to abundance, can be used to model diversification in a scenario with limits to species richness and turnover. These are the kinds of processes that are needed to explain diversity patterns at large scales, and more work connecting these process-based models to statistical inference of diversity patterns is needed (Barraclough, 2010; Morlon, 2014).

10.6 Separate limits

Section 10.5 described diversification within a single metacommunity: all individuals and species interact with each other within a single arena and therefore are jointly limited. In principle all organisms on earth are part of a big zero-sum game: if one lineage leaves descendants comprising the total biomass on the planet in the distant future, other lineages cannot. But realistically the descendants of marine algae are not likely to replace the descendants of terrestrial consumers, at least not any time soon. Section 10.3 presented evidence for limits operating separately on diversification within particular clades, but what sets those limits? How often do ecological limits apply to clades versus assemblages of species of mixed phylogenetic origins scattered across multiple clades?

The models from section 10.5 can be used to explore these questions. What happens if a single metacommunity is split into two? For example, assume the separation of a single contiguous ancestral area into two isolated areas with a dispersal barrier between them. Now individuals are limited independently in each area, and the outcome is that the metacommunities diverge over time. After a sufficiently long period of time, because of species turnover, species within each area will share a single common ancestor more recent than the time of isolation of the two areas (Fig. 10.9; Humphreys and Barraclough, 2014). The divergence time between the two metacommunities will reflect the time since the regions became isolated plus an extra amount due to sampling from the ancestral metacommunity—the radiations in each region are likely to be founded by two separate species drawn from the standing phylogenetic diversity at the time of the split.

One classic example of the effects of geographical isolation on diversity patterns is the separation of South America and Africa since the break-up of Gondwana. Many cases are known of sister clades of plants and animals indicative of vicariance between the two continents. But dating with molecular methods indicated several cases where the divergence dates were either too old or too young for the geological date of separation, despite being reciprocally monophyletic in the two regions (Lavin et al., 2004). An old date is explicable by lineage sorting of phylogenetic diversity in the ancestral region, whereas a young date can be explained by later dispersal followed by species turnover occurring separately in each region. With fast-enough species turnover,

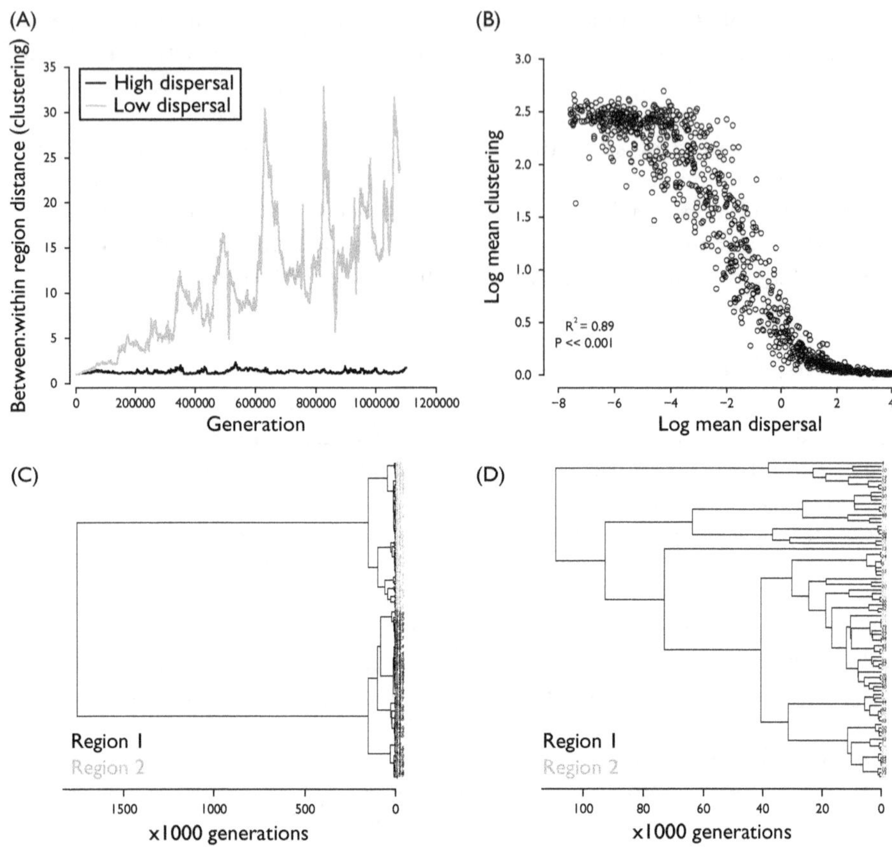

Fig. 10.9 Simulation of the model with two metacommunities representing separate geographical regions. (A) With a low rate of dispersal between regions, metacommunities diverge into distinct phylogenetic clusters over time as measured by ratio of average between-region divergence to within-region divergence among species. (B) Degree of phylogenetic clusters decreases as dispersal rate increases. (C) With low dispersal, two divergent clades restricted to each region in turn emerge. (D) With high dispersal, there is a single cohesive phylogenetic cluster with frequent shifts in region of occupancy. Equivalent patterns would be expected for metacommunities occupying different ecological adaptive zones in relation to the frequency of shifts occurring between use of each zone. (Reprinted from Humphreys and Barraclough (2014) with permission.)

monophyly can be restored in each region and the divergence time represents a date earlier than the date of dispersal (again because of sampling of phylogenetic diversity by the dispersing species versus future common ancestor in the source region), but potentially later than the date of separation of the continents (Barraclough, 2010).

Once again, a direct analogy can be made to population processes, in this case to speciation as outlined in chapter 3. The degree to which geographical isolation will generate reciprocally monophyletic clades in each region, eventually separated by long stem branches, depends on the rate of dispersal between the two regions (Fig. 10.9;

analogous to gene flow) and the rate of species turnover in each region (analogous to coalescence rate within populations). Even when dispersal is negligible, if turnover rates are slow relative to the time since isolation, paraphyly will still be apparent and there will be incomplete lineage sorting between the two regions.

An alternative mechanism for maintaining distinct metacommunities is through ecological separation (Humphreys and Barraclough, 2014). Imagine the classic neutral model, but now each grid cell provides two slots for individuals—one for a tree and one for an understorey shrub. If species specialize as either a tree or a shrub, individuals occupying tree slots compete in a separate game from individuals occupying the shrub slots. If evolutionary shifts between a tree and shrub lifestyle were common, species turnover at longer timescales would still be linked between the two ecological guilds, and tree and shrub species would be intermingled on the phylogenetic tree. But if shifts were sufficiently rare, species turnover would occur independently in each ecological guild and, after a long enough time, reciprocally monophyletic clades of shrubs and trees separated by long stem branches would emerge. In this case, the frequency of ecological shifts plays the same role as dispersal between regions in the geographical scenario.

Monophyletic clades of species with shared ecological traits is not a mind-blowing prediction because it is expected under any phylogenetic null model of slow trait changes. But there is more going on here than just ecological similarity within a clade. In this model, ecological structure due to the pattern of available resources or niches and the ease of shifting between resource or niches is determining the evolutionary fate of sets of species. Species within an interacting metacommunity share evolutionary fate, because if one of them leaves descendants that occupy all available slots in the future, the others cannot. This occurs over a longer timescale than population processes, but it still leads to evolutionary interdependence among interacting species over those timescales. In contrast, when ecological shifts between the two arenas are sufficiently rare, evolution of species in one arena is truly independent of species in the other arena. If these conditions are met and persist for long enough, they would leave a signature in phylogenetic diversity patterns—ecologically or geographically coherent clades forming clusters on relatively long stem branches (Humphreys and Barraclough, 2014). Furthermore, traits associated with membership of a metacommunity would be conserved within each clade but divergent between them, which might be detectable against a model of even changes across the tree.

The expectations are clear from the models, but do these conditions apply in reality and influence the structure of diversity patterns? Section 10.4 discussed the case of cycads and Mediterranean-climate plant groups comprising recent radiations of species on long stem branches. The models provide a hypothesis to explain that pattern—each clade represents an independently limited group that has experienced high rates of species turnover. The patterns could be formally compared to alternative models that there was a recent uniform increase in diversification rates, high rates of background extinction, or that all species belong to a single 'macroevolutionary metacommunity'. What is the probability of the observed distribution of radiations across lineages and of stem branch lengths under these alternatives?

Humphreys and Barraclough (2014, 2016) extended methods for detecting evolutionary independence at the species level (described in chapter 3), to test for higher-level patterns in representative clades of mammals and birds, and gymnosperms (including cycads). Specifically, they tested for phylogenetic clusters of closely related species separated by long stem branches, detected as a shift in branching rate from low rates deep in the tree to higher rates more recently. They termed the clusters 'higher evolutionarily significant units' (hESUs), to indicate units that are predicted to evolve if there exist sets of species that are evolutionarily interdependent over macroevolutionary timescales (Box 10.1). A model with a transition to higher branching rates was significantly preferred compared to a constant birth–death model of exponential clade growth, but an alternative model of a uniform increase in diversification rate across all lineages could not be rejected based on node age data alone (Humphreys et al., 2016). Some ecological traits displayed significant coherence within putative hESUs (e.g. body size in euungulates), as predicted by the hESU theory (Humphreys and Barraclough, 2014).

Box 10.1 The possibility of evolutionarily significant higher taxa.

The theory outlined in sections 10.5 and 10.6 opens up the intriguing possibility that there might be some evolutionary reality to the existence of higher taxa. Prior to the rise of cladistic thinking, higher taxa were routinely used as units of diversity, and they remain so in palaeontology where resolution to the species-level is rare. But since the rise of cladistics, it is widely held that no higher grouping above the level of species holds any special evolutionary significance beyond the definition of monophyly. Species are special because they represent the transition from microevolutionary processes of interbreeding, selection, and drift to macroevolutionary processes of speciation and extinction. In the terminology of cladistics, there is a distinction between tokogeny within species (i.e. parent–offspring relationships) and phylogeny between species (i.e. ancestor–descendent relationships of species). Species are defined by sharing evolutionary fate as well as history, but higher clades are defined purely based on history—they share an exclusive most recent common ancestor. Calling a particular level a 'genus' or a 'family' is simply a matter of convenience for taxonomy, rather than reflecting evolutionarily significant groupings in the same sense as species.

The alternative hypothesis discussed in section 10.6 is that some higher clades do share a measure of evolutionary interdependence because their members occupy the same geographical or ecological zone. If these conditions persist for long enough, they could result in detectable patterns of phylogenetic clustering on long stem branches and coherent trait distributions. If such entities exist, they may or may not coincide with traditionally named higher taxa—after all, there is a subjective element to deciding what to call a genus or family, and indeed these levels do vary significantly among clades in terms of their typical ages. But it is also possible that, if natural structure does exist at higher diversity levels, the eye of taxonomists might tend to be drawn to that and recognize it by application of names. A phylogenetic cluster of species at the end of a long stem branch with coherent trait variation indicative of their shared occupancy of a particular ecological zone might tend to represent a discrete and well-recognized grouping via traditional systematic methods. While there was no 1:1 match with a particular level in the taxonomic hierarchy, Humphreys and Barraclough (2014) and Humphreys et al. (2016) found that most hESUs corresponded to genera or families in mammals and genera in gymnosperms (Fig. 10.6).

Referring to higher taxa as evolutionarily significant is not to everyone's taste (although a surprisingly high proportion of evolutionary botanists surveyed in 2014 still regarded genera as more evolutionarily significant than species; Barraclough and Humphreys, 2015). Furthermore, whether currently named taxa such as families can be attributed to the theoretical entities described in section 10.6 is irrelevant both for taxonomic classification (which is a matter of practicality) and for understanding diversity patterns (which does not require particular naming conventions, although studies do tend to select named taxa to focus on). The important point is that the possibility of evolutionary interdependence at higher levels above species emerges as one prediction from theory of species turnover and limits. The alternative hypothesis is that species turnover and ecological limits do not operate in a phylogenetically structured way but rather across ecological guilds of mixed phylogenetic membership. These alternatives now need testing in order to understand observed diversity patterns.

More work is needed to determine whether evidence of evolutionary interdependence and independence over macroevolutionary timescales, which is expected when clades experience separate limits on species richness and turnover, can be detected in phylogenetic trees. There are many reasons why they might not be found: shifts between geographical regions or ecological lifestyles might be too frequent; or independently limited regions or lifestyles might not persist long enough relative to species turnover rates to generate distinct clades. In general, interactions might not be determined by, and in turn shape, the phylogeny, but instead occur between sets of distantly related species with similar lifestyles that live in the same region.

10.7 Individual selection versus species selection

A final point concerns the distinction between individual selection versus species selection. Species selection is selection acting on species as units (Stanley, 1979; Williams, 1992; Chevin, 2016). The concept assumes that there is heritable variation in the probability of diversifying and a limit on the number of species. As a result, some clades grow, others contract, and the number of species with traits conducive to speciation increases. This process would play out over much longer timescales than individual selection within populations, so it is not expected to operate in opposition to individual and gene-level selection. But if traits that are individually advantageous or neutral also affected probabilities of diversification, these would influence diversity patterns among lineages. The theory and examples described in sections 10.5 and 10.6 point to the alternative view that individual selection itself shapes diversity patterns. The way that organisms adapt to the range of conditions they experience over time and space automatically explains diversification, in terms of both promoting speciation and determining the longer-term proliferation of species.

As described for species origins in chapters 4 and 5, this shifts the focus towards trying to understand the pattern of selection pressures across species over time. For example, phenotypic disparity refers to the range of phenotypes observed across a clade of species. Traditional methods for investigating this have again focused on phylogenetic null models (e.g. Brownian motion)—similarly to those developed for species richness—and methods for detecting correlations between multiple traits (Felsenstein, 1985). An alternative perspective is to ask how selection pressures are distributed across a wider set of species—which traits are under uniform selection across those species (i.e. towards the same optimum) and which are under diversifying selection? Future availability of whole-genome data across phylogenetic scales might help to quantify selection on genes underlying phenotypic disparity.

10.8 Conclusions

Irrespective of whether species turnover and ecological opportunity/limits operate on clades or on mixed phylogenetic assemblages, there is now substantial evidence for their importance in structuring diversity patterns. Modelling diversification at broad scales in this way brings the theories for microevolution and macroevolution much closer together. The same principles of drift, coalescence, and selection operate to generate diversity patterns as those acting on genetic variation within populations, under the assumption of limits on diversification. In particular, species turnover driven by environmental change or the emergence of new opportunities is selection playing out over long timescales across sets of interacting species. This framework explains why there is little evidence for traits that consistently promote diversification—evolutionary success depends on the match between traits and environment, and therefore different traits promote diversification in different contexts.

Conclusions

Diversity is a fundamental property of life. There is not just one kind of life, but many different kinds that vary in their genes, phenotypes, and ecological roles. The foundation for cataloguing and understanding diversity is the concept of species. This book has interrogated the concept of species and its consequences for evolution. In this final chapter I summarize the main conclusions and highlight a few overarching thoughts for future work (Fig. 11.1).

11.1 Species are a model for the structure of biological diversity

The existence of species is not a self-evident fact, and represents a hypothesis or model for describing and explaining the pattern of diversity. In broad terms, the species model proposes that life does not fall into a hierarchy or continuum of forms, but instead individuals belong to discrete groups called species. Individuals interact and evolve interdependently within species, whereas individuals in different species evolve independently. The manifestation of this phenomenon is discrete clustering of genetic and phenotypic variation (i.e. the species pattern). The boundaries between species can be blurred and change over time as part of the ongoing dynamics of species formation and evolution, but the species model asserts that the units are real, and not just a convenience or artefact of human endeavour to classify diversity.

The species model generates predictions that can be tested statistically against empirical data. Current evidence supports the reality of species based on morphometric, single-locus, and multilocus data and that this pattern is ubiquitous across all forms of life. Few studies have used these methods to test the species model directly, however, as opposed to species validation or discovery assuming a priori that the model is true. Also, very few studies have considered alternative models for how diversity is structured. More work is needed therefore to apply these methods systematically across a range of taxa.

Multiple forces influence the pattern of differentiation within clades and the extent to which groups of individuals evolve cohesively versus independently. As whole-genome data come online for broad taxonomic surveys, it will be possible to test formally for arenas of recombination and selection and how these processes interact

The Evolutionary Biology of Species. Timothy G. Barraclough, Oxford University Press (2019).

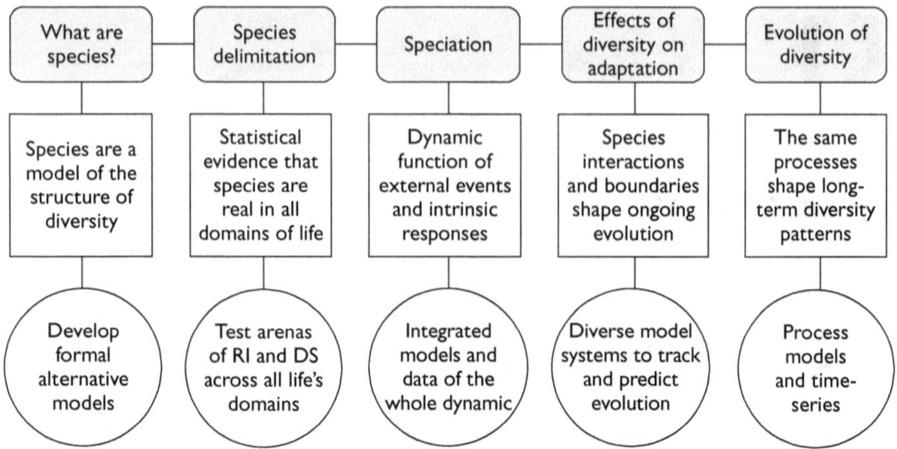

Fig. 11.1 Summary of main topics covered by the book (top row), some headline conclusions (middle row), and some important areas for future work (bottom row). RI, Reproductive isolation; DS, divergent selection.

to generate patterns of diversity, across all of life's domains. We have good understanding of the role of reproductive isolation in speciation of sexual organisms, but more work is needed on the role of barriers to recombination in bacteria. Divergent selection clearly plays a big role in driving speciation, but the dynamics of how conditions change from cohesive to diversifying forces, and how intrinsic genetic attributes influence the response of organisms to those changes, needs to be clarified to obtain a mechanistic and dynamic view of speciation. Model systems are needed where the full dynamics of speciation and its component steps can be investigated within the same clade or even experimentally in microbes.

11.2 Species, and the interactions among them, shape ongoing evolutionary dynamics

The conventional unit for studying contemporary evolution is the population. But evolution always plays out in diverse systems, never in a single population in isolation. With multiple species present, ecological interactions can lead to correlated evolution among species when aspects of their environment are shared or when species impose selection on each other via their interaction. Only in a narrow set of circumstances are the rate and/or trajectory of evolution of a local population unaffected by other species and therefore predictable without consideration of eco-evolutionary interactions. This might seem curious, as species were defined as 'evolving independently' and yet now there is correlated evolution between them. Correlated evolution does not contravene the assumption of the species model, however, if species still evolve to different phenotypic optima at least for some aspects of phenotype—it is just that those optima depend on the abundance and traits of other species.

Co-occurring species also interact if there is residual gene flow between them: that is, if the assumptions of the species model are relaxed to allow partial gene transfer. Again, this can alter evolutionary rates and trajectories compared to single-population predictions. In some organisms, notably bacteria, there are specific mechanisms to allow differential arenas of transfer for different genes, which seem to be tailored to respond to fluctuating conditions. This raises the possibility that species boundaries and mechanisms for interspecies transfer are adaptive—in other words, fine-tuned by selection to enhance adaptation to heterogeneous and fluctuating environments. Alternatively, the consequences of species boundaries for ongoing evolution might be incidental by-products of past selection and the mechanisms or agents (such as plasmids) that are involved in gene transfer. Either way, the nature of species and boundaries between them is an important determinant of contemporary evolution. More work is needed to understand the effects of both ecological and genetic interactions among species on evolution; specifically, theoretical models and empirical systems are needed that can track evolution of diverse assemblages over thousands of generations (see section 11.4).

Over longer timescales, the focus shifts to modelling the birth and death of species as a unit for understanding large-scale diversity patterns. Speciation and the build-up of diversity within clades depend on the rate of encountering new environments and opportunities, as well as on the intrinsic ability to take advantage of them (e.g. rates of evolution of reproductive isolation or of adaptive divergence). Both aspects are difficult to measure across broad taxonomic scales, which has led to an emphasis on statistical approaches looking at rates and correlates of speciation and extinction. There is a need for more mechanistic theory in which speciation and extinction emerge from processes operating within species, and for concerted efforts to assemble data at broad enough taxonomic scales to test the theory. Considering the effects of ecological limits and turnover on diversity brings the theory for macroevolution and microevolution closer together. Some processes that shape diversity at the species level, such as independent limitation and patterns of divergent selection, can also influence higher-level diversity patterns. In particular, large-scale diversity patterns and dynamics reflect the outcome of selection on individuals played out over large temporal and geographical scales.

11.3 A shared framework for all domains of life

Both theory and evidence show that the species model is useful across all domains of life, not just for sexual eukaryotes. The meme that bacteria and archaea do not form species is prevalent but can be dismissed. Yes, there are complications, as indeed there are in eukaryotes, and many interesting features (e.g. mechanisms of gene transfer) differ from eukaryotes that affect outcomes and patterns. But the same basic processes cause diversification and their outcome is broadly the same: we observe discrete genetic and phenotypic clusters indicative of independently evolving units. Bacteria do display less recombination within populations and greater levels of gene transfer between distantly related strains than in eukaryotes, but this does not appear to disrupt the species pattern for the core genome. The extent of gene transfer between species and its importance for evolutionary dynamics remains an interesting topic for study in all domains of life, not just bacteria.

I have attempted to synthesize perspectives from microbes and macrobes in every chapter, but inevitably the coverage is patchy. For example, the sheer diversity of microbes makes macroevolutionary studies of the kind discussed in chapter 10 rare, but the same principles still apply. In contrast, chapters 8 and 9 emphasized microbes because of their tractability for experimental evolution, but again the same principles apply to macrobes. It would be highly profitable to try to fill in these gaps in understanding in each domain. Of course, there are many fascinating differences between bacteria, archaea, and eukaryotes, but in terms of diversification and the utility of the species model, they share more than is often credited.

11.4 Evolutionary biology scales up

A recurrent theme in this book has been the need to bridge mechanistic and broad-scale approaches. Doing that requires systematic compilation of mechanistic data across all of the species in a wider clade or all of the species in a given local or regional assemblage. At present, the only data type really available at required scales is DNA barcodes of marker genes. Very shortly, whole-genome sequence data will be generated at similar scales. But the use of those data to answer questions posed in this book requires functional information and knowledge of phenotypes and environmental conditions. This is still a big ask.

One solution would be greater coordination among researchers to develop model systems for evolution and diversity. Work in molecular and cell biology has benefitted greatly from the concentration on a small number of model organisms. Those organisms do not represent the full range of cellular and genetic mechanisms operating across all life, but they led to deep understanding of processes and techniques that can then be applied elsewhere. Our field tends to be very different, with each laboratory developing its own study system, and knowledge accumulating from different pieces of information from different systems. There are model taxa for speciation and the origins of diversity that are used by multiple teams, such as *Heliconius* butterflies, *Anolis* lizards, sticklebacks, Darwin's finches, cichlid fish, and *Drosophila*, the wonder-model. But there are missing features to these systems, notably the lack of fossil record or time-series of evolution beyond relatively few generations. Taxa with well-resolved fossil records such as foraminifera lack detailed information on ecology and genetics. Identifying a few cases that could build a complete knowledge of diversification in one system would be highly profitable for understanding species origins. Researchers working on contemporary evolution have coalesced around long-term studies of focal species, such as the Soay sheep of the St Kilda archipelago (Hayward et al., 2018). But there is no established model system for tracking evolution 'live' in a diverse assemblage containing dozens of species.

These examples highlight another recurrent theme, which is the need for time-series of evolution in diverse systems. For contemporary evolution discussed in chapters 7 to 9, model assemblages of co-occurring species are needed that are amenable to tracking evolution over thousands of generations, together with surveys of genetic and ecological determinants of evolutionary dynamics that are scalable to multiple species. Microbes seem most suitable for this, and studies are now emerging (e.g. tracking evolution in human guts over several years; see chapter 9). For long-term dynamics of species origin, extinction, and proliferation, fossil records encompassing millions of years are needed, but with sufficient resolution to track morphological change and reconstruct branching and extinction events. Perhaps the trickiest questions lie in between these two timescales, concerning the dynamics of environmental changes and organismal distributions that result in shifts between diversifying and stabilizing conditions, which underpin speciation dynamics. These events occur at intermediate time-scales of 10,000 to 1,000,000 generations that are too short for most fossil records, which in any case lack information on genetics and population processes, but too long for direct observation. Solving these problems, for example with metagenomics of ancient DNA from sediment cores (see chapter 9), would fill a major gap in knowledge.

11.5 Genomes are coming—look busy

The big, imminent opportunity for this field is the capacity to sequence whole genomes for multiple individuals of thousands of species. This boost in technology is almost upon us and will revolutionize our field, as it is already revolutionizing more focused studies of model species, single populations, and pairs of related species. A huge amount of work is needed to make use of this new scale of data. Aside from the practicalities of wrangling and comparing thousands of genomes, these data will demand new approaches for extracting evolutionary information. The most widely used current methods will not scale to this volume of data, they are not designed for simultaneous evaluation of within- and between-species variation across thousands of samples, and they are unable to extract the information that is really needed, namely linking genome variation to phenotypic variation and fitness in non-model organisms.

A particularly interesting challenge will be to develop new methods for inferring gene interactions underlying phenotypic variation. Many current methods for evolutionary inference from genomes focus on the genome itself and explaining variation across SNPs in terms of selection pressures inferred separately on each SNP. Higher-order interactions and pathways underlying particular traits are harder to establish. In order to document selection pressures acting across clades or suites of co-occurring species, we need new methods that infer gene interaction networks, either connecting them with measured trait variation or inferring 'hidden' traits that map selection from the environment to variants in individual genes. Methods are being developed to infer gene interactions (epistasis) from sequence variation (Jerison and Desai, 2015; Otwinowski et al., 2018; Behrouzi and Wit, 2019), which could be extended for this purpose. There will be limits to our ability to map from selection on phenotypes to genotypes for non-model organisms, especially in organisms that lack practical methods for genetic manipulation. But by taking advantage of unprecedented replication within and between species, supplemented by phenotypic trait measurements at similar scales, a 10-year goal is to be able to infer whole-genome and -phenome architecture in relation to selection pressures acting across a clade of non-model organisms.

11.6 Predicting evolution in diverse systems

Evolutionary studies of species have mainly focused on the past and explaining origins of species and where diversity came from. A central theme to this book is that the species model is about fate as well as history, and that species (or an alternative model of diversity where applicable) are vital to predicting ongoing and future change. Traditionally, evolutionary biologists and ecologists have shied away from prediction for the obvious reason that our study systems are viciously complex and it is very hard to gain sufficient understanding to derive useful forward predictions. Something unexpected always occurs, such as a parasite shifting host or a mutation changing the evolutionary context.

My strong feeling is that this will no longer do. We live in an increasingly crowded world in which challenges and opportunities from natural systems are ever intensified. In order to justify the relevance and importance of our science, we need to push towards predictive knowledge that allows improved management of evolutionary phenomena. And the ability to predict evolution requires an ability to predict responses of multispecies systems. This takes us outside our natural comfort zone of tractable models and simple, controlled experiments, but this is the reality of the living world, and it is our job to deal with it.

Like evolutionary biology, astrophysics delves into wondrous origins over timescales that are hard for us to perceive. But it also uses predictive understanding to steer man-made objects to precise, far-off destinations simply using the gravitational pull of the planets. Predicting evolution is harder than this—more akin to predicting weather—but unless we attempt it, and use that knowledge to manage our interactions with the living world to our own benefit, evolutionary biology remains a half-science of post hoc explanation. There will be limits on predictability, but, just as in weather forecasting, these can be quantified and improved with advances in the underlying models, data, and computational power. It is time for evolutionary biologists to embrace diversity and to look to the future.

References

Abrams, P. A. 2000. The evolution of predator–prey interactions: theory and evidence. *Annual Review of Ecology and Systematics*, **31**, 79–105.

Abrams, P. A. and Matsuda, H. 1994. The evolution of traits that determine ability in competitive contests. *Evolutionary Ecology*, **8**, 667–86.

Acinas, S. G., Klepac-Ceraj, V., Hunt, D. E., et al. 2004. Fine-scale phylogenetic architecture of a complex bacterial community. *Nature*, **430**, 551–4.

Ackerly, D. D., Schwilk, D. W., and Webb, C. O. 2006. Niche evolution and adaptive radiation: testing the order of trait divergence. *Ecology*, **87**, S50–61.

Adams, D. C. 2010. Parallel evolution of character displacement driven by competitive selection in terrestrial salamanders. *BMC Evolutionary Biology*, **10**, 72.

Aguilée, R., Gascuel, F., Lambert, A., and Ferriere, R. 2018. Clade diversification dynamics and the biotic and abiotic controls of speciation and extinction rates. *Nature Communications*, **9**, 3013.

Alberti, M., Marzluff, J., and Hunt, V. M. 2017. Urban driven phenotypic changes: empirical observations and theoretical implications for eco-evolutionary feedback. *Philosophical Transactions of the Royal Society B: Biological Sciences*, **372**, 20160029.

Al-Ghazzewi, F. H. and Tester, R. F. 2016. Biotherapeutic agents and vaginal health. *Journal of Applied Microbiology*, **121**, 18–27.

Allmon, W. D. 1992. A causal analysis of stages in allopatric speciation. In: Futuyma, D. and Antonovics, J. (eds) *Oxford Surveys in Evolutionary Biology*. Oxford: Oxford University Press.

Alroy, J. 2008. Dynamics of origination and extinction in the marine fossil record. *Proceedings of the National Academy of Sciences of the United States of America*, **105**, 11536–42.

Alzate, A., Janzen, T., Bonte, D., Rosindell, J., and Etienne, R. S. 2017. A simple spatially explicit neutral model explains range size distribution of reef fishes. *bioRxiv*, doi: https://doi.org/10.1101/238600.

Arnegard, M. E., Mcintyre, P. B., Harmon, L. J., et al. 2010. Sexual signal evolution outpaces ecological divergence during electric fish species radiation. *The American Naturalist*, **176**, 335–56.

Aze, T., Ezard, T. H., Purvis, A. , Coxall, H. K., Stewart, D. R., Wade, B. S. and Pearson, P. N. 2011. A phylogeny of Cenozoic macroperforate planktonic foraminifera from fossil data. *Biological Reviews*, **86**, 900–927.

Baack, E., Melo, M. C., Rieseberg, L. H., and Ortiz-Barrientos, D. 2015. The origins of reproductive isolation in plants. *New Phytologist*, **207**, 968–84.

Balint, M., Pfenninger, M., Grossart, H. P., et al 2018. Environmental DNA time series in ecology. *Trends in Ecology & Evolution*, **33**, 945–57.

Bantinaki, E., Kassen, R., Knight, C. G., Robinson, Z., Spiers, A. J., and Rainey, P. B. 2007. Adaptive divergence in experimental populations of *Pseudomonas fluorescens*. III. Mutational origins of wrinkly spreader diversity. *Genetics*, **176**, 441–53.

Barley, A. J., Brown, J. M., and Thomson, R. C. 2018. Impact of model violations on the inference of species boundaries under the multispecies coalescent. *Systematic Biology*, **269**, 269–84.

Barluenga, M., Stolting, K. N., Salzburger, W., Muschick, M., and Meyer, A. 2006. Sympatric speciation in Nicaraguan crater lake cichlid fish. *Nature*, **439**, 719–23.

Barraclough, T. G. 2006. What can phylogenetics tell us about speciation in the Cape flora? *Diversity and Distributions*, **12**, 21–6.

Barraclough, T. G. 2010. Evolving entities: towards a unified framework for understanding diversity at the species and higher levels. *Philosophical Transactions of the Royal Society B: Biological Sciences*, **365**, 1801–13.

Barraclough, T. G. 2015. How do species interactions affect evolutionary dynamics across whole communities? *Annual Review of Ecology, Evolution, and Systematics*, **46**, 25–48.

Barraclough, T. G. 2019. Species matter for predicting the functioning of evolving microbial communities. Unpublished work.

Barraclough, T. G. and Herniou, E. 2003. Why do species exist? Insights from sexuals and asexuals. *Zoology*, **106**, 275–82.

Barraclough, T. G. and Humphreys, A. M. 2015. The evolutionary reality of species and higher taxa in plants: a survey of post-modern opinion and evidence. *New Phytologist*, **207**, 291–6.

Barraclough, T. G. and Nee, S. 2001. Phylogenetics and speciation. *Trends in Ecology & Evolution*, **16**, 391–9.

Barraclough, T. G. and Savolainen, V. 2001. Evolutionary rates and species diversity in flowering plants. *Evolution*, **55**, 677–83.

Barraclough, T. G. and Vogler, A. P. 2000. Detecting the geographical pattern of speciation from species-level phylogenies. *The American Naturalist*, **155**, 419–34.

Barraclough, T. G. and Vogler, A. P. 2002. Recent diversification rates in North American tiger beetles estimated from a dated mtDNA phylogenetic tree. *Molecular Biology and Evolution*, **19**, 1706–16.

Barraclough, T. G., Harvey, P. H., and Nee, S. 1995. Sexual selection and taxonomic diversity in passerine birds. *Proceedings of the Royal Society B: Biological Sciences*, **259**, 211–15.

Barraclough, T. G., Harvey, P. H., and Nee, S. 1996. Rate of rbcL gene sequence evolution and species diversification in flowering plants (angiosperms). *Proceedings of the Royal Society B: Biological Sciences*, **263**, 589–91.

Barraclough, T. G., Barclay, M. V. L., and Vogler, A. P. 1998a. Species richness: does flower power explain beetle-mania? *Current Biology*, **8**, R843–5.

Barraclough, T. G., Vogler, A. P., and Harvey, P. H. 1998b. Revealing the factors that promote speciation. *Philosophical Transactions of the Royal Society of London Series B: Biological Sciences*, **353**, 241–9.

Barraclough, T. G., Hogan, J. E., and Vogler, A. P. 1999. Testing whether ecological factors promote cladogenesis in a group of tiger beetles (Coleoptera: Cicindelidae). *Proceedings of the Royal Society B: Biological Sciences*, **266**, 1061–7.

Barraclough, T. G., Birky, C. W., and Burt, A. 2003. Diversification in sexual and asexual organisms. *Evolution*, **57**, 2166–72.

Barraclough, T. G., Fontaneto, D., Ricci, C., and Herniou, E. A. 2007. Evidence for inefficient selection against deleterious mutations in cytochrome oxidase I of asexual bdelloid rotifers. *Molecular Biology and Evolution*, **24**, 1952–62.

Barraclough, T. G., Fontaneto, D., Herniou, E. A., and Ricci, C. 2008. The evolutionary nature of diversification in sexuals and asexuals. In: Butlin, R., Bridle, J., and Schluter, D. (eds) *Speciation and Ecology*. Cambridge: Cambridge University Press.

Barraclough, T. G., Hughes, M., Ashford-Hodges, N., and Fujisawa, T. 2009. Inferring evolutionarily significant units of bacterial diversity from broad environmental surveys of single-locus data. *Biology Letters*, **5**, 425–8.

Barraclough, T. G., Balbi, K. J., and Ellis, R. J. 2012. Evolving concepts of bacterial species. *Evolutionary Biology*, **39**, 148–57.

Barroso-Batista, J., Sousa, A., Lourenco, M., Bergman, M. L., Sobral, D., Demengeot, J., Xavier, K. B., and Gordo, I. 2014. The first steps of adaptation of *Escherichia coli* to the gut are dominated by soft sweeps. *PLoS Genetics*, **10**, e1004182.

Barth, J. M. I., Berg, P. R., Jonsson, P. R., et al. 2017. Genome architecture enables local adaptation of Atlantic cod despite high connectivity. *Molecular Ecology*, **26**, 4452–66.

Bazin, E., Glemin, S., and Galtier, N. 2006. Population size does not influence mitochondrial genetic diversity in animals. *Science*, **312**, 570–2.

Behrouzi, P. and Wit, E. C. 2019. Detecting epistatic selection with partially observed genotype data by using copula graphical models. *Journal of the Royal Statistical Society Series C: Applied Statistics*, **68**, 141–60.

Bell, G. 1982. *The Masterpiece of Nature*. Berkeley: University of California Press.

Bell, G. 2007. The evolution of trophic structure. *Heredity*, **99**, 494–505.

Bell, T. 2010. Experimental tests of the bacterial distance–decay relationship. *ISME Journal*, **4**, 1357–65.

Bell, T., Newman, J. A., Silverman, B. W., Turner, S. L., and Lilley, A. K. 2005. The contribution of species richness and composition to bacterial services. *Nature*, **436**, 1157–60.

Bendall, M. L., Stevens, S. L. R., Chan, L.-K., et al. 2016. Genome-wide selective sweeps and gene-specific sweeps in natural bacterial populations. *ISME Journal*, **10**, 1589–1601.

Bergsten, J., Bilton, D. T., Fujisawa, T., et al. 2012. The effect of geographical scale of sampling on DNA barcoding. *Systematic Biology*, **61**, 851–69.

Bergstrom, C. T., Lipsitch, M., and Levin, B. R. 2000. Natural selection, infectious transfer and the existence conditions for bacterial plasmids. *Genetics*, **155**, 1505–19.

Berkhout, J., Bosdriesz, E., Nikerel, E., et al. 2013. How biochemical constraints of cellular growth shape evolutionary adaptations in metabolism. *Genetics*, **194**, 505–12.

Bestion, E., Garcia-Carreras, B., Schaum, C. E., Pawar, S., and Yvon-Durocher, G. 2018. Metabolic traits predict the effects of warming on phytoplankton competition. *Ecology Letters*, **21**, 655–64.

Betts, A., Gray, C., Zelek, M., MacLean, R. C., King, K. C. 2018. High parasite diversity accelerates host adaptation and diversification. *Science*, **360**, 907–911.

Billard, S., López-Villavicencio, M., Hood, M. E., Giraud, T. 2012. Sex, outcrossing and mating types: unsolved questions in fungi and beyond. *Biology*, **25**, 1020–1038.

Bininda-Emonds, O. R. P., Cardillo, M., Jones, K. E., et al. 2007. The delayed rise of present-day mammals. *Nature*, **446**, 507–12.

Birand, A., Vose, A., and Gavrilets, S. 2012. Patterns of species ranges, speciation, and extinction. *The American Naturalist*, **179**, 1–21.

Birky, C. W. 2010. Positively negative evidence for asexuality. *Journal of Heredity*, **101**, S42–5.

Birky, C. W., Wolf, C., Maughan, H., Herbertson, L., and Henry, E. 2005. Speciation and selection without sex. *Hydrobiologia*, **546**, 29–45.

Birky, C. W., Adams, J., Gemmel, M., and Perry, J. 2010. Using population genetic theory and DNA sequences for species detection and identification in asexual organisms. *PLoS One*, **5**, e10609.

Bishop, C. J., Adams, J., Aanensen, D. M., Jordan, G. E., Kilian, M., Hanage, W. P., and Spratt, B. G. 2009. Assigning strains to bacterial species via the internet. *BMC Biology* **7**, 3.

Blanchard, J. L., Jennings, S., Holmes, R., et al. 2012. Potential consequences of climate change for primary production and fish production in large marine ecosystems. *Philosophical Transactions of the Royal Society B: Biological Sciences, 367*, 2979–89.

Bobay, L. M. and Ochman, H. 2017. Biological species are universal across life's domains. *Genome Biology and Evolution, 9*, 491–501.

Booker, T. R., Jackson, B. C., and Keightley, P. D. 2017. Detecting positive selection in the genome. *BMC Biology, 15*, 98.

Borgonie, G., Garcia-Moyano, A., Litthauer, D., et al. 2011. Nematoda from the terrestrial deep subsurface of South Africa. *Nature, 474*, 79–82.

Boschetti, C., Carr, A., Crisp, A., et al. 2012. Biochemical diversification through foreign gene expression in bdelloid rotifers. *PLoS Genetics, 8*, e1003035.

Brandt, L. J., Aroniadis, O. C., Mellow, M., et al. 2012. Long-term follow-up of colonoscopic fecal microbiota transplant for recurrent *Clostridium difficile* infection. *American Journal of Gastroenterology, 107*, 1079–87.

Brookfield, J. F. Y. 2016. Why are estimates of the strength and direction of natural selection from wild populations not congruent with observed rates of phenotypic change? *Bioessays, 38*, 927–34.

Browne, H. P., Forster, S. C., Anonye, B. O., et al 2016. Culturing of 'unculturable' human microbiota reveals novel taxa and extensive sporulation. *Nature, 533*, 543–6.

Buckling, A., Maclean, R. C., Brockhurst, M. A., and Colegrave, N. 2009. *The Beagle* in a bottle. *Nature, 457*, 824–9.

Burger, R. and Lynch, M. 1995. Evolution and extinction in a changing environment—a quantitative-genetic analysis. *Evolution, 49*, 151–63.

Burgin, C. J., Colella, J. P., Kahn, P. L., and Upham, N. S. 2018. How many species of mammals are there? *Journal of Mammalogy, 99*, 1–14.

Burt, A. 2000. Perspective: sex, recombination, and the efficacy of selection—was Weismann right? *Evolution, 54*, 337–51.

Bush, R. M., Bender, C. A., Subbarao, K., Cox, N. J., and Fitch, W. M. 1999. Predicting the evolution of human influenza A. *Science, 286*, 1921–5.

Butlin, R., Debelle, A., Kerth, C., et al. 2012. What do we need to know about speciation? *Trends in Ecology & Evolution, 27*, 27–39.

Cadillo-Quiroz, H., Didelot, X., Held, N. L., et al. 2012. Patterns of gene flow define species of thermophilic Archaea. *PLoS Biology, 10*, e1001265.

Cadotte, M. W., Livingstone, S. W., Yasui, S. L. E., et al. 2017. Explaining ecosystem multifunction with evolutionary models. *Ecology, 98*, 3175–87.

Carroll, A. C. and Wong, A. 2018. Plasmid persistence: costs, benefits, and the plasmid paradox. *Canadian Journal of Microbiology, 64*, 293–304.

Carrolo, M., Pinto, F. R., Melo-Cristino, J., and Ramirez, M. 2009. Pherotypes are driving genetic differentiation within *Streptococcus pneumoniae*. *BMC Microbiology, 9*, 191.

Carstens, B. C., Pelletier, T. A., Reid, N. M., and Satler, J. D. 2013. How to fail at species delimitation. *Molecular Ecology, 22*, 4369–83.

Case, T. J. and Taper, M. L. 2000. Interspecific competition, environmental gradients, gene flow, and the coevolution of species' borders. *The American Naturalist, 155*, 583–605.

Castiglia, R. 2014. Sympatric sister species in rodents are more chromosomally differentiated than allopatric ones: implications for the role of chromosomal rearrangements in speciation. *Mammal Review, 44*, 1–4.

Chan, K. O., Alexander, A. M., Grismer, L. L., et al. 2017. Species delimitation with gene flow: a methodological comparison and population genomics approach to elucidate cryptic species boundaries in Malaysian Torrent frogs. *Molecular Ecology, 26*, 5435–50.

Chesson, P. 2000. Mechanisms of maintenance of species diversity. *Annual Review of Ecology and Systematics*, **31**, 343–66.

Chevin, L. M. 2016. Species selection and random drift in macroevolution. *Evolution*, **70**, 513–25.

Chevin, L.-M., Lande, R., and Mace, G. M. 2010. Adaptation, plasticity, and extinction in a changing environment: towards a predictive theory. *PLoS Biology*, **8**, e1000357.

Chira, A. M., Cooney, C. R., Bright, J. A., et al. 2018. Correlates of rate heterogeneity in avian ecomorphological traits. *Ecology Letters*, **21**, 1505–14.

Chivian, D., Brodie, E. L., Alm, E. J., et al. 2008. Environmental genomics reveals a single-species ecosystem deep within earth. *Science*, **322**, 275–8.

Claramunt, S., Derryberry, E. P., Remsen, J. V., and Brumfield, R. T. 2012. High dispersal ability inhibits speciation in a continental radiation of passerine birds. *Proceedings of the Royal Society B: Biological Sciences*, **279**, 1567–74.

Clark, A. T. and Neuhauser, C. 2018. Harnessing uncertainty to approximate mechanistic models of interspecific interactions. *Theoretical Population Biology*, **123**, 35–44.

Clarke, C. R., Karl, S. A., Horn, R. L., et al. 2015. Global mitochondrial DNA phylogeography and population structure of the silky shark, *Carcharhinus falciformis*. *Marine Biology*, **162**, 945–55.

Clarkson, C. S., Weetman, D., Essandoh, J., et al. 2014. Adaptive introgression between *Anopheles* sibling species eliminates a major genomic island but not reproductive isolation. *Nature Communications*, **5**, 4248.

Clatworthy, A. E., Pierson, E., and Hung, D. T. 2007. Targeting virulence: a new paradigm for antimicrobial therapy. *Nature Chemical Biology*, **3**, 541–8.

Cohan, F. M. 2001. Bacterial species and speciation. *Systematic Biology*, **50**, 513–24.

Cohan, F. M. and Perry, E. B. 2007. A systematics for discovering the fundamental units of bacterial diversity. *Current Biology*, **17**, R373–86.

Collins, S. and Gardner, A. 2009. Integrating physiological, ecological and evolutionary change: a Price equation approach. *Ecology Letters*, **12**, 744–57.

Condamine, F. L., Clapham, M. E., and Kergoat, G. J. 2016. Global patterns of insect diversification: towards a reconciliation of fossil and molecular evidence? *Scientific Reports*, **6**, 19208.

Connallon, T. and Clark, A. G. 2014. Evolutionary inevitability of sexual antagonism. *Proceedings of the Royal Society B: Biological Sciences*, **281**, 20132123.

Connell, J. H. 1983. On the prevalence and relative importance of interspecific competition—evidence from field experiments. *The American Naturalist*, **122**, 661–96.

Cowling, R. M. and Holmes, P. M. 1992. Endemism and speciation in a lowland flora from the Cape floristic region. *Biological Journal of the Linnean Society*, **47**, 367–83.

Coyne, J. A. and Orr, H. A. 1989. Patterns of speciation in *Drosophila*. *Evolution*, **43**, 362–81.

Coyne, J. A. and Orr, H. A. 1997. 'Patterns of speciation in *Drosophila*' revisited. *Evolution*, **51**, 295–303.

Coyne, J. A. and Orr, H. A. 1998. The evolutionary genetics of speciation. *Philosophical Transactions of the Royal Society of London Series B: Biological Sciences*, **353**, 287–305.

Coyne, J. A. and Orr, H. A. 2004. *Speciation*. Sunderland, MA: Sinauer Associates.

Coyne, J. A. and Price, T. D. 2000. Little evidence for sympatric speciation in island birds. *Evolution*, **54**, 2166–71.

Crisp, M., Cook, L., and Steane, D. 2004. Radiation of the Australian flora: what can comparisons of molecular phylogenies across multiple taxa tell us about the evolution of diversity in present-day communities? *Philosophical Transactions of the Royal Society of London Series B: Biological Sciences*, **359**, 1551–71.

Cruickshank, T. E. and Hahn, M. W. 2014. Reanalysis suggests that genomic islands of speciation are due to reduced diversity, not reduced gene flow. *Molecular Ecology*, **23**, 3133–57.

Dalquen, D. A., Zhu, T. Q., and Yang, Z. H. 2017. Maximum likelihood implementation of an isolation-with-migration model for three species. *Systematic Biology,* **66**, 379–98.

Darwin, C. R. 1859. *On the Origin of Species.* London: John Murray.

Darwin, C. 1871. *The Descent of Man and Selection in Relation to Sex.* London: John Murray.

Dávalos, L. M. and Russell, A. L. 2014. Sex-biased dispersal produces high error rates in mitochondrial distance-based and tree-based species delimitation, *Journal of Mammalogy,* **95**, 781–91.

Davies, T. J., Barraclough, T. G., Chase, M. W., Soltis, P. S., Soltis, D. E., and Savolainen, V. 2004a. Darwin's abominable mystery: insights from a supertree of the angiosperms. *Proceedings of the National Academy of Sciences of the United States of America,* **101**, 1904–9.

Davies, T. J., Barraclough, T. G., Savolainen, V., and Chase, M. W. 2004b. Environmental causes for plant biodiversity gradients. *Philosophical Transactions of the Royal Society of London Series B: Biological Sciences,* **359**, 1645–56.

Davies, T. J., Savolainen, V., Chase, M. W., Moat, J., and Barraclough, T. G. 2004c. Environmental energy and evolutionary rates in flowering plants. *Proceedings of the Royal Society B: Biological Sciences,* **271**, 2195–2200.

Davies, T. J., Savolainen, V., Chase, M. W., Goldblatt, P., and Barraclough, T. G. 2005. Environment, area, and diversification in the species-rich flowering plant family Iridaceae. *The American Naturalist,* **166**, 418–25.

Davies, T. J., Meiri, S., Barraclough, T. G., and Gittleman, J. L. 2007. Species co-existence and character divergence across carnivores. *Ecology Letters,* **10**, 146–52.

Davis, J. I. and Nixon, K. C. 1992. Populations, genetic variation, and the delimitation of phylogenetic species. *Systematic Biology,* **41**, 421–35.

Davis, M. P., Midford, P. E., and Maddison, W. 2013. Exploring power and parameter estimation of the BiSSE method for analyzing species diversification. *BMC Evolutionary Biology,* **13**, 38.

De Mazancourt, C., Johnson, E., and Barraclough, T. G. 2008. Biodiversity inhibits species' evolutionary responses to changing environments. *Ecology Letters,* **11**, 380–8.

De Paepe, M., Gaboriau-Routhiau, V., Rainteau, D., Rakotobe, S., Taddei, F., and Cerf-Bensussan, N. 2011. Trade-off between bile resistance and nutritional competence drives *Escherichia coli* diversification in the mouse gut. *PLoS Genetics,* **7**, e1002107.

De Vos, J. M., Joppa, L. N., Gittleman, J. L., Stephens, P. R., and Pimm, S. L. 2015. Estimating the normal background rate of species extinction. *Conservation Biology,* **29**, 452–62.

Debortoli, N., Li, X., Eyres, I., et al. 2016. Genetic exchange among bdelloid rotifers is more likely due to horizontal gene transfer than to meiotic sex. *Current Biology,* **26**, 723–32.

Decaestecker, E., Gaba, S., Raeymaekers, J. a. M., et al. 2007. Host–parasite 'Red Queen' dynamics archived in pond sediment. *Nature,* **450**, 870–U16.

Degnan, J. H. and Rosenberg, N. A. 2006. Discordance of species trees with their most likely gene trees. *PLoS Genetics,* **2**, 762–8.

Dellicour, S. and Flot, J. F. 2015. Delimiting species-poor data sets using single molecular markers: a study of barcode gaps, haplowebs and GMYC. *Systematic Biology,* **64**, 900–908.

Delzenne, N. M., Neyrinck, A. M., Backhed, F., and Cani, P. D. 2011. Targeting gut microbiota in obesity: effects of prebiotics and probiotics. *Nature Reviews Endocrinology,* **7**, 639–46.

Desjardins-Proulx, P. and Gravel, D. 2012. How likely is speciation in neutral ecology? *American Naturalist,* **179**, 137–144.

Dettman, J. R., Jacobson, D. J., and Taylor, J. W. 2003a. A multilocus genealogical approach to phylogenetic species recognition in the model eukaryote *Neurospora. Evolution,* **57**, 2703–20.

Dettman, J. R., Jacobson, D. J., Turner, E., Pringle, A., and Taylor, J. W. 2003b. Reproductive isolation and phylogenetic divergence in *Neurospora*: comparing methods of species recognition in a model eukaryote. *Evolution,* **57**, 2721–41.

Dettman, J. R., Sirjusingh, C., Kohn, L. M., and Anderson, J. B. 2007. Incipient speciation by divergent adaptation and antagonistic epistasis in yeast. *Nature,* **447**, 585–8.

Devaux, C. and Lande, R. 2008. Incipient allochronic speciation due to non-selective assortative mating by flowering time, mutation and genetic drift. *Proceedings of the Royal Society B: Biological Sciences,* **275**, 2723–32.

Di Giuseppe, G., Dini, F., Vallesi, A., and Luporini, P. 2015. Genetic relationships in bipolar species of the protist ciliate, *Euplotes. Hydrobiologia,* **761**, 71–83.

Diabate, A., Dao, A., Yaro, A. S., et al. 2009. Spatial swarm segregation and reproductive isolation between the molecular forms of *Anopheles gambiae. Proceedings of the Royal Society B: Biological Sciences,* **276**, 4215–22.

Diehl, S. R. and Bush, G. L. 1989. The role of habitat preference in adaptation and speciation. In: Otte, D. and Endler, J. (eds) *Speciation and its Consequences.* Sunderland, MA: Sinauer Associates.

Dimitriu, T., Misevic, D., Lotton, C., Brown, S. P., Lindner, A. B., and Taddei, F. 2016. Indirect fitness benefits enable the spread of host genes promoting costly transfer of beneficial plasmids. *PLoS Biology,* **14**, e1002478.

Dittmar, E. L., Oakley, C. G., Conner, J. K., Gould, B. A., and Schemske, D. W. 2016. Factors influencing the effect size distribution of adaptive substitutions. *Proceedings of the Royal Society B: Biological Sciences,* **283**, 20153065.

Dixit, P. D., Pang, T. Y., Studier, F. W., and Maslov, S. 2015. Recombinant transfer in the basic genome of *Escherichia coli. Proceedings of the National Academy of Sciences of the United States of America,* **112**, 9070–5.

Dobzhansky, T. 1941. *Genetics and the Origin of Species.* New York: Columbia University Press.

Dochtermann, N. A. and Matocq, M. D. 2016. Speciation along a shared evolutionary trajectory. *Current Zoology,* **62**, 507–11.

Doerder, F. P. 2014. Abandoning sex: multiple origins of asexuality in the ciliate *Tetrahymena. BMC Evolutionary Biology,* **14**, 112.

Domes, K., Norton, R. A., Maraun, M., and Scheu, S. 2007. Reevolution of sexuality breaks Dollo's law. *Proceedings of the National Academy of Sciences of the United States of America,* **104**, 7139–44.

Doyle, J. J. 1995. The irrelevance of allele tree topologies for species delimitation, and a non-topological alternative. *Systematic Botany,* **20**, 574–88.

Draghi, J. A. and Whitlock, M. C. 2012. Phenotypic plasticity facilitates mutational variance, genetic variance, and evolvability along the major axis of environmental variation. *Evolution,* **66**, 2891–902.

Dunthorn, M. and Katz, L. A. 2010. Secretive ciliates and putative asexuality in microbial eukaryotes. *Trends in Microbiology,* **18**, 183–8.

Dunthorn, M., Zufall, R. A., Chi, J. Y., Paszkiewicz, K., Moore, K., and Mahe, F. 2017. Meiotic genes in colpodean ciliates support secretive sexuality. *Genome Biology and Evolution,* **9**, 1781–7.

Dupuis, J. R., Roe, A. D., and Sperling, F. a. H. 2012. Multi-locus species delimitation in closely related animals and fungi: one marker is not enough. *Molecular Ecology,* **21**, 4422–36.

Dynesius, M. and Jansson, R. 2014. Persistence of within-species lineages: a neglected control of speciation rates. *Evolution,* **68**, 923–34.

Eberle, J., Bazzato, E., Fabrizi, S., et al. 2019. Sex-biased dispersal obscures species boundaries in integrative species delimitation approaches. *Systematic Biology,* Early Online.

Edwards, D. L. and Knowles, L. L. 2014. Species detection and individual assignment in species delimitation: can integrative data increase efficacy? *Proceedings of the Royal Society B: Biological Sciences,* **281**, 20132765.

Edwards, S. V. and Beerli, P. 2000. Perspective: gene divergence, population divergence, and the variance in coalescence time in phylogeographic studies. *Evolution,* **54**, 1839–54.

Ellegren, H. 2014. Genome sequencing and population genomics in non-model organisms. *Trends in Ecology & Evolution,* **29**, 51–63.

Ellegren, H., Smeds, L., Burri, R., et al. 2012. The genomic landscape of species divergence in Ficedula flycatchers. *Nature,* **491**, 756–60.

Enard, D., Cai, L., Gwennap, C., and Petrov, D. A. 2016. Viruses are a dominant driver of protein adaptation in mammals. *Elife,* **5**, e12469.

Evans, M. R., Norris, K. J., and Benton, T. G. 2012. Predictive ecology: systems approaches. Introduction. *Philosophical Transactions of the Royal Society B: Biological Sciences,* **367**, 163–9.

Ewens, W. J. 1979. *Mathematical Population Genetics.* Berlin: Springer.

Eyres, I., Boschetti, C., Crisp, A., et al. 2015. Horizontal gene transfer in bdelloid rotifers is ancient, ongoing and more frequent in species from desiccating habitats. *BMC Biology,* **13**, 90.

Ezard, T. H. G., Pearson, P. N., and Purvis, A. 2010. Algorithmic approaches to aid species' delimitation in multidimensional morphospace. *BMC Evolutionary Biology,* **10**, 175.

Ezard, T. H. G., Aze, T., Pearson, P. N., and Purvis, A. 2011. Interplay between changing climate and species' ecology drives macroevolutionary dynamics. *Science,* **332**, 349–51.

Farrell, B. D. 1998. 'Inordinate fondness' explained: why are there so many beetles? *Science,* **281**, 555–9.

Farrington, H. L., Lawson, L. P., Clark, C. M., and Petren, K. 2014. The evolutionary history of Darwin's finches: speciation, gene flow, and introgression in a fragmented landscape. *Evolution,* **68**, 2932–44.

Felsenstein, J. 1981. Skepticism towards Santa Rosalia, or why are there so few kinds of animals? *Evolution,* **35**, 124–38.

Felsenstein, J. 1985. Phylogenies and the comparative method. *The American Naturalist* **125**, 1–15.

Fiegna, F., Moreno-Letelier, A., Bell, T., and Barraclough, T. G. 2015a. Evolution of species interactions determines microbial community productivity in new environments. *ISME Journal,* **9**, 1235–45.

Fiegna, F., Scheuerl, T., Moreno-Letelier, A., Bell, T., and Barraclough, T. G. 2015b. Saturating effects of species diversity on life-history evolution in bacteria. *Proceedings of the Royal Society B: Biological Sciences,* **282**, 20151794.

Fisher, R. A. 1930. *The Genetical Theory of Natural Selection.* Oxford: Oxford University Press.

Fitzpatrick, B. M. and Turelli, M. 2006. The geography of mammalian speciation: mixed signals from phylogenies and range maps. *Evolution,* **60**, 601–15.

Flint, H. J., Scott, K. P., Louis, P., and Duncan, S. H. 2012. The role of the gut microbiota in nutrition and health. *Nature Reviews Gastroenterology & Hepatology,* **9**, 577–89.

Flot, J. F., Hespeels, B., Li, X., et al. 2013. Genomic evidence for ameiotic evolution in the bdelloid rotifer *Adineta vaga. Nature,* **500**, 453–7.

Foley, N. M., Springer, M. S., and Teeling, E. C. 2016. Mammal madness: is the mammal tree of life not yet resolved? *Philosophical Transactions of the Royal Society B: Biological Sciences,* **371**, 20150140.

Fontaneto, D. and Barraclough, T. G. 2015. Do species exist in asexuals? Theory and evidence from bdelloid rotifers. *Integrative and Comparative Biology,* **55**, 253–63.

Fontaneto, D., Herniou, E. A., Boschetti, C., et al. 2007. Independently evolving species in asexual bdelloid rotifers. *PLoS Biology,* **5**, 914–21.

Foster, K. R. and Bell, T. 2012. Competition, not cooperation, dominates interactions among culturable microbial species. *Current Biology,* **22**, 1845–50.

France, M. T., Mendes-Soares, H., and Forney, L. J. 2016. Genomic comparisons of *Lactobacillus crispatus* and *Lactobacillus iners* reveal potential ecological drivers of community composition in the vagina. *Applied and Environmental Microbiology,* **82**, 7063–73.

Fraser, C., Hanage, W. P., and Spratt, B. G. 2007. Recombination and the nature of bacterial speciation. *Science,* **315**, 476–80.

Friedman, J., Alm, E. J., and Shapiro, B. J. 2013. Sympatric speciation: when is it possible in bacteria? *PLoS One,* **8**, e53539.

Friesen, V. L., Smith, A. L., Gomez-Diaz, E., et al. R. 2007. Sympatric speciation by allochrony in a seabird. *Proceedings of the National Academy of Sciences of the United States of America,* **104**, 18589–94.

Frost, G. S., Walton, G. E., Swann, J. R., et al. 2014. Impacts of plant-based foods in ancestral hominin diets on the metabolism and function of gut microbiota in vitro. *MBio,* **5**, e00853-14.

Fujisawa, T. and Barraclough, T. G. 2013. Delimiting species using single-locus data and the Generalized Mixed Yule Coalescent approach: a revised method and evaluation on simulated data sets. *Systematic Biology,* **62**, 707–24.

Fujisawa, T., Vogler, A. P., and Barraclough, T. G. 2015. Ecology has contrasting effects on genetic variation within species versus rates of molecular evolution across species in water beetles. *Proceedings of the Royal Society B: Biological Sciences,* **282**, 20142476.

Fujisawa, T., Aswad, A., and Barraclough, T. G. 2016. A rapid and scalable method for multilocus species delimitation using Bayesian model comparison and rooted triplets. *Systematic Biology,* **65**, 759–71.

Fujita, M. K., Leache, A. D., Burbrink, F. T., McGuire, J. A., and Moritz, C. 2012. Coalescent-based species delimitation in an integrative taxonomy. *Trends in Ecology & Evolution,* **27**, 480–8.

Funk, D. J. and Omland, K. E. 2003. Species-level paraphyly and polyphyly: frequency, causes, and consequences, with insights from animal mitochondrial DNA. *Annual Review of Ecology Evolution and Systematics,* **34**, 397–423.

Funk, D. J., Nosil, P., and Etges, W. J. 2006. Ecological divergence exhibits consistently positive associations with reproductive isolation across disparate taxa. *Proceedings of the National Academy of Sciences of the United States of America,* **103**, 3209–13.

Garud, N. R., Good, B. H., Hallatschek, O., and Pollard, K. S. 2017. Evolutionary dynamics of bacteria in the gut microbiome within and across hosts. *bioRxiv,* doi: https://doi.org/10.1101/210955.

Gavrilets, S. 2003. Perspective: models of speciation: what have we learned in 40 years? *Evolution,* **57**, 2197–215.

Gavrilets, S. 2004. *Fitness Landscapes and the Origin of Species.* Princeton, NJ: Princeton University Press.

Gavrilets, S. and Losos, J. B. 2009. Adaptive radiation: contrasting theory with data. *Science,* **323**, 732–7.

Ghoul, M. and Mitri, S. 2016. The ecology and evolution of microbial competition. *Trends in Microbiology,* **24**, 833–45.

Gianoli, E. 2004. Evolution of a climbing habit promotes diversification in flowering plants. *Proceedings of the Royal Society B: Biological Sciences,* **271**, 2011–15.

Gillespie, R. 2004. Community assembly through adaptive radiation in Hawaiian spiders. *Science,* **303**, 356–9.

Gillooly, J. F., Allen, A. P., West, G. B., and Brown, J. H. 2005. The rate of DNA evolution: effects of body size and temperature on the molecular clock. *Proceedings of the National Academy of Sciences of the United States of America,* **102**, 140–5.

Giraud, T. and Gourbiere, S. 2012. The tempo and modes of evolution of reproductive isolation in fungi. *Heredity,* **109**, 204–14.

Gladyshev, E. A., Meselson, M., and Arkhipova, I. R. 2008. Massive horizontal gene transfer in bdelloid rotifers. *Science,* **320,** 1210–13.

Goddard, M. R. and Burt, A. 1999. Recurrent invasion and extinction of a selfish gene. *Proceedings of the National Academy of Sciences of the United States of America,* **96,** 13880–5.

Gomez, P. and Buckling, A. 2011. Bacteria–phage antagonistic coevolution in soil. *Science,* **332,** 106–9.

Gonzalez-Voyer, A. and Kolm, N. 2011. Rates of phenotypic evolution of ecological characters and sexual traits during the Tanganyikan cichlid adaptive radiation. *Journal of Evolutionary Biology,* **24,** 2378–88.

Gould, F., Brown, Z. S., and Kuzma, J. 2018. Wicked evolution: can we address the sociobiological dilemma of pesticide resistance? *Science,* **360,** 728–32.

Gourbiere, S. and Mallet, J. 2010. Are species real? The shape of the species boundary with exponential failure, reinforcement, and the 'missing snowball'. *Evolution,* **64,** 1–24.

Grandcolas, P., Murienne, J., Robillard, T., et al. 2008. New Caledonia: a very old Darwinian island? *Philosophical Transactions of the Royal Society B: Biological Sciences,* **363,** 3309–17.

Grant, P. R. and Grant, B. R. 2009. The secondary contact phase of allopatric speciation in Darwin's finches. *Proceedings of the National Academy of Sciences of the United States of America,* **106,** 20141–8.

Gray, J. C. and Goddard, M. R. 2012. Gene-flow between niches facilitates local adaptation in sexual populations. *Ecology Letters,* **15,** 955–62.

Green, J. L. and Plotkin, J. B. 2007. A statistical theory for sampling species abundances. *Ecology Letters,* **10,** 1037–45.

Grossenbacher, D. L. and Whittall, J. B. 2011. Increased floral divergence in sympatric monkey-flowers. *Evolution,* **65,** 2712–18.

Guimarães, P. R., Jordano, P., and Thompson, J. N. 2011. Evolution and coevolution in mutualistic networks. *Ecology Letters,* **14,** 877–85.

Gwynne, B. J., Gordon, T. R., and Davis, R. M. 1997. A new race of *Fusarium oxysporum* f. sp. *melonis* causing Fusarium wilt of muskmelon in the Central Valley of California. *Plant Disease,* **81,** 1095.

Hanemaaijer, M. J., Collier, T. C., Change, A., et al. 2018. The fate of genes that cross species boundaries after a major hybridization event in a natural mosquito population. *Molecular Ecology,* **27,** 4978–4990.

Hansen, T. F. 2006. The evolution of genetic architecture. *Annual Review of Ecology Evolution and Systematics,* **37,** 123–57.Hansen, T. F. and Houle, D. 2008. Measuring and comparing evolvability and constraint in multivariate characters. *Journal of Evolutionary Biology,* **21,** 1201–19.

Hanson, S. J., Schurko, A. M., Hecox-Lea, B., Mark Welch, D. B., Stelzer, C.-P., Logsdon, J. M. 2013. Inventory and phylogenetic analysis of meiotic genes in monogonont rotifers, Journal of Heredity, **104,** 357–370.

Harmon, L. J. 2018. *Phylogenetic Comparative Methods: Learning from Trees.* CreateSpace.

Harmon, L. J. and Harrison, S. 2015. Species diversity is dynamic and unbounded at local and continental scales. *The American Naturalist,* **185,** 584–93.

Harmon, L. J., Weir, J. T., Brock, C. D., Glor, R. E., and Challenger, W. 2008. GEIGER: investigating evolutionary radiations. *Bioinformatics,* **24,** 129–31.

Harris, L. W. and Davies, T. J. 2016. A complete fossil-calibrated phylogeny of seed plant families as a tool for comparative analyses: testing the 'Time for Speciation' hypothesis. *PLoS One,* **11,** e0162907.

Harrison, E. and Brockhurst, M. A. 2012. Plasmid-mediated horizontal gene transfer is a coevolutionary process. *Trends in Microbiology,* **20,** 262–7.

Harrison, E., Guymer, D., Spiers, A. J., Paterson, S., and Brockhurst, M. A. 2015. Parallel compensatory evolution stabilizes plasmids across the parasitism–mutualism continuum. *Current Biology*, **25**, 2034–9.

Harrison, R. G. and Larson, E. L. 2014. Hybridization, introgression, and the nature of species boundaries. *Journal of Heredity*, **105**, 795–809.

Harvey, M. G., Seeholzer, G. F., Smith, B. T., Rabosky, D. L., Cuervo, A. M., and Brumfield, R. T. 2017. Positive association between population genetic differentiation and speciation rates in New World birds. *Proceedings of the National Academy of Sciences of the United States of America*, **114**, 6328–33.

Hausdorf, B. and Hennig, C. 2010. Species delimitation using dominant and codominant multilocus markers. *Systematic Biology*, **59**, 491–503.

Hausdorff, W. P. and Hanage, W. P. 2016. Interim results of an ecological experiment—conjugate vaccination against the pneumococcus and serotype replacement. *Human Vaccines & Immunotherapeutics*, **12**, 358–74.

Hayward, A. D., Pemberton, J. M., Berenos, C., Wilson, A. J., Pilkington, J. G., and Kruuk, L. E. B. 2018. Evidence for selection-by-environment but not genotype-by-environment interactions for fitness-related traits in a wild mammal population. *Genetics*, **208**, 349–64.

Hazzi, N. A., Moreno, J. S., Ortiz-Movliav, C., and Palacio, R. D. 2018. Biogeographic regions and events of isolation and diversification of the endemic biota of the tropical Andes. *Proceedings of the National Academy of Sciences of the United States of America*, **115**, 7985–90.

Heard, S. B. 1992. Patterns in tree balance among cladistic, phenetic, and randomly generated phylogenetic trees. *Evolution*, **46**, 1818–26.

Hebert, P. D. N., Cywinska, A., Ball, S. L., and Dewaard, J. R. 2003. Biological identifications through DNA barcodes. *Proceedings of the Royal Society B: Biological Sciences*, **270**, 313–21.

Hein, J., Schierup, M. H., and Wiuf, C. 2005. *Gene Genealogies, Variation and Evolution: A Primer in Coalescent Theory*. Oxford: Oxford University Press.

Hendry, A. P. 2017. *Eco-evolutionary Dynamics*. Princeton, NJ: Princeton University Press.

Hendry, A. P., Huber, S. K., De Leon, L. F., Herrel, A., and Podos, J. 2009. Disruptive selection in a bimodal population of Darwin's finches. *Proceedings of the Royal Society B: Biological Sciences*, **276**, 753–9.

Hennig, W. 1966. *Phylogenetic Systematics*. Urbana: University of Illinois Press.

Herniou, E., Theze, J., Cory, J., and Vaamonde, C. L. 2015. Diversity, species delimitation, and evolution of insect viruses. *Genome*, **58**, 226.

Herron, M. D. and Doebeli, M. 2013. Parallel evolutionary dynamics of adaptive diversification in *Escherichia coli*. *PLoS Biology*, **11**, e1001490.

Hey, J. 1991. The structure of genealogies and the distribution of fixed differences between DNA sequence samples from natural populations. *Genetics*, **128**, 831–40.

Hey, J. 2001. The mind of the species problem. *Trends in Ecology & Evolution*, **16**, 326–9.

Higgs, P. G. and Derrida, B. 1992. Genetic distance and species formation in evolving populations. *Journal of Molecular Evolution*, **35**, 454–65.

Hillis, D. M., Bull, J. J., White, M. E., Badgett, M. R., and Molineux, I. J. 1992. Experimental phylogenetics—generation of a known phylogeny. *Science*, **255**, 589–92.

Holmes, E. C. 2013. What can we predict about viral evolution and emergence? *Current Opinion in Virology*, **3**, 180–4.

Holmes, M. W., Hammond, T. T., Wogan, G. O. U., et al. 2016. Natural history collections as windows on evolutionary processes. *Molecular Ecology*, **25**, 864–81.

Holt, B. G. and Jonsson, K. A. 2014. Reconciling hierarchical taxonomy with molecular phylogenies. *Systematic Biology*, **63**, 1010–17.

Hooper, D. M. and Price, T. D. 2017. Chromosomal inversion differences correlate with range overlap in passerine birds. *Nature Ecology & Evolution,* **1**, 1526–34.

Hopkins, R. 2013. Reinforcement in plants. *New Phytologist,* **197**, 1095–103.

Hou, J., Friedrich, A., De Montigny, J., and Schacherer, J. 2014. Chromosomal rearrangements as a major mechanism in the onset of reproductive isolation in *Saccharomyces cerevisiae. Current Biology,* **24**, 1153–9.

Huang, Y. J. and LiPuma, J. J. 2016. The microbiome in cystic fibrosis. *Clinics in Chest Medicine,* **37**, 59–67.

Hubbell, S. P. 2001. *The Unified Neutral Theory of Biodiversity and Biogeography.* Princeton, NJ: Princeton University Press.

Hudson, L. N., Blagoderov, V., Heaton, A., et al. 2015. Inselect: automating the digitization of natural history collections. *PLoS One,* **10**, e0143402.

Hudson, R. R. 1991. Gene genealogies and the coalescent process. *Oxford Surveys in Evolutionary Biology,* **7**, 1–44.

Hudson, R. R. and Coyne, J. A. 2002. Mathematical consequences of the genealogical species concept. *Evolution,* **56**, 1557–65.

Hufford, K. M. and Mazer, S. J. 2003. Plant ecotypes: genetic differentiation in the age of ecological restoration. *Trends in Ecology & Evolution,* **18**, 147–55.

Hughes, C. and Eastwood, R. 2006. Island radiation on a continental scale: exceptional rates of plant diversification after uplift of the Andes. *Proceedings of the National Academy of Sciences of the United States of America,* **103**, 10334–9.

Humphreys, A. M. and Barraclough, T. G. 2014. The evolutionary reality of higher taxa in mammals. *Proceedings of the Royal Society B: Biological Sciences,* **281**, 20132750.

Humphreys, A. M., Rydin, C., Jonsson, K. A., Alsop, D., Callender-Crowe, L. M., and Barraclough, T. G. 2016. Detecting evolutionarily significant units above the species level using the Generalised Mixed Yule Coalescent method. *Methods in Ecology and Evolution,* **7**, 1366–75.

Hunt, T., Bergsten, J., Levkanicova, Z., et al. 2007. A comprehensive phylogeny of beetles reveals the evolutionary origins of a superradiation. *Science,* **318**, 1913–16.

Hunter, D. C., Pemberton, J. M., Pilkington, J. G., and Morrissey, M. B. 2018. Quantification and decomposition of environment–selection relationships. *Evolution,* **72**, 851–66.

Hunter, J. P. 1998. Key innovations and the ecology of macroevolution. *Trends in Ecology & Evolution,* **13**, 31–6.

Huttenhower, C., Gevers, D., Knight, R., et al. 2012. Structure, function and diversity of the healthy human microbiome. *Nature,* **486**, 207–14.

Igea, J., Bogarin, D., Papadopulos, A. S. T., and Savolainen, V. 2015. A comparative analysis of island floras challenges taxonomy-based biogeographical models of speciation. *Evolution,* **69**, 482–91.

Ingram, T. 2011. Speciation along a depth gradient in a marine adaptive radiation. *Proceedings of the Royal Society B: Biological Sciences,* **278**, 613–18.

Jackson, N. D., Carstens, B. C., Morales, A. E., and O'Meara, B. C. 2017. Species delimitation with gene flow. *Systematic Biology,* **66**, 799–812.

James, J. E., Piganeau, G., and Eyre-Walker, A. 2016. The rate of adaptive evolution in animal mitochondria. *Molecular Ecology,* **25**, 67–78.

Jansson, R. and Davies, T. J. 2008. Global variation in diversification rates of flowering plants: energy vs. climate change. *Ecology Letters,* **11**, 173–83.

Jeltsch, A. 2003. Maintenance of species identity and controlling speciation of bacteria: a new function for restriction/modification systems? *Gene,* **317**, 13–16.

Jerison, E. R. and Desai, M. M. 2015. Genomic investigations of evolutionary dynamics and epistasis in microbial evolution experiments. *Current Opinion in Genetics & Development*, **35**, 33–9.

Jetz, W., Thomas, G. H., Joy, J. B., Hartmann, K., and Mooers, A. O. 2012. The global diversity of birds in space and time. *Nature*, **491**, 444–8.

Jiggins, C. D., Mallarino, R., Willmott, K. R., and Bermingham, E. 2006. The phylogenetic pattern of speciation and wing pattern change in neotropical *Ithomia* butterflies (Lepidoptera: Nymphalidae). *Evolution*, **60**, 1454–66.

Johansson, J. 2008. Evolutionary responses to environmental changes: how does competition affect adaptation? *Evolution*, **62**, 421–35.

Johnson, L. P., Walton, G. E., Swann, J. R., Frost, G., Gibson, G. R., and Barraclough, T. G. 2019. Ecological and evolutionary dynamics of human faecal communities in vitro in response to inulin provision as a sole carbon source. Unpublished work.

Jones, G., Aydin, Z., and Oxelman, B. 2015. DISSECT: an assignment-free Bayesian discovery method for species delimitation under the multispecies coalescent. *Bioinformatics*, **31**, 991–8.

Jonsson, H., Schubert, M., Seguin-Orlando, A., et al. 2014. Speciation with gene flow in equids despite extensive chromosomal plasticity. *Proceedings of the National Academy of Sciences of the United States of America*, **111**, 18655–60.

Jordan, D. S. 1905. The origin of species through isolation. *Science*, **22**, 545–62.

Jörger, K. M. and Schrödl, M. 2013. How to describe a cryptic species? Practical challenges of molecular taxonomy. *Frontiers in Zoology*, **10**, 59.

Keller, I., Wagner, C. E., Greuter, L., et al. 2013. Population genomic signatures of divergent adaptation, gene flow and hybrid speciation in the rapid radiation of Lake Victoria cichlid fishes. *Molecular Ecology*, **22**, 2848–63.

Kerr, K. C. R. 2011. Searching for evidence of selection in avian DNA barcodes. *Molecular Ecology Resources*, **11**, 1045–55.

Kerr, K. C. R., Stoeckle, M. Y., Dove, C. J., Weigt, L. A., Francis, C. M., and Hebert, P. D. N. 2007. Comprehensive DNA barcode coverage of North American birds. *Molecular Ecology Notes*, **7**, 535–43.

Kettle, H., Louis, P., Holtrop, G., Duncan, S. H., and Flint, H. J. 2015. Modelling the emergent dynamics and major metabolites of the human colonic microbiota. *Environmental Microbiology*, **17**, 1615–30.

Kingsolver, J. G., Diamond, S. E., Siepielski, A. M., and Carlson, S. M. 2012. Synthetic analyses of phenotypic selection in natural populations: lessons, limitations and future directions. *Evolutionary Ecology*, **26**, 1101–18.

Kinnison, M. T. and Hendry, A. P. 2001. The pace of modern life. II: From rates of contemporary microevolution to pattern and process. *Genetica*, **112**, 145–64.

Kisel, Y. and Barraclough, T. G. 2010. Speciation has a spatial scale that depends on levels of gene flow. *The American Naturalist*, **175**, 316–34.

Kisel, Y., Moreno-Letelier, A. C., Bogarin, D., Powell, M. P., Chase, M. W., and Barraclough, T. G. 2012. Testing the link between population genetic differentiation and clade diversification in Costa Rican orchids. *Evolution*, **66**, 3035–52.

Kleindorfer, S., O'Connor, J. A., Dudaniec, R. Y., Myers, S. A., Robertson, J., and Sulloway, F. J. 2014. Species collapse via hybridization in Darwin's tree finches. *The American Naturalist*, **183**, 325–41.

Knowles, L. L. and Carstens, B. C. 2007. Delimiting species without monophyletic gene trees. *Systematic Biology*, **56**, 887–95.

Koch, C., Konieczka, J., Delorey, T., et al. 2017. Inference and evolutionary analysis of genome-scale regulatory networks in large phylogenies. *Cell Systems*, **4**, 543–58.

Koeppel, A., Perry, E. B., Sikorski, J., et al. 2008. Identifying the fundamental units of bacterial diversity: a paradigm shift to incorporate ecology into bacterial systematics. *Proceedings of the National Academy of Sciences of the United States of America*, **105**, 2504–9.

Kopp, M. and Gavrilets, S. 2006. Multilocus genetics and the coevolution of quantitative traits. *Evolution*, **60**, 1321–36.

Koraimann, G. and Wagner, M. A. 2014. Social behavior and decision making in bacterial conjugation. *Frontiers in Cellular and Infection Microbiology*, **4**, 54.

Koskella, B. and Brockhurst, M. A. 2014. Bacteria-phage coevolution as a driver of ecological and evolutionary processes in microbial communities. *FEMS Microbiology Reviews*, **38**, 916–31.

Koufopanou, V., Burt, A., and Taylor, J. W. 1997. Concordance of gene genealogies reveals reproductive isolation in the pathogenic fungus *Coccidioides immitis*. *Proceedings of the National Academy of Sciences of the United States of America*, **94**, 5478–82.

Kovatcheva-Datchary, P., Egert, M., Maathuis, A., et al. 2009. Linking phylogenetic identities of bacteria to starch fermentation in an in vitro model of the large intestine by RNA-based stable isotope probing. *Environmental Microbiology*, **11**, 914–26.

Kozak, K. H. and Wiens, J. J. 2006. Does niche conservatism promote speciation? A case study in North American salamanders. *Evolution*, **60**, 2604–21.

Kraus, R. H. S., Kerstens, H. H. D., Van Hooft, P., et al. 2012. Widespread horizontal genomic exchange does not erode species barriers among sympatric ducks. *BMC Evolutionary Biology*, **12**, 45.

Krausmann, F., Erb, K. H., Gingrich, S., et al. 2013. Global human appropriation of net primary production doubled in the 20th century. *Proceedings of the National Academy of Sciences of the United States of America*, **110**, 10324–9.

Kubatko, L. S., Gibbs, H. L., and Bloomquist, E. W. 2011. Inferring species-level phylogenies and taxonomic distinctiveness using multilocus data in *Sistrurus* rattlesnakes. *Systematic Biology*, **60**, 393–409.

Kuehne, H. A., Murphy, H. A., Francis, C. A., and Sniegowski, P. D. 2007. Allopatric divergence, secondary contact and genetic isolation in wild yeast populations. *Current Biology*, **17**, 407–11.

Lackey, A. C. R. and Boughman, J. W. 2017. Evolution of reproductive isolation in stickleback fish. *Evolution*, **71**, 357–72.

Lahaye, R., Van Der Bank, M., Bogarin, D., et al. 2008. DNA barcoding the floras of biodiversity hotspots. *Proceedings of the National Academy of Sciences of the United States of America*, **105**, 2923–8.

Lahr, D. J. G., Parfrey, L. W., Mitchell, E. a. D., Katz, L. A., and Lara, E. 2011. The chastity of amoebae: re-evaluating evidence for sex in amoeboid organisms. *Proceedings of the Royal Society B: Biological Sciences*, **278**, 2081–90.

Lamanna, F., Kirschbaum, F., Ernst, A. R. R., Feulner, P. G. D., Mamonekene, V., Paul, C., and Tiedemann, R. 2016. Species delimitation and phylogenetic relationships in a genus of African weakly-electric fishes (Osteoglossiformes, Mormyridae, *Campylomormyrus*). *Molecular Phylogenetics and Evolution*, **101**, 8–18.

Lamsdell, J. C. 2016. Horseshoe crab phylogeny and independent colonizations of fresh water: ecological invasion as a driver for morphological innovation. *Palaeontology*, **59**, 181–94.

Lande, R. 1976. Natural selection and random genetic drift in phenotypic evolution. *Evolution*, **30**, 314–34.

Lande, R. 1979. Quantitative genetic-analysis of multivariate evolution, applied to brain–body size allometry. *Evolution*, **33**, 402–16.

Lande, R. 1981. Models of speciation by sexual selection on polygenic characters. *Proceedings of the National Academy of Sciences of the United States of America,* **78**, 3721–5.

Lande, R. 2014. Evolution of phenotypic plasticity and environmental tolerance of a labile quantitative character in a fluctuating environment. *Journal of Evolutionary Biology,* **27**, 866–75.

Lassig, M., Mustonen, V., and Walczak, A. M. 2017. Predicting evolution. *Nature Ecology & Evolution,* **1**, 0077.

Lavin, M., Schrire, B. P., Lewis, G., et al. 2004. Metacommunity process rather than continental tectonic history better explains geographically structured phylogenies in legumes. *Philosophical Transactions of the Royal Society B: Biological Sciences,* **359**, 1509–22.

Lawrence, D. and Barraclough, T. G. 2016. Evolution of resource use along a gradient of stress leads to increased facilitation. *Oikos,* **125**, 1284–95.

Lawrence, D., Fiegna, F., Behrends, V., et al. 2012. Species interactions alter evolutionary responses to a novel environment. *PLoS Biology,* **10**, e1001330.

Lawrence, D., Bell, T., and Barraclough, T. G. 2016. The effect of immigration on the adaptation of microbial communities to warming. *The American Naturalist,* **187**, 236–48.

Le Corre, V. and Kremer, A. 2012. The genetic differentiation at quantitative trait loci under local adaptation. *Molecular Ecology,* **21**, 1548–66.

Leache, A. D., Koo, M. S., Spencer, C. L., Papenfuss, T. J., Fisher, R. N., and McGuire, J. A. 2009. Quantifying ecological, morphological, and genetic variation to delimit species in the coast horned lizard species complex (*Phrynosoma*). *Proceedings of the National Academy of Sciences of the United States of America,* **106**, 12418–23.

Leinonen, T., O'Hara, R. B., Cano, J. M., and Merila, J. 2008. Comparative studies of quantitative trait and neutral marker divergence: a meta-analysis. *Journal of Evolutionary Biology,* **21**, 1–17.

Lessios, H. A. 1998. The first stage of speciation as seen in organisms separated by the Isthmus of Panama. In: Howard, D. J. and Berlocher, S. H. (eds) *Endless Forms: Species and Speciation.* New York: Oxford University Press.

Li, W., Ortiz, G., Fournier, P.-E., et al. 2010. Genotyping of human lice suggests multiple emergences of body lice from local head louse populations. *PLoS Neglected Tropical Diseases,* **4**, e641.

Lieberman, E., Hauert, C., and Nowak, M. A. 2005. Evolutionary dynamics on graphs. *Nature,* **433**, 312–16.

Linder, H. P. 2003. The radiation of the Cape flora, southern Africa. *Biological Reviews,* **78**, 597–638.

Linder, H. P. 2005. Evolution of diversity: the Cape flora. *Trends in Plant Science,* **10**, 536–41.

Liti, G., Barton, D. B. H., and Louis, E. J. 2006. Sequence diversity, reproductive isolation and species concepts in *Saccharomyces. Genetics,* **174**, 839–50.

Liu, L. and Yu, L. L. 2010. Phybase: an R package for species tree analysis. *Bioinformatics,* **26**, 962–3.

Lobato, E., Geraldes, M., Melo, M., Doutrelant, C., and Covas, R. 2017. Diversity and composition of cultivable gut bacteria in an endemic island bird and its mainland sister species. *Symbiosis,* **71**, 155–64.

Loeuille, N. and Loreau, M. 2005. Evolutionary emergence of size-structured food webs. *Proceedings of the National Academy of Sciences of the United States of America,* **102**, 5761–6.

Lohse, K. 2009. Can mtDNA barcodes be used to delimit species? A response to Pons et al. (2006). *Systematic Biology,* **58**, 439–42; discussion 442–4.

Losos, J. B. and Schluter, D. 2000. Analysis of an evolutionary species–area relationship. *Nature,* **408**, 847–50.

Lynch, J. D. 1989. The gauge of speciation: on the frequencies of modes of speciation. In: Otte, D. and Endler, J. A. (eds) *Speciation and its Consequences.* Sunderland, MA: Sinauer Associates.

Lynch, M. 1990. The rate of morphological evolution in mammals from the standpoint of the neutral expectation. *The American Naturalist,* **136**, 727–41.

Lynch, M. and Hill, W. G. 1986. Phenotypic evolution by neutral mutation. *Evolution,* **40**, 915–35.

Lynch, M. and Marinov, G. K. 2015. The bioenergetic costs of a gene. *Proceedings of the National Academy of Sciences of the United States of America,* **112**, 15690–5.

Ma, L. J., Van Der Does, H. C., Borkovich, K. A., et al. 2010. Comparative genomics reveals mobile pathogenicity chromosomes in Fusarium. *Nature,* **464**, 367–73.

Maclean, R. C., Dickson, A., and Bell, G. 2005. Resource competition and adaptive radiation in a microbial microcosm. *Ecology Letters,* **8**, 38–46.

Maclean, R. C., Hall, A. R., Perron, G. G., and Buckling, A. 2010. The population genetics of antibiotic resistance: integrating molecular mechanisms and treatment contexts. *Nature Reviews Genetics,* **11**, 405–14.

Maddison, W. P., Midford, P. E., and Otto, S. P. 2007. Estimating a binary character's effect on speciation and extinction. *Systematic Biology,* **56**, 701–10.

Magallon, S. and Sanderson, M. J. 2001. Absolute diversification rates in angiosperm clades. *Evolution,* **55**, 1762–80.

Magnuson-Ford, K. and Otto, S. P. 2012. Linking the investigations of character evolution and species diversification. *The American Naturalist,* **180**, 225–45.

Maheshwari, S. and Barbash, D. A. 2011. The genetics of hybrid incompatibilities. *Annual Review of Genetics,* **45**, 331–55.

Majewski, J. and Cohan, F. M. 1999. Adapt globally, act locally: the effect of selective sweeps on bacterial sequence diversity. *Genetics,* **152**, 1459–74.

Mallet, J. 2008. Hybridization, ecological races and the nature of species: empirical evidence for the ease of speciation. *Philosophical Transactions of the Royal Society B: Biological Sciences,* **363**, 2971–86.

Mallet, J., Besansky, N., and Hahn, M. W. 2016. How reticulated are species? *Bioessays,* **38**, 140–9.

Martin, C. H., Cutler, J. S., Friel, J. P., Touokong, C. D., Coop, G., and Wainwright, P. C. 2015. Complex histories of repeated gene flow in Cameroon Crater Lake cichlids cast doubt on one of the clearest examples of sympatric speciation. *Evolution,* **69**, 1406–22.

Martin, S. H., Dasmahapatra, K. K., Nadeau, N. J., et al. 2013. Genome-wide evidence for speciation with gene flow in *Heliconius* butterflies. *Genome Research,* **23**, 1817–28.

Martin, S. H., Davey, J. W., Salazar, C., and Jiggins, C. D. 2019. Recombination rate variation shapes barriers to introgression across butterfly genomes. *PLoS Biology,* **17**, e2006288.

Marvig, R. L., Sommer, L. M., Molin, S., and Johansen, H. K. 2015. Convergent evolution and adaptation of *Pseudomonas aeruginosa* within patients with cystic fibrosis. *Nature Genetics,* **47**, 57–64.

Maynard Smith, J. and Szathmary, E. 1995. *The Major Transitions in Evolution.* New York: W.H. Freeman.

Mayr, E. 1963. *Animal Species and Evolution.* Cambridge, MA: Harvard University Press.

McGinty, S. E., Lehmann, L., Brown, S. P., and Rankin, D. J. 2013. The interplay between relatedness and horizontal gene transfer drives the evolution of plasmid-carried public goods. *Proceedings of the Royal Society B: Biological Sciences,* **280**, 20130400.

McDaniel, S. F. and Shaw, A. J. 2005. Selective sweeps and intercontinental migration in the cosmopolitan moss *Ceratodon purpureus* (Hedw.) Brid. *Molecular Ecology,* **14**, 1121–32.

McDonald, J. H. and Kreitman, M. 1991. Adaptive protein evolution at the adh locus in *Drosophila. Nature,* **351**, 652–4.

Melendrez, M. C., Becraft, E. D., Wood, J. M., et al. 2016. Recombination does not hinder formation or detection of ecological species of *Synechococcus* inhabiting a hot spring cyanobacterial mat. *Frontiers in Microbiology,* **6**, 1540.

Melian, C. J., Vilas, C., Baldo, F., Gonzalez-Ortegon, E., Drake, P., and Williams, R. J. 2011. Eco-evolutionary dynamics of individual-based food webs. *Advances in Ecological Research: The Role of Body Size in Multispecies Systems,* **45**, 225–68.

Merila, J. and Hendry, A. P. 2014. Climate change, adaptation, and phenotypic plasticity: the problem and the evidence. *Evolutionary Applications,* **7**, 1–14.

Messer, P. W., Ellner, S. P., and Hairston, N. G. 2016. Can population genetics adapt to rapid evolution? *Trends in Genetics,* **32**, 408–18.

Meyer, C. P. and Paulay, G. 2005. DNA barcoding: error rates based on comprehensive sampling. *PLoS Biology,* **3**, 2229–38.

Meyer, J. R., Dobias, D. T., Medina, S. J., Servilio, L., Gupta, A., and Lenski, R. E. 2016. Ecological speciation of bacteriophage lambda in allopatry and sympatry. *Science,* **354**, 1301–4.

Millanes, A. M., Truong, C., Westberg, M., Diederich, P., and Wedin, M. 2014. Host switching promotes diversity in host-specialized mycoparasitic fungi: uncoupled evolution in the biatoropsis–usnea system. *Evolution,* **68**, 1576–93.

Miller, E. A., Beasley, D. E., Dunn, R. R., and Archie, E. A. 2016. Lactobacilli dominance and vaginal pH: why is the human vaginal microbiome unique? *Frontiers in Microbiology,* **7**, 1936.

Mindell, D. P. 2013. The tree of life: metaphor, model, and heuristic device. *Systematic Biology,* **62**, 479–89.

Minot, S., Bryson, A., Chehoud, C., Wu, G. D., Lewis, J. D., and Bushman, F. D. 2013. Rapid evolution of the human gut virome. *Proceedings of the National Academy of Sciences of the United States of America,* **110**, 12450–5.

Mirarab, S., Reaz, R., Bayzid, M. S., Zimmermann, T., Swenson, M. S., and Warnow, T. 2014. ASTRAL: genome-scale coalescent-based species tree estimation. *Bioinformatics,* **30**, i541–8.

Mitter, C., Farrell, B., and Wiegmann, B. 1988. The phylogenetic study of adaptive zones: has phytophagy promoted insect diversification? *The American Naturalist,* **132**, 107–28.

Monaghan, M. T., Wild, R., Elliot, M., et al. 2009. Accelerated species inventory on Madagascar using coalescent-based models of species delineation. *Systematic Biology,* **58**, 298–311.

Moritz, C. C., Pratt, R. C., Bank, S., et al. 2018. Cryptic lineage diversity, body size divergence, and sympatry in a species complex of Australian lizards (*Gehyra*). *Evolution,* **72**, 54–66.

Morjan, C. L. and Rieseberg, L. H. 2004. How species evolve collectively: implications of gene flow and selection for the spread of advantageous alleles. *Molecular Ecology,* **13**, 1341–56.

Morlon, H. 2014. Phylogenetic approaches for studying diversification. *Ecology Letters,* **17**, 508–25.

Moya-Larano, J., Verdeny-Vilalta, O., Rowntree, J., Melguizo-Ruiz, N., Montserrat, M., and Laiolo, P. 2012. Climate change and eco-evolutionary dynamics in food webs. *Advances in Ecological Research: Global Change in Multispecies Systems,* **47**, 1–80.

Nachman, M. W. and Payseur, B. A. 2012. Recombination rate variation and speciation: theoretical predictions and empirical results from rabbits and mice. *Philosophical Transactions of the Royal Society B: Biological Sciences,* **367**, 409–21.

Nee, S. 2005. The neutral theory of biodiversity: do the numbers add up? *Functional Ecology,* **19**, 173–6.

Nee, S., Mooers, A. O., and Harvey, P. H. 1992. Tempo and mode of evolution revealed from molecular phylogenies. *Proceedings of the National Academy of Sciences of the United States of America,* **89**, 8322–6.

Nee, S., May, R. M., and Harvey, P. H. 1994. The reconstructed evolutionary process. *Philosophical Transactions of the Royal Society of London Series B: Biological Sciences,* **344**, 305–11.

Neher, R. A., Bedford, T., Daniels, R. S., Russell, C. A., and Shraiman, B. I. 2016. Prediction, dynamics, and visualization of antigenic phenotypes of seasonal influenza viruses. *Proceedings of the National Academy of Sciences of the United States of America*, **113**, E1701–9.

Nicholson, J. K., Holmes, E., Kinross, J., Burcelin, R., Gibson, G., Jia, W. and Petttersen, S. 2012. Host-gut microbiota metabolic interactions. *Science*, **336**, 1262–1267.

Niehus, R., Mitri, S., Fletcher, A. G., and Foster, K. R. 2015. Migration and horizontal gene transfer divide microbial genomes into multiple niches. *Nature Communications*, **6**, 8924.

Nijman, V. and Aliabadian, M. 2013. DNA barcoding as a tool for elucidating species delineation in wide-ranging species as illustrated by owls (Tytonidae and Strigidae). *Zoological Science*, **30**, 1005–9.

Northfield, T. D. and Ives, A. R. 2013. Coevolution and the effects of climate change on interacting species. *PLoS Biology*, **11**, e1001685.

Nosil, P. 2012. *Ecological Speciation*. Oxford: Oxford University Press.

Nosil, P., Funk, D. J., and Ortiz-Barrientos, D. 2009. Divergent selection and heterogeneous genomic divergence. *Molecular Ecology*, **18**, 375–402.

Nowell, R. W., Almeida, P., Wilson, C. G., et al. 2018. Comparative genomics of bdelloid rotifers: insights from desiccating and nondesiccating species. *PLoS Biology*, **16**, e2004830.

Nowell, R. W., Green, S., Laue, B. E. & Sharp, P. M, 2014. The extent of genome flux and its role in the differentiation of bacterial lineages. *Genome Biology and Evolution* **6**, 1514–1529.

Obst, M., Faurby, S., Bussarawit, S., and Funch, P. 2012. Molecular phylogeny of extant horseshoe crabs (Xiphosura, Limulidae) indicates Paleogene diversification of Asian species. *Molecular Phylogenetics and Evolution*, **62**, 21–6.

O'Brien, S., and Fothergill, J.L. 2017. The role of multispecies social interactions in shaping *Pseudomonas aeruginosa* pathogenicity in the cystic fibrosis lung. *FEMS Microbiology Letters*, **364**, fnx128

O'Connor, M. I., Piehler, M. F., Leech, D. M., Anton, A., and Bruno, J. F. 2009. Warming and resource availability shift food web structure and metabolism. *PLoS Biology*, **7**, e1000178.

O'Donnell, K., Kistler, H. C., Cigelnik, E., and Ploetz, R. C. 1998. Multiple evolutionary origins of the fungus causing Panama disease of banana: concordant evidence from nuclear and mitochondrial gene genealogies. *Proceedings of the National Academy of Sciences of the United States of America*, **95**, 2044–9.

Omelchenko, M. V., Makarova, K. S., Wolf, Y. I., Rogozin, I. B., and Koonin, E. V. 2003. Evolution of mosaic operons by horizontal gene transfer and gene displacement in situ. *Genome Biology*, **4**, R55.

Osmond, M. M. and De Mazancourt, C. 2013. How competition affects evolutionary rescue. *Philosophical Transactions of the Royal Society B: Biological Sciences*, **368**, 20120085.

Ostevik, K. L., Andrew, R. L., Otto, S. P., and Rieseberg, L. H. 2016. Multiple reproductive barriers separate recently diverged sunflower ecotypes. *Evolution*, **70**, 2322–35.

Otto, S. P., and Michalakis, Y. 1998. The evolution of recombination in changing environments. *Trends in Ecology and Evolution*, **13**, 145–151.

Otwinowski, J., McCandlish, D. M., and Plotkin, J. B. 2018. Inferring the shape of global epistasis. *Proceedings of the National Academy of Sciences of the United States of America*, **115**, E7550–8.

Ozkilinc, H., Rotondo, F., Pryor, B. M., and Peever, T. L. 2018. Contrasting species boundaries between sections Alternaria and Porri of the genus *Alternaria*. *Plant Pathology*, **67**, 303–14.

Papadopoulou, A., Bergsten, J., Fujisawa, T., Monaghan, M. T., Barraclough, T. G., and Vogler, A. P. 2008. Speciation and DNA barcodes: testing the effects of dispersal on the formation of discrete sequence clusters. *Philosophical Transactions of the Royal Society B: Biological Sciences*, **363**, 2987–96.

Papadopoulou, A., Monaghan, M. T., Barraclough, T. G., and Vogler, A. P. 2009. Sampling error does not invalidate the Yule-Coalescent model for species delimitation. A response to Lohse (2009). *Systematic Biology,* **58**, 442–4.

Papadopulos, A. S. T., Baker, W. J., Crayn, D., et al. 2011. Speciation with gene flow on Lord Howe Island. *Proceedings of the National Academy of Sciences of the United States of America,* **108**, 13188–93.

Papadopulos, A. S. T., Price, Z., Devaux, C., et al. 2013. A comparative analysis of the mechanisms underlying speciation on Lord Howe Island. *Journal of Evolutionary Biology,* **26**, 733–45.

Papke, R. T., Ramsing, N. B., Bateson, M. M., and Ward, D. M. 2003. Geographical isolation in hot spring cyanobacteria. *Environmental Microbiology,* **5**, 650–9.

Paradis, E., Claude, J., and Strimmer, K. 2004. APE: analyses of phylogenetics and evolution in R language. *Bioinformatics,* **20**, 289–90.

Paun, O., Turner, B., Trucchi, E., Munzinger, J., Chase, M. W., and Samuel, R. 2016. Processes driving the adaptive radiation of a tropical tree (*Diospyros*, Ebenaceae) in New Caledonia, a biodiversity hotspot. *Systematic Biology,* **65**, 212–27.

Payseur, B. A., Presgraves, D. C., and Filatov, D. A. 2018. Introduction: sex chromosomes and speciation. *Molecular Ecology,* **27**, 3745–8.

Peccoud, J., Ollivier, A., Plantegenest, M., and Simon, J.-C. 2009. A continuum of genetic divergence from sympatric host races to species in the pea aphid complex. *Proceedings of the National Academy of Sciences of the United States of America,* **106**, 7495–500.

Pentinsaari, M., Salmela, H., Mutanen, M., and Roslin, T. 2016. Molecular evolution of a widely-adopted taxonomic marker (COI) across the animal tree of life. *Scientific Reports,* **6**, 35275.

Perret, M., Chautems, A., Spichiger, R., Barraclough, T. G., and Savolainen, V. 2007. The geographical pattern of speciation and floral diversification in the neotropics: the tribe Sinningieae (Gesneriaceae) as a case study. *Evolution,* **61**, 1641–60.

Perron, G. G., Gonzalez, A., and Buckling, A. 2007. Source-sink dynamics shape the evolution of antibiotic resistance and its pleiotropic fitness cost. *Proceedings of the Royal Society B: Biological Sciences,* **274**, 2351–6.

Petchey, O. L., Pontarp, M., Massie, T. M., et al. 2015. The ecological forecast horizon, and examples of its uses and determinants. *Ecology Letters,* **18**, 597–611.

Phillimore, A. B. and Price, T. D. 2008. Density-dependent cladogenesis in birds. *PLoS Biology,* **6**, 483–9.

Phillimore, A. B., Freckleton, R. P., Orme, C. D. L., and Owens, I. P. F. 2006. Ecology predicts large-scale patterns of phylogenetic diversification in birds. *The American Naturalist,* **168**, 220–9.

Phillimore, A. B., Orme, C. D. L., Thomas, G. H., et al. 2008. Sympatric speciation in birds is rare: insights from range data and simulations. *The American Naturalist,* **171**, 646–57.

Phillimore, A. B., Hadfield, J. D., Jones, O. R., and Smithers, R. J. 2010. Differences in spawning date between populations of common frog reveal local adaptation. *Proceedings of the National Academy of Sciences of the United States of America,* **107**, 8292–7.

Pigot, A. L. and Tobias, J. A. 2013. Species interactions constrain geographic range expansion over evolutionary time. *Ecology Letters,* **16**, 330–8.

Pigot, A. L., Phillimore, A. B., Owens, I. P. F., and Orme, C. D. L. 2010. The shape and temporal dynamics of phylogenetic trees arising from geographic speciation. *Systematic Biology,* **59**, 660–73.

Pinsky, M. L., Saenz-Agudelo, P., Salles, O. C., et al. 2017. Marine dispersal scales are congruent over evolutionary and ecological time. *Current Biology,* **27**, 149–54.

Pitchers, W., Wolf, J. B., Tregenza, T., Hunt, J., and Dworkin, I. 2014. Evolutionary rates for multivariate traits: the role of selection and genetic variation. *Philosophical Transactions of the Royal Society B: Biological Sciences,* **369**, 1–13.

Polz, M. F., Alm, E. J., and Hanage, W. P. 2013. Horizontal gene transfer and the evolution of bacterial and archaeal population structure. *Trends in Genetics,* **29**, 170–5.

Pons, J., Barraclough, T. G., Gomez-Zurita, J., et al. 2006. Sequence-based species delimitation for the DNA taxonomy of undescribed insects. *Systematic Biology,* **55**, 595–609.

Popa, O., Landan, G., and Dagan, T. 2017. Phylogenomic networks reveal limited phylogenetic range of lateral gene transfer by transduction. *ISME Journal,* **11**, 543–54.

Popp, J., Peto, K., and Nagy, J. 2013. Pesticide productivity and food security. A review. *Agronomy for Sustainable Development,* **33**, 243–55.

Price, T. 2008. *Speciation in Birds.* Greenwood Village, CO: Roberts & Company.

Price, T. D., Hooper, D. M., Buchanan, C. D., et al. 2014. Niche filling slows the diversification of Himalayan songbirds. *Nature,* **509**, 222–5.

Proulx, S. R., Promislow, D. E. L., and Phillips, P. C. 2005. Network thinking in ecology and evolution. *Trends in Ecology & Evolution,* **20**, 345–53.

Puillandre, N., Lambert, A., Brouillet, S., and Achaz, G. 2012. ABGD, automatic barcode gap discovery for primary species delimitation. *Molecular Ecology,* **21**, 1864–77.

Qin, J., Li, R., Raes, J., et al. 2010. A human gut microbial gene catalogue established by metagenomic sequencing. *Nature,* **464**, 59–65.

R Core Team 2018. R: A Language and Environment for Statistical Computing. Vienna: R Foundation for Statistical Computing. URL https://www.R-project.org/.

Rabosky, D. L. 2010. Extinction rates should not be estimated from molecular phylogenies. *Evolution,* **64**, 1816–24.

Rabosky, D. L. 2013. Diversity-dependence, ecological speciation, and the role of competition in macroevolution. *Annual Review of Ecology, Evolution, and Systematics,* **44**, 481–502.

Rabosky, D. L. and Goldberg, E. E. 2015. Model inadequacy and mistaken inferences of trait-dependent speciation. *Systematic Biology,* **64**, 340–55.

Rabosky, D. L. and Matute, D. R. 2013. Macroevolutionary speciation rates are decoupled from the evolution of intrinsic reproductive isolation in *Drosophila* and birds. *Proceedings of the National Academy of Sciences of the United States of America,* **110**, 15354–9.

Rabosky, D. L., Chang, J., Title, P. O., et al. 2018. An inverse latitudinal gradient in speciation rate for marine fishes. *Nature,* **559**, 392–5.

Rainey, P. B. and Travisano, M. 1998. Adaptive radiation in a heterogeneous environment. *Nature,* **394**, 69–72.

Ravel, J., Gajer, P., Abdo, Z., et al. 2011. Vaginal microbiome of reproductive-age women. *Proceedings of the National Academy of Sciences of the United States of America,* **108**, 4680–7.

Renaut, S., Grassa, C. J., Yeaman, S., et al. 2013. Genomic islands of divergence are not affected by geography of speciation in sunflowers. *Nature Communications,* **4**, 1827.

Renner, S. S. and Zohner, C. M. 2018. Climate change and phenological mismatch in trophic interactions among plants, insects, and vertebrates. *Annual Review of Ecology, Evolution, and Systematics,* **49**, 165–82.

Revell, L. J. 2012. Phytools: an R package for phylogenetic comparative biology (and other things). *Methods in Ecology and Evolution,* **3**, 217–23.

Ribas, C. C., Aleixo, A., Nogueira, A. C. R., Miyaki, C. Y., and Cracraft, J. 2012. A palaeobiogeographic model for biotic diversification within Amazonia over the past three million years. *Proceedings of the Royal Society B: Biological Sciences,* **279**, 681–9.

Ricci, C., Melone, G., Santo, N., Caprioli, M. 2003. Morphological response of a bdelloid rotifer to desiccation. *Journal of Morphology,* **257**, 246–253.

Rice, W. R. and Hostert, E. E. 1993. Laboratory experiments on speciation: what have we learned in the last forty years? *Evolution,* **47**, 1637–53.

Richards, E. J., Poelstra, J. W., and Martin, C. H. 2018. Don't throw out the sympatric speciation with the crater lake water: fine-scale investigation of introgression provides equivocal support for causal role of secondary gene flow in one of the clearest examples of sympatric speciation. *Evolution Letters, 2,* 524–40.

Ricklefs, R. E. 2003. Global diversification rates of passerine birds. *Proceedings of the Royal Society of London Series B: Biological Sciences, 270,* 2285–91.

Ricklefs, R. E. 2006. The unified neutral theory of biodiversity: do the numbers add up? *Ecology, 87,* 1424–31.

Ricklefs, R. E. 2007. Estimating diversification rates from phylogenetic information. *Trends in Ecology & Evolution, 22,* 601–10.

Rieseberg, L. H. 2001. Chromosomal rearrangements and speciation. *Trends in Ecology & Evolution, 16,* 351–8.

Rieseberg, L. H., Wood, T. E., and Baack, E. J. 2006. The nature of plant species. *Nature, 440,* 524–7.

Ritchie, M. G. 2007. Sexual selection and speciation. *Annual Review of Ecology Evolution and Systematics, 38,* 79–102.

Rivett, D. W., Scheuerl, T., Culbert, C. T, Mombrikotb, S. B., Johnstone, E., Barraclough, T. G., and Bell, T. 2016. Diversity-dependent attenuation of species interactions during bacterial succession. *ISME* Journal, **10,** 2259–2268.

Robinson, C. J., Bohannan, B. J. M., and Young, V. B. 2010. From structure to function: the ecology of host-associated microbial communities. *Microbiology and Molecular Biology Reviews, 74,* 453–76.

Roderick, G. K., Hufbauer, R., and Navajas, M. 2012. Evolution and biological control. *Evolutionary Applications, 5,* 419–23.

Roemhild, R., Barbosa, C., Beardmore, R. E., Jansen, G., and Schulenburg, H. 2015. Temporal variation in antibiotic environments slows down resistance evolution in pathogenic *Pseudomonas aeruginosa. Evolutionary Applications, 8,* 945–55.

Romiguier, J., Gayral, P., Ballenghien, M., et al. 2014. Comparative population genomics in animals uncovers the determinants of genetic diversity. *Nature, 515,* 261–3.

Rosenberg, N. A. 2003. The shapes of neutral gene genealogies in two species: probabilities of monophyly, paraphyly, and polyphyly in a coalescent model. *Evolution, 57,* 1465–77.

Rosenberg, N. A. and Nordborg, M. 2002. Genealogical trees, coalescent theory and the analysis of genetic polymorphisms. *Nature Reviews Genetics, 3,* 380–90.

Rosindell, J. and Cornell, S. J. 2007. Species–area relationships from a spatially explicit neutral model in an infinite landscape. *Ecology Letters, 10,* 586–95.

Rosindell, J., Harmon, L. J., and Etienne, R. S. 2015. Unifying ecology and macroevolution with individual-based theory. *Ecology Letters, 18,* 472–82.

Rosser, N., Kozak, K. M., Phillimore, A. B., and Mallet, J. 2015. Extensive range overlap between heliconiine sister species: evidence for sympatric speciation in butterflies? *BMC Evolutionary Biology, 15,* 125.

Roughgarden, J. 1976. Resource partitioning among competing species—a coevolutionary approach. *Theoretical Population Biology, 9,* 388–424.

Ryan, P. G., Bloomer, P., Moloney, C. L., Grant, T. J., and Delport, W. 2007. Ecological speciation in South Atlantic Island finches. *Science, 315,* 1420–3.

Samuk, K., Owens, G. L., Delmore, K. E., Miller, S. E., Rennison, D. J., and Schluter, D. 2017. Gene flow and selection interact to promote adaptive divergence in regions of low recombination. *Molecular Ecology, 26,* 4378–90.

San Millan, A. S., Pena-Miller, R., Toll-Riera, M., et al 2014. Positive selection and compensatory adaptation interact to stabilize non-transmissible plasmids. *Nature Communications,* **5**, 5208.

Sanderson, M. J. and Donoghue, M. J. 1994. Shifts in diversification rate with the origin of angiosperms. *Science,* **264**, 1590–3.

Sasaki, A. and Godfray, H. C. J. 1999. A model for the coevolution of resistance and virulence in coupled host–parasitoid interactions. *Proceedings of the Royal Society B: Biological Sciences,* **266**, 455–63.

Savolainen, V., Anstett, M. C., Lexer, C., et al. 2006. Sympatric speciation in palms on an oceanic island. *Nature,* **441**, 210–13.

Scheuerl, T., Hopkins, M., Nowell, R. W., Rivett, D. W., Barraclough, T. G., and Bell, T. 2019. Bacterial adaptation is constrained in complex communities. In prep.

Schiffman, J. S. and Ralph, P. L. 2018. System drift and speciation. *bioRxiv,* doi: https://doi. org/10.1101/231209.

Schliep, K. P. 2011. phangorn: phylogenetic analysis in R. *Bioinformatics,* **27**, 592–3.

Schliewen, U. K., Tautz, D., and Paabo, S. 1994. Sympatric speciation suggested by monophyly of Crater Lake cichlids. *Nature,* **368**, 629–32.

Schluter, D. 1996. Adaptive radiation along genetic lines of least resistance. *Evolution,* **50**, 1766–74.

Schluter, D. 2000. *The Ecology of Adaptive Radiation.* Oxford: Oxford University Press.

Schluter, D. 2001. Ecology and the origin of species. *Trends in Ecology & Evolution,* **16**, 372–80.

Schluter, D. 2009. Evidence for ecological speciation and its alternative. *Science,* **323**, 737–41.

Schluter, D. 2016. Speciation, ecological opportunity, and latitude. *American Naturalist,* **187**, 1–18.

Schluter, D. and Conte, G. L. 2009. Genetics and ecological speciation. *Proceedings of the National Academy of Sciences of the United States of America,* **106**, 9955–62.

Schluter, D. and Pennell, M. W. 2017. Speciation gradients and the distribution of biodiversity. *Nature,* **546**, 48–55.

Schmutzer, M. and Barraclough, T. G. 2019. The role of recombination, niche-specific gene pools and flexible genomes in the ecological speciation of bacteria. *Ecology and Evolution.* **9**, 4544–4556.

Schnitzler, J., Barraclough, T. G., Boatwright, J. S., et al. 2011. Causes of plant diversification in the Cape biodiversity hotspot of South Africa. *Systematic Biology,* **60**, 343–57.

Schoch, C. L., Seifert, K. A., Huhndorf, S., et al. 2012. Nuclear ribosomal internal transcribed spacer (ITS) region as a universal DNA barcode marker for fungi. *Proceedings of the National Academy of Sciences of the United States of America,* **109**, 6241–6.

Scopece, G., Cozzolino, S., and Bateman, R. M. 2010. Just what is a genus? Comparing levels of postzygotic isolation to test alternative taxonomic hypotheses in Orchidaceae subtribe Orchidinae. *Taxon,* **59**, 1367–74.

Seehausen, O., Vanalphen, J. J. M., and Witte, F. 1997. Cichlid fish diversity threatened by eutrophication that curbs sexual selection. *Science,* **277**, 1808–11.

Seehausen, O., Van Alphen, J. J. M., and Lande, R. 1999. Color polymorphism and sex ratio distortion in a cichlid fish as an incipient stage in sympatric speciation by sexual selection. *Ecology Letters,* **2**, 367–78.

Seehausen, O., Butlin, R. K., Keller, I., et al. 2014. Genomics and the origin of species. *Nature Reviews Genetics,* **15**, 176–92.

Sepp, S. and Paal, J. 1998. Taxonomic continuum of *Alchemilla* (Rosaceae) in Estonia. *Nordic Journal of Botany,* **18**, 519–35.

Shapiro, B. J. and Polz, M. F. 2014. Ordering microbial diversity into ecologically and genetically cohesive units. *Trends in Microbiology,* **22**, 235–47.

Shapiro, B. J., Friedman, J., Cordero, O. X., et al. 2012. Population genomics of early events in the ecological differentiation of bacteria. *Science,* **336**, 48–51.

Sheppard, R., Barraclough, T. G., and Jansen, V. A. 2019. The coevolution and control of conjugative plasmid transfer rate. Unpublished work.

Shoaie, S., Karlsson, F., Mardinoglu, A., Nookaew, I., Bordel, S., and Nielsen, J. 2013. Understanding the interactions between bacteria in the human gut through metabolic modeling. *Scientific Report,* **3**, 2532.

Signorovitch, A., Hur, J., Gladyshev, E., and Meselson, M. 2015. Allele sharing and evidence for sexuality in a mitochondrial clade of bdelloid rotifers. *Genetics,* **200**, 581–90.

Simpson, G. G. 1961. *Principles of Animal Taxonomy.* New York: Columbia University Press.

Singhal, S., Huang, H. T., Grundler, M. R., et al. 2018. Does population structure predict the rate of speciation? A comparative test across Australia's most diverse vertebrate radiation. *The American Naturalist,* **192**, 432–47.

Siren, J., Ovaskainen, O., and Merila, J. 2017. Structure and stability of genetic variance-covariance matrices: a Bayesian sparse factor analysis of transcriptional variation in the three-spined stickleback. *Molecular Ecology,* **26**, 5099–113.

Sites, J. W. and Marshall, J. C. 2003. Delimiting species: a Renaissance issue in systematic biology. *Trends in Ecology & Evolution,* **18**, 462–70.

Skeels, A. and Cardillo, M. 2019. Reconstructing the geography of speciation from contemporary biodiversity data. *The American Naturalist,* **193**, 240–255.

Skulason, S. and Smith, T. B. 1995. Resource polymorphisms in vertebrates. *Trends in Ecology & Evolution,* **10**, 366–70.

Slatkin, M. 1984. Ecological causes of sexual dimorphism. *Evolution,* **38**, 622–30.

Slatkin, M. 1985. Gene flow in natural populations. *Annual Review of Ecology and Systematics,* **16**, 393–430.

Slatkin, M. 1991. Inbreeding coefficients and coalescence times. *Genetical Research,* **58**, 167–75.

Slatkin, M. and Smith, J. M. 1979. Models of coevolution. *Quarterly Review of Biology,* **54**, 233–63.

Sleeth, M. L., Thompson, E. L., Ford, H. E., Zac-Varghese, S. E. K., and Frost, G. 2010. Free fatty acid receptor 2 and nutrient sensing: a proposed role for fibre, fermentable carbohydrates and short-chain fatty acids in appetite regulation. *Nutrition Research Reviews,* **23**, 135–45.

Smillie, C., Garcillan-Barcia, M. P., Francia, M. V., Rocha, E. P. C., and De La Cruz, F. 2010. Mobility of plasmids. *Microbiology and Molecular Biology Reviews,* **74**, 434–52.

Smillie, C. S., Smith, M. B., Friedman, J., Cordero, O. X., David, L. A., and Alm, E. J. 2011. Ecology drives a global network of gene exchange connecting the human microbiome. *Nature,* **480**, 241–4.

Sobel, J. M., Chen, G. F., Watt, L. R., and Schemske, D. W. 2010. The biology of speciation. *Evolution,* **64**, 295–315.

Solis-Lemus, C., Knowles, L. L., and Ane, C. 2015. Bayesian species delimitation combining multiple genes and traits in a unified framework. *Evolution,* **69**, 492–507.

Soria-Carrasco, V., Gompert, Z., Comeault, A. A., et al. 2014. Stick insect genomes reveal natural selection's role in parallel speciation. *Science,* **344**, 738–42.

Speirs, D. C., Greenstreet, S. P. R., and Heath, M. R. 2016. Modelling the effects of fishing on the North Sea fish community size composition. *Ecological Modelling,* **321**, 35–45.

Stadler, T., Rabosky, D. L., Ricklefs, R. E., and Bokma, F. 2014. On age and species richness of higher taxa. *The American Naturalist,* **184**, 447–55.

Stanley, S. M. 1979. *Macroevolution: Pattern and Process.* San Francisco: W.H. Freeman.

Steiner, C. F. 2012. Environmental noise, genetic diversity and the evolution of evolvability and robustness in model gene networks. *PLoS One,* **7**, e52204.

Stewart, F. M. and Levin, B. R. 1973. Partitioning of resources and the outcome of interspecific competition: a model and some general considerations. *The American Naturalist,* **107**, 171–98.

Stewart, F. M. and Levin, B. R. 1977. Population biology of bacterial plasmids—a priori conditions for existence of conjugationally transmitted factors. *Genetics,* **87,** 209–28.

Stoelting, K. N., Nipper, R., Lindtke, D., et al. 2013. Genomic scan for single nucleotide polymorphisms reveals patterns of divergence and gene flow between ecologically divergent species. *Molecular Ecology,* **22,** 842–55.

Strauss, S. Y., Sahli, H., and Conner, J. K. 2005. Toward a more trait-centered approach to diffuse (co)evolution. *New Phytologist,* **165,** 81–9.

Strayer, D. L. 2012. Eight questions about invasions and ecosystem functioning. *Ecology Letters,* **15,** 1199–210.

Sukumaran, J. and Knowles, L. L. 2017. Multispecies coalescent delimits structure, not species. *Proceedings of the National Academy of Sciences of the United States of America,* **114,** 1607–12.

Sun, J. M., Ni, X. J., Bi, S. D., et al. 2014. Synchronous turnover of flora, fauna, and climate at the Eocene–Oligocene boundary in Asia. *Scientific Reports,* **4,** 7463.

Surette, M. G. 2014. The cystic fibrosis lung microbiome. *Annals of the American Thoracic Society,* **11,** S61–5.

Swain, D. P., Sinclair, A. F., and Hanson, J. M. 2007. Evolutionary response to size-selective mortality in an exploited fish population. *Proceedings of the Royal Society B: Biological Sciences,* **274,** 1015–22.

Tamminen, M., Virta, M., Fani, R., and Fondi, M. 2012. Large-scale analysis of plasmid relationships through gene-sharing networks. *Molecular Biology and Evolution,* **29,** 1225–40.

Tang, C. Q., Humphreys, A. M., Fontaneto, D., and Barraclough, T. G. 2014a. Effects of phylogenetic reconstruction method on the robustness of species delimitation using single-locus data. *Methods in Ecology and Evolution,* **5,** 1086–94.

Tang, C. Q., Obertegger, U., Fontaneto, D., and Barraclough, T. G. 2014b. Sexual species are separated by larger genetic gaps than asexual species in rotifers. *Evolution,* **68,** 2901–16.

Tautz, D., Arctander, P., Minelli, A., Thomas, R. H., and Vogler, A. P. 2003. A plea for DNA taxonomy. *Trends in Ecology & Evolution,* **18,** 70–4.

Taylor, J. W., Jacobson, D. J., Kroken, S., et al. 2000. Phylogenetic species recognition and species concepts in fungi. *Fungal Genetics and Biology,* **31,** 21–32.

Templeton, A. 1989. The meaning of species and speciation: a population genetics approach. In: Otte, D. and Endler, J. (eds) *Speciation and its Consequences.* Sunderland, MA: Sinauer Associates.

TerHorst, C. P., Zee, P. C., Heath, K. D., et al. 2018. Evolution in a community context: trait responses to multiple species interactions. *The American Naturalist,* **191,** 368–80.

Thompson, L. R., Sanders, J. G., McDonald, D., et al. 2017. A communal catalogue reveals Earth's multiscale microbial diversity. *Nature,* **551,** 457–63.

Thorpe, R. B., Dolder, P. J., Reeves, S., Robinson, P., and Jennings, S. 2016. Assessing fishery and ecological consequences of alternate management options for multispecies fisheries. *ICES Journal of Marine Science,* **73,** 1503–12.

Tilman, D. 1982. *Resource Competition and Community Structure.* Princeton, NJ: Princeton University Press.

Tilman, D. and Lehman, C. 2001. Human-caused environmental change: impacts on plant diversity and evolution. *Proceedings of the National Academy of Sciences of the United States of America,* **98,** 5433–40.

Tobias, J. A., Seddon, N., Spottiswoode, C. N., Pilgrim, J. D., Fishpool, L. D. C., and Collar, N. J. 2010. Quantitative criteria for species delimitation. *Ibis,* **152,** 724–46.

Tognon, M., Köhler, T., Gdaniec, B.G., Hao, Y., Lam, J.S., Beaume, M., Luscher, A., Buckling, A., van Delden, C. 2017. Co-evolution with *Staphylococcus aureus* leads to lipopolysaccharide alterations in *Pseudomonas aeruginosa. ISME Journal,* **11,** 2233–2243.

Tomasetto, F., Tylianakis, J. M., Reale, M., Wratten, S., and Goldson, S. L. 2017. Intensified agriculture favors evolved resistance to biological control. *Proceedings of the National Academy of Sciences of the United States of America*, **114**, 3885–90.

Torresen, O. K., Star, B., Jentoft, S., et al. 2017. An improved genome assembly uncovers prolific tandem repeats in Atlantic cod. *BMC Genomics*, **18**, 95.

Turissini, D. A., McGuire, J. A., Patel, S. S., David, J. R., and Matute, D. R. 2018. The rate of evolution of postmating–prezygotic reproductive isolation in *Drosophila*. *Molecular Biology and Evolution*, **35**, 312–34.

Turner, B., Paun, O., Munzinger, J., Duangjai, S., Chase, M. W., and Samuel, R. 2013. Analyses of amplified fragment length polymorphisms (AFLP) indicate rapid radiation of *Diospyros* species (Ebenaceae) endemic to New Caledonia. *BMC Evolutionary Biology*, **13**, 269.

Urwin, R. and Maiden, M. C. J. 2003. Multi-locus sequence typing: a tool for global epidemiology. *Trends in Microbiology*, **11**, 479–87.

Valente, L. M., Reeves, G., Schnitzler, J., et al. 2010. Diversification of the African genus *Protea* (Proteaceae) in the Cape biodiversity hotspot and beyond: equal rates in different biomes. *Evolution*, **64**, 745–59.

Valente, L. M., Britton, A. W., Powell, M. P., Papadopulos, A. S. T., Burgoyne, P. M., and Savolainen, V. 2014. Correlates of hyperdiversity in southern African ice plants (Aizoaceae). *Botanical Journal of the Linnean Society*, **174**, 110–29.

Van Dam, P., Fokkens, L., Ayukawa, Y., et al. 2017. A mobile pathogenicity chromosome in *Fusarium oxysporum* for infection of multiple cucurbit species. *Scientific Reports*, **7**, 9042.

Van Moorsel, S. J., Hahl, T., Wagg, C., et al. 2018. Community evolution increases plant productivity at low diversity. *Ecology Letters*, **21**, 128–37.

Venturelli, O. S., Carr, A. V., Fisher, G., et al. 2018. Deciphering microbial interactions in synthetic human gut microbiome communities. *Molecular Systems Biology*, **14**, e8157.

Vieira, C., D'Hondt, S., De Clerck, O., and Payri, C. E. 2014. Toward an inordinate fondness for stars, beetles and *Lobophora*? Species diversity of the genus *Lobophora* (Dictyotales, Phaeophyceae) in New Caledonia. *Journal of Phycology*, **50**, 1101–19.

Voje, K. L., Holen, O. H., Liow, L. H., and Stenseth, N. C. 2015. The role of biotic forces in driving macroevolution: beyond the Red Queen. *Proceedings of the Royal Society B: Biological Sciences*, **282**, 20150186.

Vonlanthen, P., Bittner, D., Hudson, A. G., et al. 2012. Eutrophication causes speciation reversal in whitefish adaptive radiations. *Nature*, **482**, 357–62.

Vos, M. and Didelot, X. 2009. A comparison of homologous recombination rates in bacteria and archaea. *ISME Journal*, **3**, 199–208.

Wade, M. J. and Kalisz, S. 1990. The causes of natural-selection. *Evolution*, **44**, 1947–55.

Wagner, C. E., Harmon, L. J., and Seehausen, O. 2012. Ecological opportunity and sexual selection together predict adaptive radiation. *Nature*, **487**, 366–9.

Wagner, C. E., Harmon, L. J., and Seehausen, O. 2014. Cichlid species–area relationships are shaped by adaptive radiations that scale with area. *Ecology Letters*, **17**, 583–92.

Wallace, S. J., Morris-Pocock, J. A., Gonzalez-Solis, J., Quillfeldt, P., and Friesen, V. L. 2017. A phylogenetic test of sympatric speciation in the Hydrobatinae (Aves: Procellariiformes). *Molecular Phylogenetics and Evolution*, **107**, 39–47.

Wang, I. J., Glor, R. E., and Losos, J. B. 2013. Quantifying the roles of ecology and geography in spatial genetic divergence. *Ecology Letters*, **16**, 175–82.

Waterman, R. J., Bidartondo, M. I., Stofberg, J., et al. 2011. The effects of above- and belowground mutualisms on orchid speciation and coexistence. *The American Naturalist*, **177**, E54–68.

Webster, A. J., Payne, R. J. H., and Pagel, M. 2003. Molecular phylogenies link rates of evolution and speciation. *Science*, **301**, 478.

Weiner, A., Aurahs, R., Kurasawa, A., Kitazato, H., and Kucera, M. 2012. Vertical niche partitioning between cryptic sibling species of a cosmopolitan marine planktonic protist. *Molecular Ecology,* **21**, 4063–73.

Weir, J. T. and Schluter, D. 2007. The latitudinal gradient in recent speciation and extinction rates of birds and mammals. *Science,* **315**, 1574–6.

Wiedenbeck, J. and Cohan, F. M. 2011. Origins of bacterial diversity through horizontal genetic transfer and adaptation to new ecological niches. *FEMS Microbiology Reviews,* **35**, 957–76.

Wiens, J. J. and Servedio, M. R. 2000. Species delimitation in systematics: inferring diagnostic differences between species. *Proceedings of the Royal Society B: Biological Sciences,* **267**, 631–6.

Wiens, J. J., Engstrom, T. N., and Chippindale, P. T. 2006. Rapid diversification, incomplete isolation, and the 'speciation clock' in North American salamanders (genus *Plethodon*): testing the hybrid swarm hypothesis of rapid radiation. *Evolution,* **60**, 2585–603.

Williams, G. C. 1992. *Natural Selection: Domains, Levels and Challenges.* Oxford: Oxford University Press.

Wilson, C. G., Nowell, R. W., and Barraclough, T. G. 2018. Cross-contamination explains 'inter and intraspecific horizontal genetic transfers' between asexual bdelloid rotifers. *Current Biology,* **28**, 2436–444.

Winstanley, C., O'Brien, S., and Brockhurst, M. A. 2016. *Pseudomonas aeruginosa* evolutionary adaptation and diversification in cystic fibrosis chronic lung infections. *Trends in Microbiology,* **24**, 327–37.

Wood, J. L. A., Yates, M. C., and Fraser, D. J. 2016. Are heritability and selection related to population size in nature? Meta-analysis and conservation implications. *Evolutionary Applications,* **9**, 640–57.

Wright, S. 1931. Evolution in Mendelian populations. *Genetics,* **16**, 97–159.

Yablonovitch, A. L., Fu, J., Li, K. X., et al. 2017. Regulation of gene expression and RNA editing in *Drosophila* adapting to divergent microclimates. *Nature Communications,* **8**, 1570.

Yang, Z. H. and Rannala, B. 2010. Bayesian species delimitation using multilocus sequence data. *Proceedings of the National Academy of Sciences of the United States of America,* **107**, 9264–9.

Yin, Y. B., Zhang, H., Olman, V., and Xu, Y. 2010. Genomic arrangement of bacterial operons is constrained by biological pathways encoded in the genome. *Proceedings of the National Academy of Sciences of the United States of America,* **107**, 6310–15.

Zhang, C., Zhang, D. X., Zhu, T. Q., and Yang, Z. H. 2011. Evaluation of a Bayesian coalescent method of species delimitation. *Systematic Biology,* **60**, 747–61.

Zhang, J. J., Kapli, P., Pavlidis, P., and Stamatakis, A. 2013. A general species delimitation method with applications to phylogenetic placements. *Bioinformatics,* **29**, 2869–76.

Zhao, L. P., Zhang, F., Ding, X. Y., et al. 2018. Gut bacteria selectively promoted by dietary fibers alleviate type 2 diabetes. *Science,* **359**, 1151–6.

Zhao, S., Lieberman, T. D., Poyet, M., et al. 2017. Adaptive evolution within the gut microbiome of individual people. *bioRxiv,* doi: https://doi.org/10.1101/208009.

Zuckerkandl, E. and Pauling, L. B. 1962. Molecular disease, evolution, and genic heterogeneity. In: Kasha, M. and Pullman, B. (eds) *Horizons in Biochemistry.* New York: Academic Press.

Zuppinger-Dingley, D., Schmid, B., Petermann, J. S., Yadav, V., De Deyn, G. B., and Flynn, D. F. B. 2014. Selection for niche differentiation in plant communities increases biodiversity effects. *Nature,* **515**, 108–11.

Index

The manufacturer's authorised representative in the EU for product safety is Oxford University Press España S.A. of el Parque Empresarial San Fernando de Henares, Avenida de Castilla, 2 – 28830 Madrid (www.oup.es/en or product. safety@oup.com). OUP España S.A. also acts as importer into Spain of products made by the manufacturer.

www.ingramcontent.com/pod-product-compliance
Lightning Source LLC
Chambersburg PA
CBHW051552030726
47592CB00001B/248